De Gruyter Graduate

Blümich, Haber-Pohlmeier, Zia • Compact NMR

Also of Interest

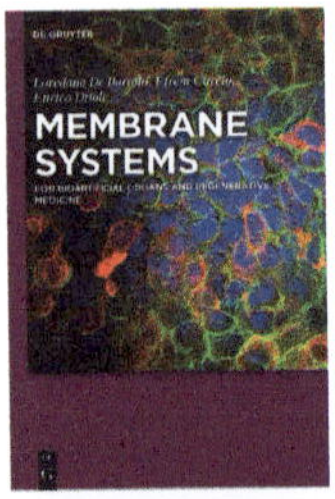

Membrane Systems.
For Bioartificial Organs and Regenerative Medicine
Loredana De Bartolo, Efrem Curcio, Enrico Drioli 2014
ISBN 978-3-11-026798-3, e-ISBN 978-3-11-026801-0

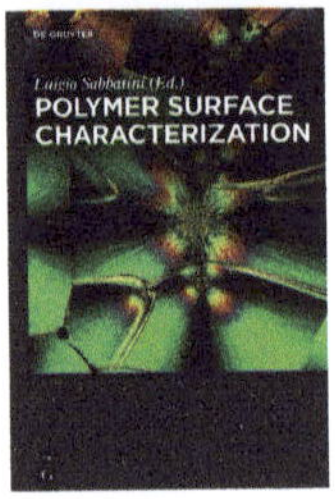

Polymer Surface Characterization
Luigia Sabbatini (Ed.) 2015
ISBN 978-3-11-027508-7, e-ISBN 978-3-11-028811-7

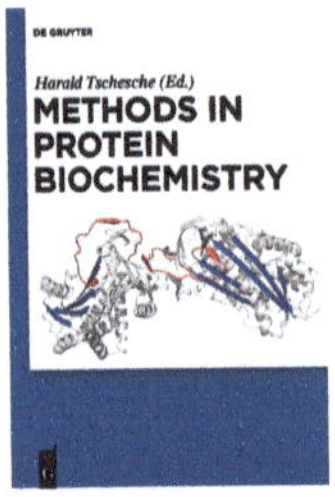

Methods in Protein Biochemistry
Harald Tschesche (Ed.) 2011
ISBN 978-3-11-025233-0, e-ISBN 978-3-11-025236-1

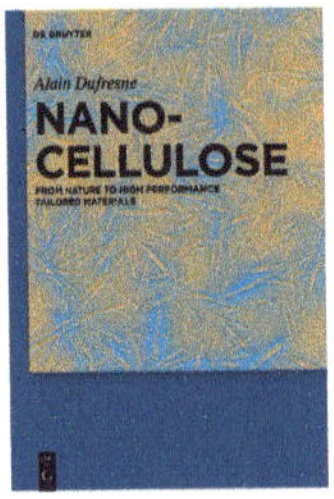

Nanocellulose.
From Nature to High Performance Tailored Materials
Alain Dufresne 2012
ISBN 978-3-11-025456-3, e-ISBN 978-3-11-025460-0

Reviews in Analytical Chemistry
Israel Schechter (Editor-in-Chief)
ISSN 0793-0135, e-ISBN 2191-0189

Bernhard Blümich, Sabina Haber-Pohlmeier,
Wasif Zia

Compact NMR

DE GRUYTER

Authors
Professor Dr. Bernhard Blümich
RWTH Aachen University
Institut für Technische und
Makromolekulare Chemie (ITMC)
Worringerweg 1
52074 Aachen
bluemich@itmc.rwth-aachen.de

Dr. Sabina Haber-Pohlmeier
RWTH Aachen University
Institut für Technische und
Makromolekulare Chemie (ITMC)
Worringerweg 1
52074 Aachen
haber-pohlmeier@itmc.rwth-aachen.de

Wasif Zia
RWTH Aachen University
Institut für Technische und
Makromolekulare Chemie (ITMC)
Worringerweg 1
52074 Aachen
zia@itmc.rwth-aachen.de

ISBN 978-3-11-026628-3
e-ISBN 978-3-11- 026671-9

Library of Congress Cataloging-in-Publication Data
A CIP catalog record for this book has been applied for at the Library of Congress.

Bibliographic information published by the Deutsche Nationalbibliothek
The Deutsche Nationalbibliothek lists this publication in the Deutsche Nationalbibliografie; detailed bibliographic data are available in the Internet at http://dnb.dnb.de.

Cover image: Measurement of an NMR depth profile of a Roman mosaic in Herculaneum during a study within the Herculaneum Conservation Project, an initiative of the Packard Humanities Institute in collaboration with the Soprintendenza Speciale per i Beni Archeologici di Napoli e Pompei and the British School at Rome.
Typesetting: PTP-Berlin, Protago-TEX-Production GmbH, Berlin
Printing and binding: Hubert & Co. GmbH & Co. KG, Göttingen
♾Printed on acid-free paper
Printed in Germany

www.degruyter.com

Foreword

NMR is a complicated business. When we want to learn the trade, being fascinated by all these wonderful experiments the NMR experts tell us are simple to do, we get confronted with mountains of information about pulses, flip angles, the rotating coordinate frame, the density matrix, and more. How is it, that we cannot do NMR in the same manner as we are used to operating a DVD player or some other sophisticated technical gadget like a cell phone. Clearly, an NMR spectrometer is a bit more complicated than a coffee machine, and so the instructions for operating a spectrometer are not expected to be trivial. But is an NMR spectrometer really more complicated than a cell phone? Probably not! Nevertheless we can operate a mobile phone without knowing the technical details about its function and the electronics inside. So why can we not operate an NMR spectrometer and get decent results on a level of expertise similar to that required using a cell phone?

Today one of the most challenging tasks is the miniaturization of NMR machines for use as dedicated and personalized devices. Because the use of such compact devices is comparatively new and these NMR devices become more popular, this book intends to be a guide to their use and applications by providing the basic knowledge and operating instructions to perform successful NMR measurements. The content focuses on compact and mobile NMR machines for the analysis of materials and processes, because this technology is most likely to be in demand by investigators who have little to no knowledge about NMR. Following a basic introduction to NMR and compact NMR equipment in Chapter 1, the experimental set-up is discussed in Chapter 2. Several general types of NMR experiments are described in Chapter 3. Representative applications of these experiments to liquids, polymers, biological tissue, porous media, and objects from cultural heritage are collected in Chapters 4 to 8. For each case, a description of the measurement and data evaluation procedures is given with reference to the principles and procedures explained in the preceding chapters. Each description follows the same pattern including the objective, the theoretical background, the pulse sequences and parameters, beginners-level measurements, advanced-level measurements, and data processing. It is hoped, the information given will help the NMR novice to successfully conduct measurements with compact NMR equipment although it is not quite yet at the level of operating a cell phone.

This book has benefitted from the help of many friends and members of the Aachen NMR research group. They contributed earlier versions of some sections of the book, helped with proof reading, checked the pulse sequences and phase tables, and most importantly provided many of the experimental data given as examples in the text. We thank Alina Adams, Sophia Anferova, Vladimir Anferov, Stephan Appelt, Juliane Arnold, Maria Baias, Peter Blümler, Federico Casanova, Ernesto Danieli, Vasiliki Demas, Dan Demco, Gunnar Eidmann, Ralf Eymael, Stefan Glöggler, Nicolae Goga, Andreas Guthausen, Gisela Guthausen, Agnes Haber, Rolf Haken,

Song-I Han, Christian Hedesiu, Volker Herrmann, Jürgen Kolz, Kidist Hailu, Bharatam Jagadeesh, Martin Klein, Kai Kremer, Rance Kwamen, Maxime Van Landeghem, Dirk Oligschläger, Eva Paciok, Josefina Perlo, Juan Perlo, Pablo Prado, Gabriel Rata Doru, Ralf Savelsberg, Udo Schmitz, Andrea Schweiger, Shatrughan Sharma, Siegfried Stapf, Oscar Sucre, Yadoallah Teimouri, Jochen Vieß, Alexandra Voda, and Anette Wiesmath for their scientific contributions to the contents of this book. Eiichi Fukushima, Tia Ishi, Eva Paciok and Lutz Weihermüller helped with proof reading early versions of the book and demanding chapters. Individual figures were kindly supplied by Gisela Guthausen, Burkhard Luy, Antonio Marchi, Peter McDonald, and Frank Rühli. Special thanks are from Bernhard to Joanie for accepting endless evenings of an absent minded husband, from Sabina to Andreas for many discussions even before breakfast, and from Wasif to his mother Rifqua Ejaz for listening to him everyday and to Aroosa Ijaz for standing by him.

Aachen, 31st August 2013

Bernhard Blümich
Sabina Haber-Pohlmeier
Wasif Zia

Contents

List of Symbols

a_{fast}	amplitude of a rapidly decaying signal component
a_{slow}	amplitude of a slowly decaying signal component
a_{ij}	component amplitudes
A	absorption signal, cross sectional area
b	power exponent in the Kohlrausch function
b_1, b_2	diffusion attenuation coefficient
B	magnitude of the magnetic field
$\boldsymbol{B}$	magnetic field vector
B_0	magnitude of the homogeneous magnetic field
B_1	amplitude of the radio-frequency magnetic field
$\boldsymbol{B}_1$	vector of the rf magnetic field
$\boldsymbol{B}_{\mathrm{eff}}$	effective field
$\boldsymbol{B}_{\mathrm{fic}}$	fictive field
B_z	z-component of the magnetic field vector
C	constant
C_1, C_2	constants
C_{M}	spin-density contrast
C_{w}	contrast of the weight parameter, water capacity function
dB	decibel
d_{i}	diameter of the interface domain
d_{m}	diameter of the mobile domain
d_{r}	diameter of the rigid domain
D	diffusion coefficient, dispersion signal, soil water diffusivity
D_0	self-diffusion coefficient of the pure substance
D_{eff}	effective diffusion coefficient
exp	exponential function
E	energy
$E_{\uparrow}$	energy of spins pointing up
$E_{\downarrow}$	energy of spins pointing down
f	scaling factor
FT	Fourier transformation
G	magnitude of the magnetic field gradient, shear modulus
$\boldsymbol{G}$	magnetic field gradient vector
G_{eff}	effective field gradient
G_{int}	internal field gradient
G_{max}	maximum gradient value
G_{phase}	phase-encoding gradient
G_{read}	read gradient
G_{slice}	slice-selection gradient
G_x	x-component of the gradient vector
$G_{x,\mathrm{max}}$	maximum gradient in x-direction
h	hydraulic head, Planck's constant
$^1\mathrm{H}$	hydrogen nucleus
i	square root of −1
J	water flow velocity
$\boldsymbol{k}$	wave vector
k_{B}	Boltzmann constant
k_{ij}	exchange rate
k_y	wave number in y-direction
$k_{y,\mathrm{max}}$	maximum wave number in y-direction
k_z	wave number in z-direction
$k_{z,\mathrm{max}}$	maximum wave number in z-direction
$\mathbf{K}$	matrix of exchange rates
K	hydraulic conductivity
K_{S}	hydraulic conductivity at full saturation
l_{d}	diffusion length
ln	natural logarithm
log	logarithm of base 10
L	length
L_0	initial length
m_{s}	mass of solid phase
m_{t}	total mass
m_{w}	mass of water phase
$\boldsymbol{M}$	nuclear magnetization vector
M_0	thermodynamic equilibrium magnetization, spin density
$M_{0\mathrm{ref}}$	spin density of a reference compound
M_2	second moment
$M(t)$	complex transverse magnetization
M_{c}	molecular weight between cross-links in an elastomer
M_{n}	number-average molecular weight
M_x	component of the magnetization vector in x-direction
M_y	component of the magnetization vector in y-direction
M_z	component of the magnetization vector in z-direction
n	number, cross-link density, van Genuchten parameter
$n_{\uparrow}$	number of spins pointing up

$n_\downarrow$	number of spins pointing down
n_1, n_2	numbers
n_{acq}	number of acquired data points
n_E	number of echoes
$n_{f,max}$	maximum number of frequency encoding steps
n_G	number of gradient steps
$n_{p,max}$	maximum number of phase encoding steps
n_s	number of scans
p	Laplace variable
P	probability density
$P(r)$	droplet size distribution
q	fraction
Q	volume flow rate
r	radius, magnitude of space vector
$\mathbf{r}$	space vector
$\langle r \rangle$	mean radius
R	ideal gas constant, constant, magnitude of the displacement vector, hydrodynamic radius
$\mathbf{R}$	relaxation matrix
RX	receiver
s	signal
s_{2Q}	double quantum signal
s_d	signal of a distribution
s_{ref}	reference signal
S	spectrum, Fourier transform of s, surface area
SFC	solid fat content
t	time
t_{acq}	acquisition time
t_d	dead time
t_E	echo time, dito t_{E1}, t_{E2}, t_{E2}
$t_{E,eff}$	effective echo time
t_f	filter time
t_m	mixing time
t_{MQ}	multiquantum evolution time
t_p	pulse width
t_{phase}	duration of G_{phase}
t_{read}	duration of G_{read}
t_R	repetition time, recycle delay
t_z	storage time for a longitudinal magnetization
t_0	recovery time
t_1	evolution time, evolution period
t_2	detection time, detection period
t_{90}	duration of the 90° pulse
t_{180}	duration of the 180° pulse
T	absolute temperature
TX	transmitter
T_1	longitudinal relaxation time
$1/T_1$	longitudinal relaxation rate
$1/T_2$	transverse relaxation rate
$1/T_{1bulk}$	longitudinal relaxation rate of the bulk
$1/T_{2bulk}$	transverse relaxation rate of the bulk
$1/T_{2diff}$	transverse relaxation rate for diffusion in gradients
$1/T_{2surf}$	surface transverse relaxation rate
T_0	time constant
T_{1o}	longitudinal relaxation time of the oil phase
T_{1w}	longitudinal relaxation time of the water phase
T_2	transverse relaxation time in homogeneous field
T_2^*	transverse relaxation time including field inhomogeneity
T_{2A}	transverse relaxation time of component A
T_{2B}	T_2 of component B
T_{2C}	T_2 of component C
T_{2bulk}	transverse relaxation time of the bulk
T_{2eff}	effective transverse relaxation time
$T_{2eff,0}$	reference effective transverse relaxation time
$T_{2eff,aniso}$	anisotropic T_{2eff}
$T_{2eff,inter}$	T_{2eff} of the interfacial component
$T_{2eff,iso}$	isotropic T_{2eff}
$T_{2eff,long}$	T_{2eff} of the mobile component
$T_{2eff,short}$	T_{2eff} of the rigid component
T_{2lm}	logarithmic mean value of the relaxation distribution
T_{2long}	long relaxation time
T_{2short}	short relaxation time
u	real part of the NMR signal recorded in the time domain
U	real part of the spectrum
v	imaginary part of the NMR signal recorded in the time domain
V	imaginary part of the spectrum, volume
V_g	volume of gas phase
V_s	volume of solid phase
V_t	total volume
V_w	volume of water phase
w	weight parameter, relaxation-weighted spin density

x, y, z	space coordinates
$\mathbf{x}, \mathbf{y}, \mathbf{z}$	unit vectors along the Cartesian space coordinates
x_A	number fraction of component A, crystallinity
x_B	number fraction of component B
x_C	number fraction of component C
x_i	number fraction of the interfacial component
x_m	number fraction of the mobile component
x_{max}	field of view in x-direction
x_r	number fraction of the rigid component
z_{max}	field of view in z-direction
α	flip angle of an rf pulse, van Genuchten parameter
α_E	Ernst angle
γ	gyromagnetic ratio
Δ	diffusion time
ΔG_{phase}	increment of G_{phase}
ΔG_z	increment of gradient in z-direction
Δt	dwell time, sampling interval
Δt_1	increment of the evolution time
Δt_E	increment of the echo time
$1/\Delta x$	spatial resolution in x-direction
$\Delta\phi$	phase increment
$\Delta\nu_{1/2}$	line width in Hz
$\Delta\omega_{1/2}$	line width in rad/s
δ	chemical shift, duration of a gradient pulse
δ_{max}	maximum chemical shift
ϵ	strain
φ	precession angle
φ_1	transmitter phase of pulse 1
φ_{acq}	receiver phase for data acquisition
φ_{RX}	receiver phase
φ_{TX}	transmitter phase
ϕ	phase error
ϕ_0	constant phase correction parameter
ϕ_1	linear phase correction parameter
Φ	porosity
η	viscosity
κ	fluid permeablity
Λ	elongation ratio
ν	frequency in Hz
ν_0	Larmor or precession frequency in Hz
ν_{max}	maximum frequency
ν_r	reference frequency
ν_{rf}	NMR frequency for transmitter and receiver in Hz
θ_m	water content of mass fraction
θ_r	bound water content
θ_V	volumetric water content
Θ	normalized water saturation
Θ_S	water saturation
ω	frequency in rad/s
ω_0	Larmor frequency in rad/s
ω_1	nutation frequency in rad/s
ω_f	final frequency in rad/s
ω_i	initial frequency in rad/s
ω_{rf}	NMR frequency for transmitter and receiver in rad per second
Ω	frequency offset from the rf transmitter frequency
π	half the circumference of the unit circle
ϖ_D	residual dipole-dipole interaction
θ	polar angle
ρ	density
$\rho_{1,2}$	surface relaxivities
ρ_s	density of solid particles
ρ_t	density of bulk soil
σ	standard deviation
τ	delay between the first and the second pulse in an echo sequence
τ_c	correlation time of molecular motion
τ_{ij}	exchange time between sites i and j
ψ	total water potential
ψ_m	matrix potential
ψ_o	osmotic potential
ψ_p	additional air pressure potential
ψ_z	gravitational potential
ψ_H	hydraulic potential
ψ_Ω	overburden potential

List of Acronyms

1D	one-dimensional
2D	two-dimensional
2Q	double quantum
3D	three-dimensional
ADC	Analog-to-Digital Converter
AOCS	American Oil Chemists' Society
ASCII	Amercian Standard Code for Information Interchange
CLI	Command Line Interface
COSY	COrrelation SpectroscopY
CPMG	Carr, Purcell, Meiboom, Gill
C-S-H	calcium silicate hydrate
CT	Computer Tomography
CUFF	Cut-open, Uniform, Force Free
CYCLOPS	CYClically Ordered Phase Sequence
DOSY	Diffusion Ordered SpectroscopY
EXSY	EXchange SpectroscopY
FID	Free Induction Decay
FLASH	Fast Low Angle Shot
FSP	Fiber Saturation Point
FT	Fourier Transformation
GARfield	Gradient At Right angle to the magnetic field
GPC	Gel Permeation Chromatography
HetCor	Hetero-nuclear Correlation
HDPE	High-Density PolyEthylene
IR	InfraRed
ISO	International Organization for Standardization
LDPE	Low-Density PolyEthylene
LLDPE	Linear Low-Density PolyEthylene
ln	logarithmus naturalis
MAS	Magic Angle Spinning
MIP	Mercury Intrusion Porosimetry
MOUSE	Mobile Universal Surface Explorer
MRI	Magnetic Resonance Imaging
MRT	Magnetic Resonance Tomography
MQ	MultiQuantum
NMR	Nuclear Magnetic Resonance
NOESY	Nuclear Overhauser Effect SpectroscopY
o/w	oil-in-water
PE	PolyEthylene
PFG	Pulsed Field Gradient
Prospa	Processing package
PVC	PolyVinylChloride
PTFE	PolyTetraFluoroEthylene
rf	radio frequency
RARE	Rapid Acquisition with Relaxation Enhancement
RMS	Root Mean Square
ROSY	Relaxation Ordered Spectoscopy
RPA	Rubber Process Analyzer
RX	Receiver
SEC	Size-Exclusion Chromatography
SFC	Solid Fat Content
SNR	Signal-to-Noise Ratio
SPAC	Soil Plant Atmosphere Continuum
S/V	Surface-to-Volume ratio
SQUID	Superconducting Quantum Interference Device
TX	Transmitter
USB	Universal Serial Bus
UV	Ultra Violet
w/c	water/cement ratio
w/o	water-in-oil

1 Introduction to NMR

The chemist calls it *NMR* and the medical doctor *MRI*. *Nuclear magnetic resonance* (NMR) is the most popular tool in chemistry to analyze molecular structures, and *magnetic resonance imaging* (MRI) is a non-invasive diagnostic tool in the hospital that provides high-contrast *images* of tissues depicting the brain functions and the beating heart. In both cases large and expensive superconducting magnets are employed (Fig. 1.0.1), which magnetize the object by aligning the atomic nuclei inside the magnet. The resulting magnetization can be triggered by radio frequency waves to rotate around the direction of the magnetic field. Depending on the operating mode, the frequency *spectrum* of the rotating nuclear magnetization provides the chemist with molecular information and the medical doctor with anatomical images, while the materials scientist may be interested in the decay of the *impulse response* to learn about physical properties of a solid object like a wet wall.

1.1 NMR: Nuclear Magnetic Resonance

NMR can be defined as a physical phenomenon which is utilized to investigate molecular properties of matter by irradiating atomic nuclei in a *magnetic field* with electromagnetic *radio waves*. Many nuclear isotopes possess an angular momentum called *spin*. In classical terms, spins appear to rotate around an axis like a bicycle wheel (Fig. 1.1.2a). For atomic nuclei, however, the somewhat unusual laws of *quantum mechanics* apply. For example, every spin is associated with a *magnetic moment* like the

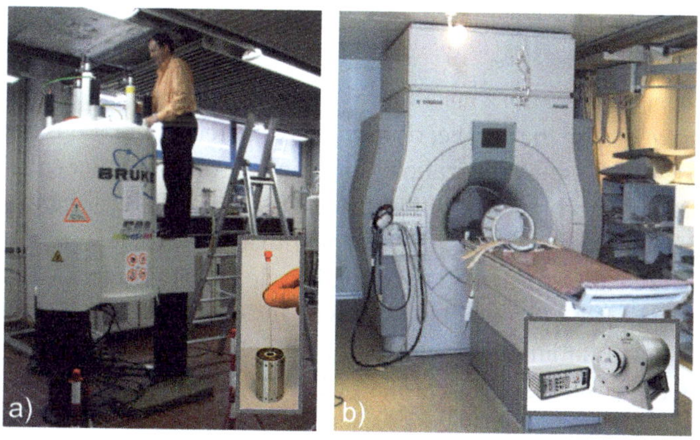

Figure 1.0.1. High-field NMR machines with superconducting magnets and compact, low-field machines with permanent magnets (insets). (a) Magnets for chemical analysis. (b) Magnets for magnetic resonance imaging (MRI). The patient or the object is positioned in the center of the magnet hole. The bulky electronics of medical MRI machines are typically hidden in a separate room.

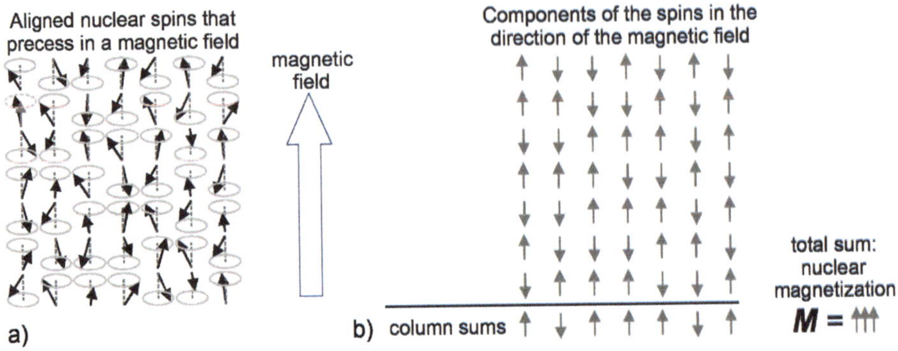

Figure 1.1.1. Schematic drawing of 49 out of 10^{23} proton spins which are aligned in a magnetic field. (a) Each spin appears to rotate or 'precess' around the direction of the magnetic field in a manner similar to a spinning bicycle wheel, which precesses around the direction of the gravitational field (Fig. 1.1.2a). (b) The up and down states of the spins are more easily recognized when only the part of the spin vector parallel to the direction of the magnetic field is drawn. Because each spin is a magnet, each of these arrows represents a magnet. The nuclear magnetization $\boldsymbol{M}$ is the sum of the magnetizations from each component magnet.

needle of a compass. Depending on the magnitude of the spin, it can align with a magnetic field in different stable orientations, which differ in their inclination angles with respect to the magnetic field and therefore also differ in their energies (Fig. 1.1.1a). *Protons*, the most abundant nuclear spins in organic matter, align in two states, called up (↑) and down (↓). The relative numbers $n_\uparrow$ and $n_\downarrow$ of spins for the two states with energies $E_\uparrow$ and $E_\downarrow$ follow the *Boltzmann distribution*, where k_B is the Boltzmann constant and T is the temperature in Kelvin,

$$\frac{n_\downarrow}{n_\uparrow} = \exp\{-(E_\downarrow - E_\uparrow)/(k_B T)\}. \tag{1.1.1}$$

The *nuclear magnetization* $\boldsymbol{M}$ of a macroscopic sample with some 10^{23} spins is formed by the difference $n_\uparrow - n_\downarrow$ of the number of spins with different orientations (Fig. 1.1.1b).

Because the resulting magnetization $\boldsymbol{M}$ is composed of an unimaginably large number of quantum mechanical entities, it behaves like a classical *magnet*, which spins around its magnetization axis. It interacts with a magnetic field $\boldsymbol{B}_0$ in the same way as a *gyroscope* e.g., in the way a spinning bicycle wheel interacts with the gravitational field (Fig. 1.1.2a): When not aligned with the direction of the field, the magnetization axis rotates around the direction of the field (Fig. 1.1.2b). This rotation is called precession. The *precession frequency* or *Larmor frequency* ω_0 is proportional to the strength B_0 of the applied field,

$$\omega_0 = 2\pi\nu_0 = \gamma B_0 \tag{1.1.2}$$

where the *gyromagnetic ratio* γ is a constant specific to the type of atomic nucleus, and $\nu_0 = (E_\downarrow - E_\uparrow)/h$, where h is Planck's constant. For example, the frequency ν_0 for protons ^{1}H is 21.29 MHz in a magnetic field with $B_0 = 0.5$ T, which is a typical value for compact permanent magnets of *mobile NMR* instruments.

When a sample is exposed to a magnetic field, it takes time for the spins to align with the field and establish the nuclear magnetization along the field direction (cf. Fig. 1.1.1). This time is characterized by the *longitudinal relaxation time* T_1. The resultant magnetization can be tipped away from the direction of the magnetic field $\boldsymbol{B}_0$ by a radio frequency (rf) magnetic field *impulse* (Fig. 1.1.2c). The impulse is applied to the sample by means of a current pulse through an rf *coil* (Fig. 1.1.2b), which usually is part of an rf resonance circuit. This coil produces an oscillating magnetic field $\boldsymbol{B}_1$ perpendicular to the static magnetic field $\boldsymbol{B}_0$ (Fig. 1.1.2b), and the magnetization also rotates around this field with frequency $\omega_1 = \gamma B_1$. The *flip angle* α by which the magnetization is tipped away from the direction of the magnetic field $\boldsymbol{B}_0$ is adjusted by the amplitude B_1 of the rf field and the duration t_p of the rf impulse according to

$$\alpha = \gamma B_1 t_p. \tag{1.1.3}$$

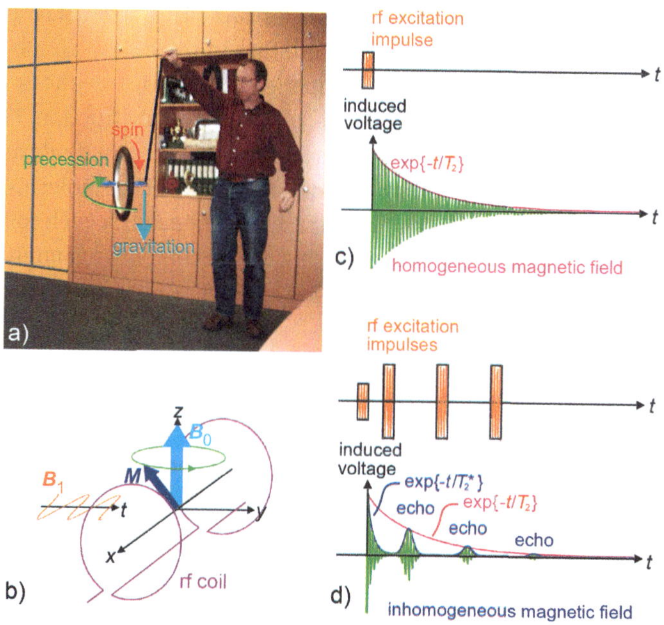

Figure 1.1.2. Precession and impulse response. (a) A bicycle wheel spins around its axis and precesses in the gravitational field. (b) The nuclear magnetization $\boldsymbol{M}$ follows the same motion in a magnetic field $\boldsymbol{B}_0$. (c) A magnetic rf impulse with its frequency ν_{rf} tuned to the precession frequency can excite this precession of nuclear magnetization. The impulse response decays with the time constant T_2 in a homogeneous magnetic field $\boldsymbol{B}_0$. (d) In an inhomogeneous field the impulse response decays with the time constant T_2^*. Echoes can be generated by two and more excitation impulses. Ideally, the echo amplitudes stroboscopically sample the envelope of the impulse decay from a homogeneous magnetic field.

With a single impulse for excitation, maximum signal is generated when the flip angle is 90°, so that the magnetization after the impulse is perpendicular to the $\boldsymbol{B}_0$ field direction.

In conventional NMR and MRI the sample rests inside the coil, which is inside the magnet; in *single-sided NMR*, the sample is placed outside in the *stray field* of the coil and the magnet. Following the excitation impulse, the precessing magnetization induces an oscillating voltage in the coil in the same way as a bicycle dynamo generates electricity. The induced voltage decays with a time constant T_2 when the magnetic field is homogeneous (Fig. 1.1.2c). The decay results from destructive interference of different magnetization components in the sample, because fluctuating magnetic fields caused by the motion of the molecules randomly change the precession frequencies of the components so that fewer and fewer magnetization components are aligned in the same direction as they precess, and the resulting magnetization amplitude decreases.

The *impulse response* contains all the information that can be measured. It provides oscillation frequencies, decay times, and amplitudes. The chemist evaluates the oscillating part of the impulse response by *Fourier transformation* to analyze components with small shifts of precession frequencies caused by weak magnetic fields associated with the currents of the electrons orbiting in the chemical bonds of the molecules. These shifts and the associated component amplitudes are the fingerprints of chemical groups and their concentrations, i.e. of the molecular structure. The physicist determines the decay time of the impulse response, the so-called *relaxation time* T_2 to study the physics of matter associated with the mobility of the molecules. In heterogeneous matter the impulse response is the sum of responses for each component. When relaxation times differ for different components, the component amplitudes can be determined from the *distribution of relaxation times*. The relative component amplitudes are proportional to *concentrations*.

In inhomogeneous fields the precession frequencies differ in each voxel of the sample because the applied magnetic field differs from voxel to voxel. Then the impulse response decays faster with a relaxation time T_2^* which is shorter than T_2 (Fig. 1.1.2d). The fraction of the signal amplitude that is only affected by the static inhomogeneity of the applied field and not by the fluctuating fields can be recovered in a so-called *spin echo* after effectively reverting the sense of precession of the magnetization components by a second rf impulse. This fundamental phenomenon was discovered by Erwin Hahn. In his honor the spin echo is also called the *Hahn echo*. The echo-generating process can be repeated many times to generate a train of echoes. The only signal loss is then due to T_2 *relaxation* invoked by the ever present fluctuating fields so that the echo train envelope ideally corresponds to the signal decay in a homogeneous magnetic field. Following the initials of the inventors Carr, Purcell, Meiboom and Gill, a train of many Hahn echoes is also called a *CPMG echo train*. The generation of *echo trains* is the fundamental principle followed when measuring signals in the inhomogeneous fields of the permanent magnets frequently used for portable *stray-field NMR* devices like the *NMR-MOUSE*.

1.2 Mobile NMR

Today NMR with portable and compact machines is experiencing a period of rapid growth. This process started when scientists learned how to measure NMR signals with simple magnets and with inhomogeneous magnetic fields from objects outside the magnet. A further push came when it was demonstrated that magnets with extremely homogeneous magnetic fields could be assembled from blocks of permanent magnets that show large variations in their magnetic properties.

There are now portable NMR magnets for all three types of basic NMR measurements. These are NMR *relaxometry* for analysis of material properties by relaxation measurements, NMR *imaging* for measuring 2D and 3D images, and NMR *spectroscopy* for chemical analysis (Fig. 1.2.1). While in high-field NMR, the object is always nearly placed inside the magnet (Fig. 1.0.1), mobile low-field NMR also uses magnets that are placed near the object so that the object is exposed to the *stray field* of the magnet. Here the object can be arbitrarily large, because its size is not restricted by the magnet bore, and the volume inside the object from which the signal is collected is well defined. These two magnet geometries are often referred to as open and closed. In either case, the demands on magnetic field homogeneity increase from relaxometry to imaging to spectroscopy. Relaxation measurements can be performed by *echo* techniques in arbitrarily shaped magnetic fields, images are best acquired in linearly varying magnetic fields with constant *field gradients*, and spectra are best acquired in extremely *homogeneous fields*.

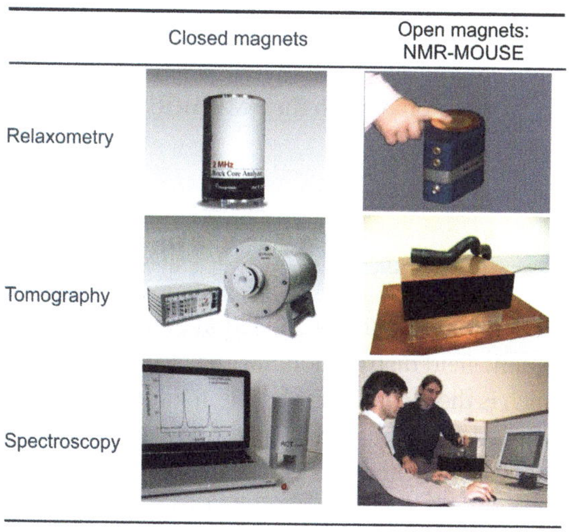

Figure 1.2.1. Compact magnets for mobile NMR. Left: Closed magnets that accommodate the sample in their center. Right: Open magnets that are placed close to the object to measure NMR signals from selected volume regions of the object. An early small stray-field NMR device of this kind is the NMR-MOUSE®.

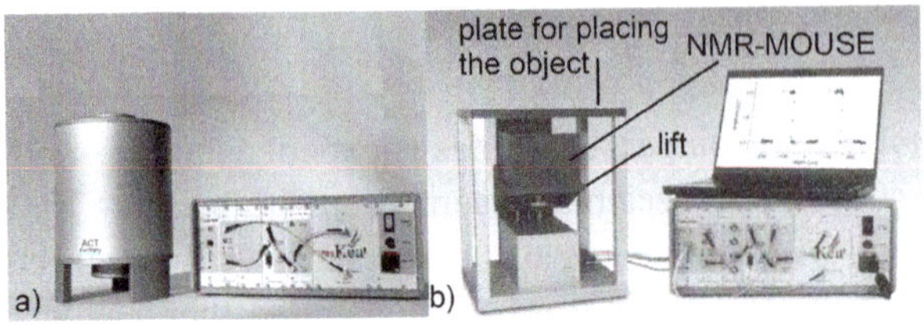

Figure 1.2.2. Mobile NMR magnets and spectrometers. (a) Halbach magnet for NMR spectroscopy. (b) Profile NMR-MOUSE® on a computer-controlled lift for depth profiling.

The first, small device with an open design was named *NMR-MOUSE* for Nuclear Magnetic Resonance *MObile Universal Surface Explorer*. It can be shifted around like a computer mouse to scan large objects (Figs. 1.2.1, top right and 1.2.2b). The magnet of the NMR-MOUSE can be designed in different ways so that the field is either parallel or perpendicular to the sensor surface. There are also different ways of building closed magnets from permanent magnet pieces. A particularly ingenious way is due to Klaus *Halbach* and many closed NMR magnets follow the original Halbach design (Fig. 1.2.2a). They possess maximal magnetic field strength inside and ideally zero stray field outside. While the NMR-MOUSE is good for testing material properties of large objects, *closed magnets* are suitable for analysis of substances transported in tubes through their centers and for *process control* by imaging of objects passing through the magnet.

The compact NMR magnets most important for practical applications are the NMR-MOUSE for relaxometry and the *Halbach magnet* for imaging and spectroscopy. Because the field strength of permanent magnets varies with temperature, many applications require *temperature control*. This is essential when measuring data with high resolution, for example, depth profiles with the NMR-MOUSE and spectra with a Halbach magnet.

1.2.1 Halbach magnets

Halbach discovered the principle of arranging a collection of magnet blocks in such a way that the resulting magnetic-field distribution has the desired shape. One shape is the *homogeneous field*, which is the most desirable field for any NMR measurement. The magnet blocks are positioned on a circle forming a ring with the polarization of each block aligned in the plane of the circle but rotated from block to block in such a way that the orientation of the magnetization performs two rotations when passing full circle around the ring (Fig. 1.2.3). Ideally, the field is uniform and maximal inside the ring and zero outside. Several such rings can be stacked to form a cylinder magnet that accommodates the sample inside just like a conventional superconducting magnet.

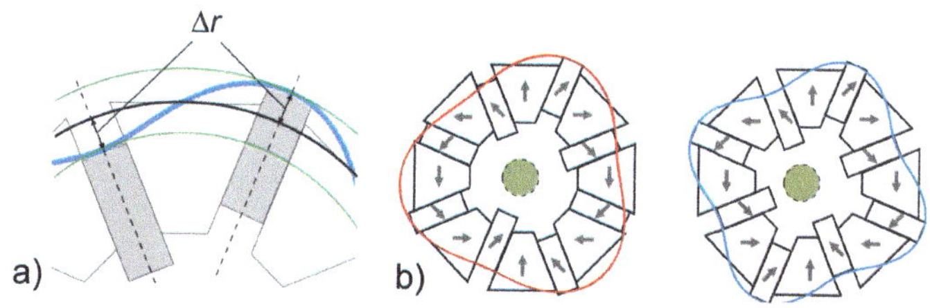

Figure 1.2.3. Halbach magnets. (a) Section of a Halbach ring with trapezoidal and rectangular magnet pieces. The rectangular blocks in the gaps can be shifted to eliminate the field distortions resulting from magnet imperfections. (b) Depending on the displacements Δr of the rectangular magnet blocks, different field correction terms can be generated.

The field inside a simple Halbach magnet is not sufficiently homogeneous to measure *spectra* or *images* but quite sufficient to measure NMR relaxation. The residual field inhomogeneity originates from the use of a finite number of blocks in place of polarization varying continuously within the ring, the statistics of the magnetic domains within the magnet blocks, which are made from sintered granular material, and the finite precision of their dimensions and magnetizations. A clever way to compensate for this *inhomogeneity* is to *shim* the field by radially displacing rectangular magnet blocks positioned in gaps between the trapezoidal magnet pieces (Fig. 1.2.3b). Depending on the relative diameter of the hole, typical field strengths of such Halbach magnets are 0.5 T to 2 T.

1.2.2 Magnets for stray-field NMR: The NMR-MOUSE

To measure a sample positioned in the *stray field* outside a magnet is a well-known idea. The oil industry started to use it with *well-logging sensors* to measure the NMR *relaxation* of fluids confined to the pores of the borehole wall. Early materials studies employed this principle to measure moisture in soil, bridge decks, and other building materials. Subsequently the stray field of superconducting spectroscopy and imaging magnets was used to measure *images* and *diffusion coefficients*, because the high *field gradient* of 50 T/m and more, together with the high field strength, promised superior resolution and sensitivity. Similarly strong gradients of some 20 T/m at lower field strengths can also be obtained with small permanent magnets. However, rather than the gradient it was the simplicity of the device that led to the construction of the *NMR-MOUSE*, a small stray-field NMR sensor, which turned out to be well suited to test all kinds of hydrogen-containing materials such as rubber, polymers, wood, food, and moisture in porous media.

The NMR-MOUSE is obtained from a classical C-shaped magnet that produces a largely homogeneous magnetic field between its poles (Fig. 1.2.4a) by opening up the C (Fig. 1.2.4b). A sample positioned near the surface experiences the *stray field* of this

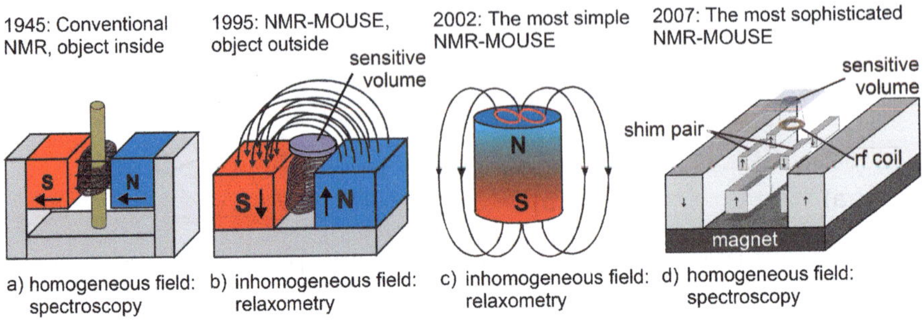

Figure 1.2.4. Permanent magnet assemblies for stray-field NMR. (a) The NMR magnet used in the early days of NMR produces a homogeneous magnetic field in a gap between two magnet poles. (b) The NMR-MOUSE® is obtained from this design by opening it up. The stray fields of both the magnet and the rf coil are inhomogeneous. (c) The most simple stray-field NMR magnet is just a bar magnet. (d) With additional magnet blocks, the stray field can be shimmed locally in a sensitive volume to good enough homogeneity to resolve the small frequency differences in proton NMR spectra.

magnet and the stray field of the rf *coil* between the two magnet poles (Fig. 1.2.4b). The sensitive volume of such a simple sensor is oddly shaped more like an upside-down soup dish than a flat slice. A different shape for the sensitive volume is obtained with a simple *bar magnet* (Fig. 1.2.4c). The best homogeneity of the magnetic field in a plane above the rf coil is produced by the *Profile NMR-MOUSE*, which looks like two of the u-shaped magnets in Fig. 1.2.4b placed side by side. In this case the magnetic field varies within the sensitive volume only in the direction perpendicular to the sensor surface. Signals could be measured in this way from slices as thin as 2.3 micrometers. The ultra-high magnetic-field homogeneity required to measure NMR spectra is typically only obtained inside the magnet, but by shimming the magnetic field profile of the u-shaped NMR-MOUSE with additional magnets, field homogeneity sufficient to resolve the *chemical shift* of protons could be obtained (Fig. 1.3.4b).

1.3 Measuring methods

The measuring methods can be grouped into two types (Fig. 1.3.1). Those in one group (Fig. 1.3.1a) require a *homogeneous* field to resolve the oscillations in the measured signal that result from the *precession* of the transverse magnetization generated by an rf *impulse*. Those in the other group (Fig. 1.3.1b) only measure the signal decay from the envelope of several echoes generated at different times in an inhomogeneous field. In the following, the oscillating impulse response measured in a homogeneous or linearly and weakly space-dependent magnetic field is called the *free induction decay* or *FID*. The train of echoes measured in an *inhomogeneous field* is called the *CPMG echo train* because it is most often measured in a way that follows the pioneering work of Carr, Purcell, Meiboom, and Gill.

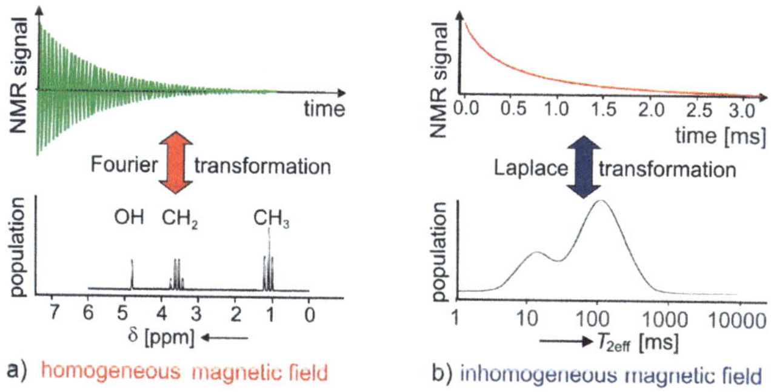

Figure 1.3.1. NMR signals and processing. (a) Free induction decay acquired in a homogeneous magnetic field. Its Fourier transform is a distribution of frequencies known as the NMR spectrum, which is used to derive molecular structures. When acquired in a linearly varying magnetic field, the NMR spectrum is a projection of the object, i.e. a one-dimensional image. (b) Multi-echo envelope measured with the CPMG sequence. Its inverse Laplace transform is a distribution of relaxation times.

Either signal is further analyzed to extract information in terms of parameters or distributions of parameters (Fig. 1.3.1). The FID is almost always converted into a *distribution of frequencies* by *Fourier transformation*. This distribution is the *NMR spectrum* when measured in a homogeneous magnetic field or a 1D projection *image* of an object when measured in a linearly space-dependent field. The CPMG train is either fitted with model functions like an exponential or a bi-exponential decay to extract amplitudes and relaxation times, or it is converted into a *distribution of relaxation times* by operations similar to an inverse *Laplace transformation*.

Both schemes can be expanded to acquire multi-dimensional data sets (Fig. 1.3.2). The underlying idea is to start the data acquisition not with the nuclear magnetization in the *thermodynamic equilibrium* established when the samples have been exposed sufficiently long and undisturbed to the magnetic field, but rather in a non-equilibrium state that is generated by manipulation of the initial magnetization prior to acquisition. Such manipulations are achieved by further rf impulses and waiting times. When carried out in a systematic manner, the evolution of the nuclear spin system before acquisition is also oscillatory or follows some type of multi-exponential curve. By measuring many FIDs or CPMG trains for systematically changing initial conditions, multidimensional data sets are acquired that can be analyzed by multidimensional Fourier and Laplace transformations and combinations thereof to obtain *multi-dimensional distributions* of frequencies and relaxation times. Such multidimensional parameter distributions reveal more detailed information about the system under study than one-dimensional distributions.

A well-known *two-dimensional distribution* of signal amplitude versus frequency is an NMR *image*. In magnetic fields with a constant gradient, i.e., in magnetic fields that vary linearly in space, the NMR frequency in a particular voxel is proportional to the position of the object. Other important examples are correlation and exchange

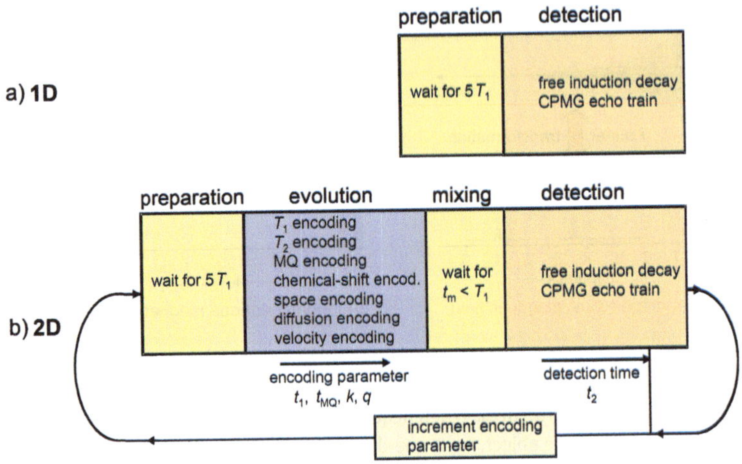

Figure 1.3.2. NMR data acquisition schemes. (a) One-dimensional NMR. Following a waiting time during which the magnetization reaches thermodynamic equilibrium, an FID or a CPMG echo train are acquired depending on the magnetic field homogeneity. (b) Two-dimensional NMR. The initial state of the spin system is varied from acquisition to acquisition in a systematic fashion so that encoding parameters are changed in a systematic fashion. This introduces a second dimension to the acquired data set, which can also be processed by Fourier or Laplace transformations. Which encoding parameter is varied depends on the particular setup of the rf impulses and gradient impulses which is defined in the NMR experiment.

NMR spectra where *cross peaks* in the plane of a *two-dimensional spectrum* identify interactions between magnetization components that precess with different frequencies in the evolution and detection periods of the measurement scheme. Frequency distributions or NMR spectra can have hundreds of discrete resonance lines in *high-resolution NMR spectroscopy*, and *two-dimensional Fourier NMR* helps to identify the structures and dynamics of the molecules investigated. On the other hand, *relaxation time distributions* and *distributions of diffusion coefficients* may show only up to about five different peaks, and *two-dimensional Laplace NMR* helps to identify the nature of relaxation centers and the geometry of the pore spaces explored by fluid molecules.

1.3.1 Relaxometry

Relaxometry denotes the measurement of relaxation curves. Such measurements are the least demanding on *magnetic field homogeneity*. Compared to *spectroscopy*, the magnetic field can be highly inhomogeneous as in the case of the *stray field* of an ordinary magnet and the fields inside an unshimmed Halbach magnet. In fact, the use of *well-logging NMR* instruments and of the *NMR-MOUSE* relies mostly on measurements of transverse and longitudinal relaxation curves. For example, water saturating the pores of rock may produce the *CPMG* decay shown in Fig. 1.3.1b when measured with

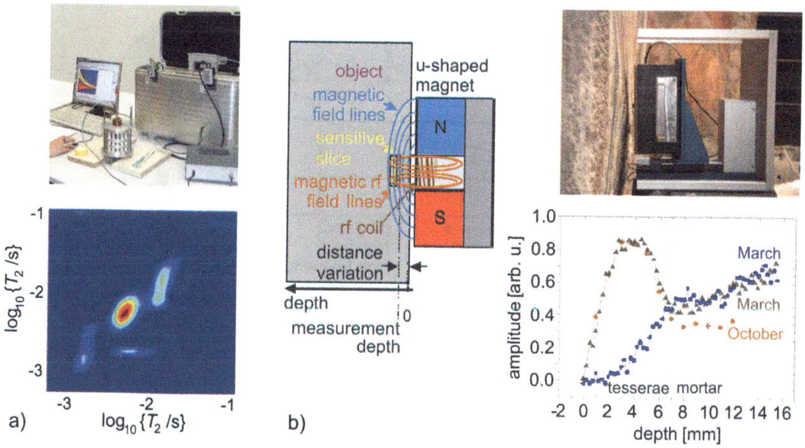

Figure 1.3.3. Applications of NMR relaxometry. (a) 2D relaxation-exchange NMR of water saturated soil with a simple Halbach magnet operated by a Bruker Minispec spectrometer enclosed in a pilot's carrying case. The 2D Laplace exchange map identifies different relaxation sites by peaks on the diagonal. Cross peaks indicate spatial proximity of the sites that can be accessed by water molecules diffusing between them during a mixing time of 10 ms. (b) Depth profiling with the Profile NMR-MOUSE. By varying the distance between the sensor and the object, the sensitive slice is displaced inside the object. The amplitude of the CPMG signal from the sensitive slice defines the amplitude in proton-density profiles shown here through a mosaic at Herculaneum measured at different times and different parts of the mosaic.

a simple Halbach magnet, which transforms into the distribution of relaxation times shown along with it.

Such simple instrumentation can deliver detailed information on the pore space when measuring *2D relaxation maps* (Fig. 1.3.3a) of the fluid molecules moving inside the porous matrix. For example, following the 2D scheme of Fig. 1.3.2, the distribution of transverse relaxation times is measured twice in a *relaxation exchange* experiment, once at an initial time and once at a final time separated by a time delay or *mixing time* t_m during which the magnetization is carried by the fluid molecules from one relaxation site to another. In many cases the fluid molecules are water molecules moving by *self-diffusion* in the pores of soil or rock where the different relaxation sites mark different positions in the complex pore space formed by compacted irregular grains. From a computer simulation of the experimental exchange map, the *exchange rates* can be extracted and from those the distances between the relaxation centers can be calculated.

Much simpler 1D relaxation curves like CPMG trains (Fig. 1.3.1b) are routinely measured with the *NMR-MOUSE*. When using the *Profile NMR-MOUSE*, the *sensitive volume* is a thin slice at a well-defined position which can be shifted through the object by varying the distance between the sensor and the object (Fig. 1.3.3b). The amplitude of the *CPMG* signal from the sensitive slice is proportional to the number of protons in

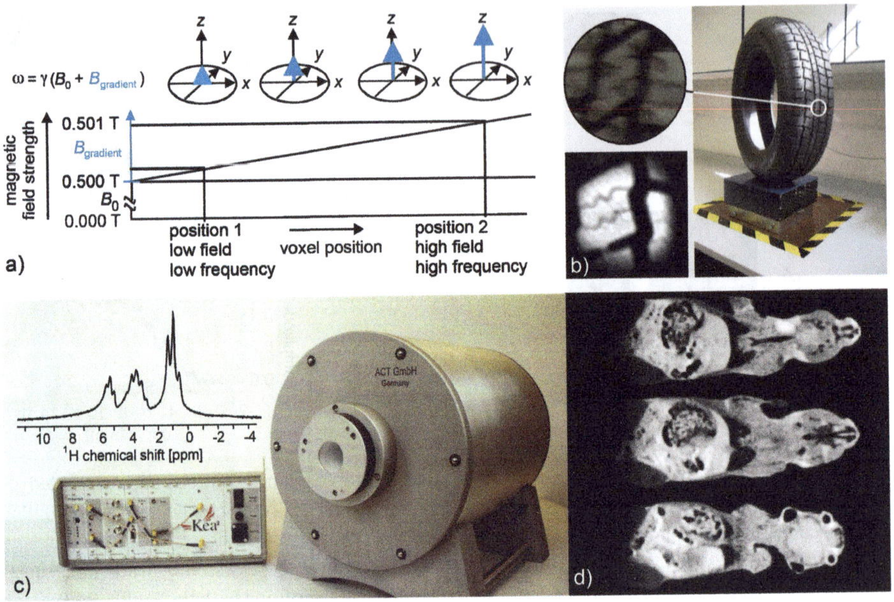

Figure 1.3.4. NMR imaging. (a) Different voxels are discriminated by means of their NMR frequencies. Because the field varies linearly across the object, the NMR frequency is proportional to the voxel position. The linear field variation is achieved with current-driven coils inside the magnet. Amplitude and timing of the currents are determined by the particular sequence of commands to the spectrometer hardware when acquiring the image data. (b) Stray-field NMR imaging of tire tread with the NMR-MOUSE: Photo of a tread section, NMR image, and setup. (c) Kea NMR spectrometer and 0.5 T Halbach imaging magnet. The field of the Halbach magnet is so homogeneous that the fine structure of the resonances in the ^{1}H ethanol spectrum can be resolved within a sample volume of 1 cm^3. (d) NMR images through a mouse measured with the 0.5 T Halbach magnet.

the sensitive slice, which may correspond to the water or wax content in some porous building material like a concrete wall or a Roman mosaic. The amplitude values as a function of the sensor displacement define a 1D image of the proton content versus depth into the object. Instead of the signal amplitude, the relaxation time or another parameter of the measured signal can be plotted depending on how one wants to define the *contrast*.

To measure *depth profiles* by scanning the object voxel by voxel is a slow process. This had already been observed in the early days of medical MRI. A much faster way is to measure the impulse response in a magnetic field with a weak gradient, where the positions of the voxels can be identified from their different NMR frequencies in the direction in which the magnetic field strength changes, i.e., in the *gradient direction* (Fig. 1.3.4a). For the NMR-MOUSE this is the direction away from the surface. Indeed, the signal from a narrow distance range covering the depth of the sensitive slice can be measured by a CPMG train without displacing the sensor. The depth profile across the sensitive slice is then obtained in terms of the Fourier transform of an NMR *echo*. This approach to *imaging* is basic to almost all imaging schemes in diagnostic medicine, but the field gradients used there are much smaller than those produced by the NMR-MOUSE, so that larger objects can be imaged albeit at lower resolution.

1.3.2 Imaging

Imaging requires magnetic field profiles to be better defined than for *relaxometry*. Ideally the magnetic field varies linearly across the object and the direction of field variation should be adjustable during the measurement (Fig. 1.3.4a). The linear field variation corresponds to a constant spatial derivative or *gradient*. In practice these demands are met by a magnet that consists of two parts. One part produces a *homogeneous field* across the extension of the object. This would be a superconducting magnet in clinical MRI (Fig. 1.0.1b) or a *Halbach magnet* (Fig. 1.2.1 left, middle) in compact NMR. The switchable gradient field is produced in either case by currents through sets of electrical coils inside the magnet or on top of the NMR-MOUSE. Depending on which of the so-called gradient coils is powered up, the *field gradient* is applied in a different direction but with the direction of the *gradient field* always aligned with that of the homogeneous field.

Because the magnetic field strength B is proportional to the NMR frequency ω (eqn. (1.1.2)), each group of spins at different positions along the direction of the field gradient experiences a different frequency, so that the frequency distribution or the *spectrum* of the NMR signal measured in a magnetic field with constant gradient is a 1D *image* of the magnetization distribution along the gradient direction. The art of magnetic resonance imaging consists of orchestrating the sequence of radio frequency impulses and current impulses through the gradient coils in such a way that the information about the signal distribution inside the object is measured in the optimum way with the best *contrast*. Typically this distribution or image is obtained by *Fourier transformation* of the measured signal, because the Fourier transformation disentangles the different frequency components of the *impulse response*. The contrast is achieved by further manipulations in the measuring schemes that result in weighting the signal from each voxel with functions of the *relaxation times* and *diffusion coefficients* of the molecules in the object. There are numerous ways of doing this, and the wide range of contrast values that can be accessed by NMR makes NMR imaging superior to any other method of imaging when applied to soft matter like fluids, moisture in porous materials, biological tissue, food, plants, rubber, and polymers.

Compact, single-sided *MRI* was first realized by an *NMR-MOUSE* fitted with gradient coils to supply the gradient fields needed to measure Fourier images of slices through objects parallel to the sensor surface (Fig. 1.3.4b). The strong field gradient in the direction perpendicular to the scanner surface, however, results in a very thin slice, so that measurement times are long. With today's technology, the field gradient of the NMR-MOUSE could be lowered to boost the sensitivity by measuring signal from thicker slices. Moreover, Halbach magnets can be built with outstanding homogeneity over extended volumes for imaging of objects a few centimeters in diameter (Fig. 1.3.4c). Within seconds and up to a few minutes, 2D NMR images can be acquired from various soft-matter objects such as a mouse (Fig. 1.3.4d) but also rheological properties of fluids passing through the magnet can be studied with it by *flow imaging*.

1.3.3 Spectroscopy

NMR spectroscopy requires extreme *homogeneity* of the magnetic field across the sample, because the differences in NMR frequency for protons that are usually caused by differences in chemical structure are in the range of one part per million and less. It is far less difficult to achieve this level of homogeneity within small volumes inside big magnets. But as the homogeneous volume increases and the magnet size decreases, such field homogeneity is more and more difficult to obtain. In this respect, the homogeneity of the *Halbach magnet* in Fig. 1.3.4c is already outstanding, because typical sample diameters are less than 5 mm in high-resolution NMR spectroscopy, while the *high-resolution spectrum* shown in Fig. 1.3.4c is from a 10 mm diameter sample. The worst-case scenario is to measure a ^{1}H NMR spectrum in the *stray field* outside the magnet. While for a long time this was believed to be an impossible task, an NMR-MOUSE could nevertheless be shimmed well enough by placing additional magnets in the gap (Fig. 1.2.4d) so that the field becomes sufficiently homogeneous in a finite volume region outside the magnet assembly to measure ^{1}H NMR spectra that resolve the *chemical shift* (Fig. 1.3.5). It should be noted that this is a proof-of-principle demonstration only. Before this technology can be used in routine applications, the drift of the magnetic field from temperature fluctuations has to be eliminated.

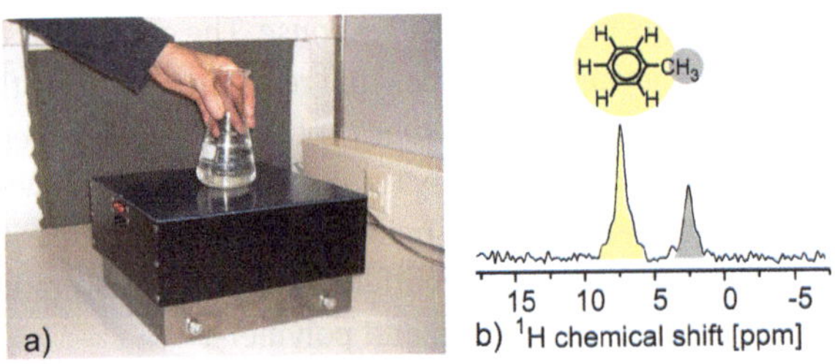

Figure 1.3.5. Stray-field NMR spectroscopy. The stray field of the NMR-MOUSE can be shimmed in the sensitive volume to such extreme homogeneity that the chemical shift can be resolved to measure ^{1}H NMR spectra. (a) For measurement, a flask with the solution to be measured is positioned on top of the magnet so that the sensitive volume resides inside the flask. (b) NMR spectrum of toluene measured in this way with a single excitation impulse.

1.4 Hardware

Essentially an NMR instrument is a device that enables two-way radio communication between the operator and the atomic nuclei in a magnetic field (Fig. 1.4.1). It consists of a *console*, which is connected to the *probe* that communicates with the sample and is positioned in a *magnetic field*. The console is operated by a computer, which times all events of the measurement and is often also used for data processing. A *transmitter* TX sends the excitation impulses, a *receiver* RX detects the NMR signal, and the computer inside the console controls the currents that power the gradient field coils in imaging experiments. The *magnet* and the probe usually form an entity separate from the console. This entity is either an *NMR-MOUSE*, which consists of a *stray-field magnet* and an rf resonant circuit, or it is a *Halbach magnet* or another magnet which encloses the probe with the rf coil containing the sample.

The probe is shared between the transmitter and the receiver and connected on demand to one or the other by a transmit/receive switch, which is also called a *duplexer*. The probe can typically be exchanged to fit different types of samples. Its resonant circuit needs to be tuned to the NMR frequency of the sample and matched to 50 Ω impedance. From time to time this tuning and matching may need to be checked and adjusted for optimum performance of the instrument.

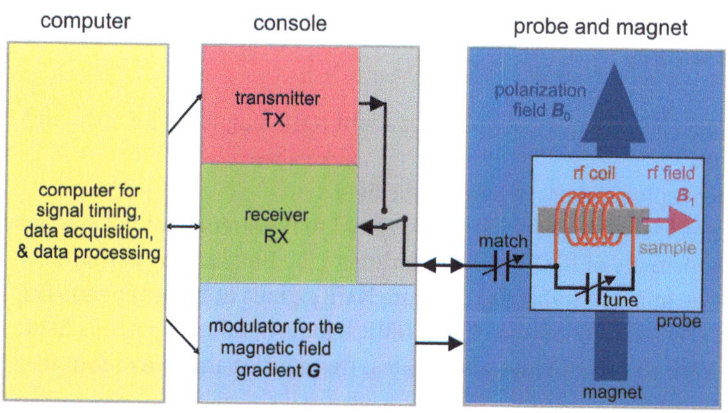

Figure 1.4.1. The three main components of an NMR machine are the computer, the console, and the magnet with the probe. When setting up the NMR equipment, these three components often need to be physically and electrically connected and the proper communication between them established. The magnet produces the constant magnetic field $\boldsymbol{B}_0$ to which the sample is exposed in close proximity to the probe. The main component of the probe is an rf resonant circuit which produces the magnetic rf field $\boldsymbol{B}_1$ of the excitation impulses and receives the NMR signal. The console contains a transmitter TX, a receiver RX, and a gradient modulator. The transmitter is connected to the rf resonant circuit only when the rf impulse is sent. Otherwise the receiver is connected to the resonant circuit for detection of the NMR signal. A current modulator is needed to turn gradient fields on and off. TX, RX and the gradient modulator are timed by a computer, which often also processes the data, for example, by executing the Fourier and Laplace transformations.

In the receiver, the voltage induced in the coil by the precessing nuclear magnetization is amplified and converted from the radio frequency regime to the audio frequency regime in the same way a radio converts the radio waves received via the antenna to audio signals. In NMR rotating waves are processed which, when written in complex notation, consist of a real and an imaginary part. Essentially each part passes through one receiver channel. The signal from both channels is referred to as the *quadrature signal*. The quadrature audio signal is digitized in an *analog-to-digital converter*, stored in the computer, and further processed, for example, by Fourier and Laplace transformation.

Often a *gradient controller* is also part of the spectrometer. It produces time-dependent currents to drive gradient coils located near the rf coil that generate additional magnetic fields in the sensitive volume of the object parallel to the main magnetic field and with well-defined *gradients*. Such gradient fields are used to encode the positions of magnetization components within the sensitive volume in imaging experiments and to destroy unwanted transverse magnetization by spoiling the homogeneity of the magnetic field. In the latter case the gradients are called *homospoil gradients* or *crusher gradients*. While the magnet and the spectrometer are specific to the demands of NMR, the computer can typically be a general one, which is capable of running the NMR software package that comes along with the NMR spectrometer.

1.5 Summary

- Compact NMR employs small permanent NMR magnets with open and closed geometry
- Nuclear spins precess in a magnetic field
- The precession frequency is proportional to the strength of the magnetic field
- The sensitivity of NMR is low according to the Boltzmann distribution
- A single impulse suffices to measure NMR signals in homogeneous fields
- Spin echoes are used to measure NMR signals in inhomogeneous fields
- The NMR signal provides amplitudes, frequencies and relaxation times of signal components
- Longitudinal and transverse relaxation times are determined by molecular mobility
- Distributions of relaxation times can be measured with simple magnets
- NMR imaging requires linear magnetic field profiles
- NMR spectroscopy needs homogeneous magnetic fields
- Depth profiles of different NMR parameters can be measured with open magnets

1.6 Further reading

Casanova F, Perlo P, Blümich B, editors. Single-Sided NMR. Berlin: Springer; 2011.

Soltner H, Blümler P. Dipolar Halbach magnet stacks made from identically shaped permanent magnets for magnetic resonance. Conc Magn Reson 2010; 36A:211–222.

Kose GK, Haishi T, Handa S. Applications of Permanent-Magnet Compact NMR Systems. In: Codd S, Seymour JD, editors. Magnetic Resonance Microscopy. Weinheim: Wiley-VCH; 2009, pp. 365–380.

Blümich B, Casanova F, Perlo J. Mobile single-sided NMR. Prog Nucl Magn Reson Spectr 2008; 52: 197–26.

Mitchell J, Blümler P, McDonald PJ. Spatially resolved nuclear magnetic resonance studies of planar samples. Prog Nucl Magn Reson Spectr 2006; 48: 161–181.

Stapf S, Han SI, editors. NMR Imaging in Chemical Engineering. Weinheim: Wiley-VCH; 2006.

Blümich B. Essential NMR. Berlin: Springer; 2005.

Freeman R. Magnetic Resonance in Chemistry and Medicine. Oxford: Oxford University Press; 2003.

Fukushima E, Roeder SBW. Experimental Pulse NMR: A Nuts and Bolts Approach. Reading: Addison Wesley; 1981.

2 Hardware setup and operation

Depending on the type of compact NMR equipment, individual hardware components need to be connected and tuned for optimum performance before measuring the object of interest. Traditionally these are three basic components, the *magnet* with the radio frequency *probe*, the spectrometer *console*, and the computer (Fig. 1.4.1). Whereas in the past, most NMR-literate scientists were expected to be trained in setting up the hardware, modern compact NMR machines will at best require to be connected to a few power and data lines, and automatic setup routines will take care of any tuning steps required to optimize the performance of the equipment. Today fully integrated compact NMR spectrometers are about to become routine, which do not require any hardware setup or maintenance operations by the user and can be operated by apps dedicated to particular types of measurements. At present, such compact NMR machines are not yet widely available, so that the different steps in setting up the hardware are described in the following. Depending on the instrument manufacturer, the appearance of the components differs as well as the names of pulse sequence parameters and spectrometer commands. The overall procedures, however, are similar with most producers like Bruker, Magritek, Oxford, Tecmag, and others. The following description refers largely to the `Kea` spectrometer by Magritek. `Kea` specific names are identified by Courier font.

2.1 Connecting the spectrometer

The different functional units of the console (Fig. 1.4.1) are mounted in different slots in the `Kea` spectrometer (Fig. 2.1.1). They are connected at the front panel. The rf pulses are generated by a synthesizer in the *transmitter*. In combination with a *receiver* this unit is called a *transceiver*. The transmitter output (1) is fed to the rf amplifier input (2), which raises the excitation power to the required power level. The rf amplifier output

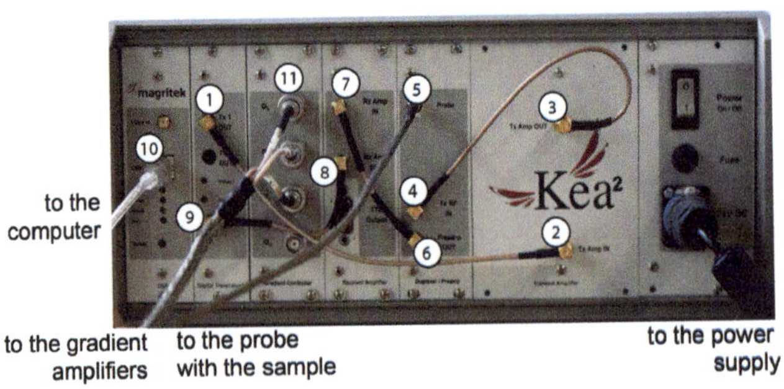

Figure 2.1.1. The front panel of the Kea spectrometer. Each unit (cf. Fig. 1.4.1) of the spectrometer occupies one slot. The signal passes from one unit to the next through connecting cables. The connections are described in the text.

(3) connects via a transmit-receive switch or duplexer (4) to the probe with the sample (5) which is located in the magnetic field, e.g. inside the magnet. For an NMR-MOUSE (Fig. 1.2.2b) the components of the probe electronics are mounted underneath a plate that covers the magnet, and the rf coil that excites and receives the NMR signal is located in the center as close to the object as possible.

From the probe the NMR signal is returned to the duplexer and passed through a preamplifier when the transmitter is inactive. From the preamplifier output (6) the signal is passed to the receiver amplifier input (7) and from the receiver amplifier output (8) to the transceiver input (9) where it is digitized and fed to the computer via the computer-interface connector (10). For *imaging* and *diffusion* measurements magnetic *field gradients* are needed which can be turned on and off and modulated in time. To this end three *gradient modulators* (11) are part of the spectrometer, which are connected to the gradient coils in the magnet assembly via *external gradient amplifiers*.

2.2 Test samples

To test the hardware and to set up NMR experiments, suitable *test samples* are needed. Readily available test samples are a piece of soft rubber like silicone rubber and natural rubber, e.g. a pencil eraser, a stabilized emulsion like *skin cream*, and some common fluids like water. Because pure water has a long T_1 relaxation time, the magnetization builds up slowly, requiring several seconds, so that the water should be doped with a relaxation agent such as *copper sulfate* ($CuSO_4$) to avoid long delays between successive scans. A highly concentrated copper sulfate solution has a T_1 of about 10 ms so that a recycle delay of 50 ms can be used. When measuring liquid samples in a magnetic field with a strong gradient like that of the NMR-MOUSE, the signal decays rapidly within a few milliseconds because the molecules permanently change their resonance frequencies by diffusing between regions with different magnetic field strength. In *emulsions*, their diffusion distance is confined by the droplet size. This is why skin cream is a better test sample for the NMR-MOUSE than water, because the signal does not decay as rapidly.

2.3 Starting the operating software

Once the hardware components are properly connected and a test sample is available, NMR experiments can be set up. Although every NMR experiment is different, some basic steps are common to most of them. They are executed with the *operating software* of the *spectrometer*. For the `Kea` spectrometer the software for the data processing package is called `Prospa`. It is started by clicking on the `Prospa` icon on the screen of the computer, which is connected to the spectrometer via one of its USB ports. A main window will appear which contains additional sub-windows named for example `1D`

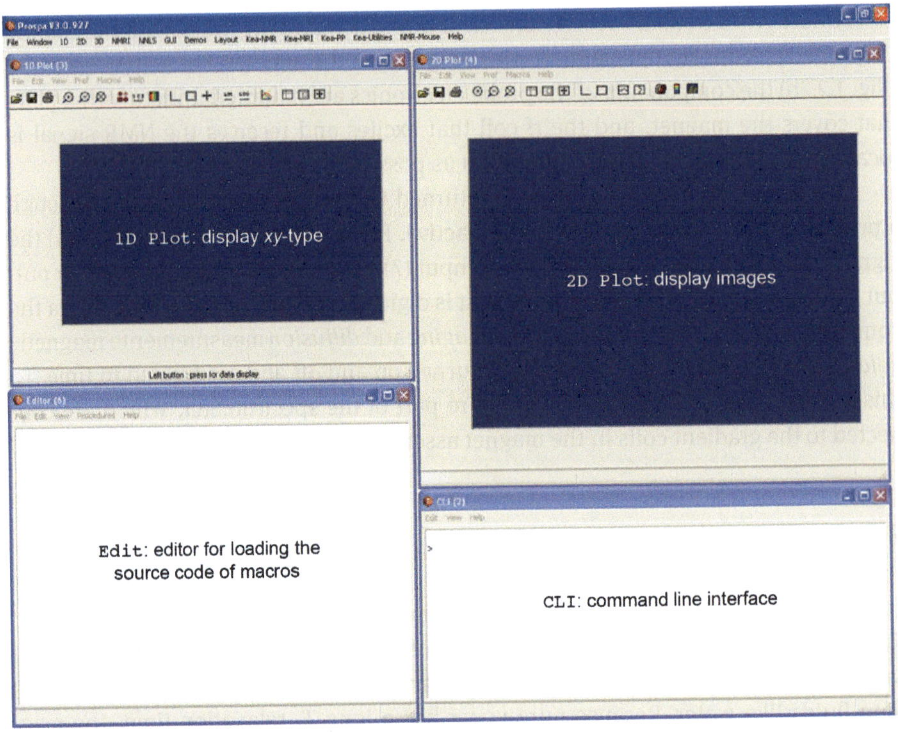

Figure 2.3.1. The main `Prospa` screen with windows.

`Plot`, `2D Plot`, `CLI`, and `Edit` (Fig. 2.3.1). `1D Plot` displays acquired *xy* data such as an `FID`, an echo, or an echo train. `2D Plot` displays images and other two-dimensional data sets. `CLI` is short for `command line interface` and shows the values of relevant parameters like pulse length, information flags, and error messages during a measurement. In the `Edit` window the source code of the currently active macro command can be visible.

Like most office programs, `Prospa` has a main menu bar with different menus such as `Windows` or `Kea-MRI`. These menus contain a series of commands or macros for controlling the `Prospa` interface and for running the experiments. If the use of the spectrometer is shared between different operators, it is helpful to define one's own measurement environment with a personal entry in the menu bar. Prior to running the first experiment, the proper function of the spectrometer has to be assured by executing the following steps:

- Checking the external *noise level*
- Checking the *tuning* and *matching* of the probe
- Calibrating the *pulse length*
- Defining the parameters of the *acquisition program*, e.g. for *single-pulse excitation* or the *CPMG sequence*

2.4 Noise level

When closed magnets are used, for example for NMR spectroscopy, the probe and the sample reside inside the magnet where they are shielded from the electromagnetic rf noise of the environment. This is different in single-sided NMR, because the probe and the magnet are open, and the NMR detection circuit may receive *noise* from external sources (Fig. 2.4.1a). Strategies for reducing such noise are: (1) partial shielding of the probe together with the object by placing the equipment on a grounded copper plate, (2) grounding the magnet to zero electric potential, (3) covering the whole setup with conducting tissues such as copper mesh or parachute silk which are electrically connected to ground.

The noise level can be determined by measuring the ratio of the root mean square (RMS) of the *noise* from the rf coil containing the sample and connected to the spectrometer and the RMS of the noise produced by the spectrometer only (Fig. 2.4.1b). Note that noise originates not only from the coil but also from the sample, especially if the sample is electrically conducting like a moist wall or a human limb. The noise of the spectrometer is the noise detected when the console is connected to a 50 Ω resistor instead of the probe. Ideally the noise ratio is 1, and the probe noise is negligible. But in practice this ratio is larger. An acceptable value is 1.5. For values higher than this, the external noise needs to be reduced. Figure 2.4.1b demonstrates how the noise level may change when shielding and grounding the sample and the NMR-MOUSE.

2.5 Tuning and matching

The probe consists of a tunable rf *oscillator* circuit composed of a coil and two adjustable capacitors to *tune* the oscillation frequency ν_{rf} to the precession frequency ν_0 of the nuclear magnetization and to *match* the electrical impedance to 50 Ω (Fig. 1.4.1).

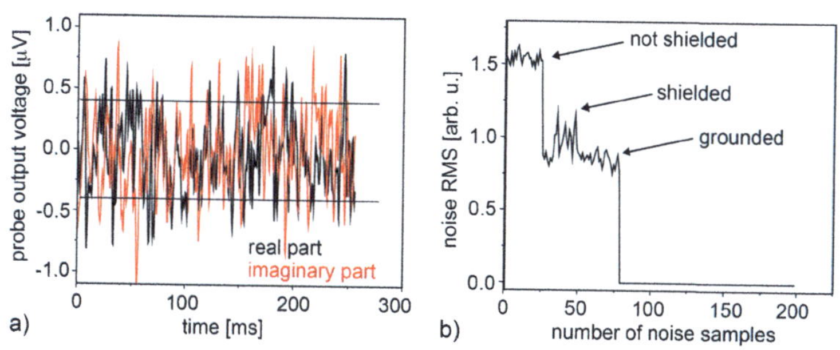

Figure 2.4.1. Measuring the noise level. (a) Actual noise signal recorded by the probe. The horizontal lines show the reference voltage due to noise using a 50 Ω resistor instead of the rf coil. (b) Change of the noise level of a 13.42 MHz NMR-MOUSE from unacceptable to acceptable values by shielding and grounding of the magnet and the object.

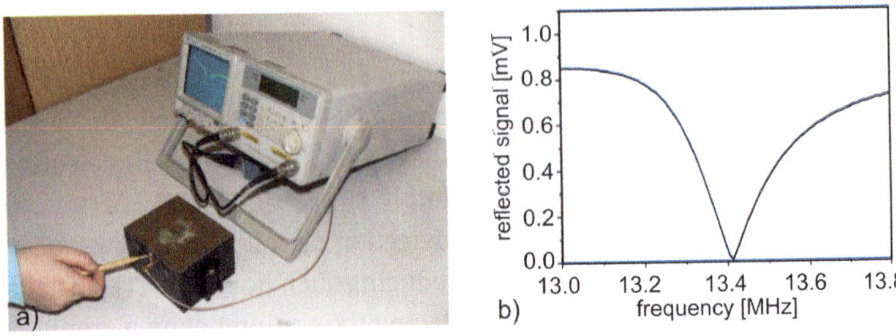

Figure 2.5.1. Tuning and matching the rf circuit of the NMR-MOUSE. (a) Use of a frequency analyzer. (b) Reflected transmitter signal as a function of frequency. The minimum at 13.41 MHz indicates optimum tuning.

The precession frequency ν_0 is determined by the field strength B_0 at the position of the sensitive volume according to $\nu_0 = \gamma B_0/(2\pi)$ (eqn. (1.1.2)). The frequency ν_{rf} of the oscillator circuit is tuned by changing the capacitance of the capacitor parallel to the coil. The *impedance* is matched by variation of the series capacitance (Figs. 1.4.1 and 2.5.1a). The capacitances of both capacitors cannot be varied independently, so that tuning and matching has to be repeated a number of times. Matching the impedance of the oscillator circuit to the 50 Ω impedance of the spectrometer is important to deliver the maximum rf power from the transmitter to the probe and the maximum signal power from the probe to the receiver.

For tuning and matching, the *transmitter power* reflected by the probe needs to be observed over a frequency range wider than that of the resonance circuit. The *tuning* and *matching* procedure can be executed in two ways, either with the help of an external *spectrum analyzer* (Fig. 2.5.1a) or with the wobble function of the spectrometer. In both cases, an rf signal is sent to the probe at low power, and its frequency is repetitively swept or wobbled across the frequency range of interest (Fig. 2.5.1b). At the same time, the reflected power is measured. The frequency dependence of the reflected power shows a dip at the resonance frequency. The reflected signal is minimal when the impedance of the resonance circuit matches that of the transmitter. Typical parameters for operating the Kea spectrometer to tune and match an *NMR-MOUSE* at a Larmor frequency of 13.76 MHz are listed in Tab. 2.5.1. Important are the lower and upper frequencies ν_0, which define the *wobble* range and the high *transmitter amplitude* attenuation.

Table 2.5.1. Default instrument settings for tuning and matching an NMR-MOUSE® PM25

Parameter	Value
lower frequency ν_0	13.4 MHz
upper frequency ν_0	14.1 MHz
receiver gain	1 dB
transmitter attenuation	−44 dB at 300 W

Although many sensors do not require frequent tuning and matching it is recommended to check the quality of tuning and matching when a sensor is used for the first time or samples with very different properties are investigated. Conducting samples may couple to the rf circuit and require tuning and matching values different from those of non-conducting samples. Operating the equipment with severe deviations from the correct tuning and matching values may damage the spectrometer electronics.

2.6 Calibration of the excitation pulse flip angle

The *flip angle* that gives maximum transverse magnetization starting from thermodynamic equilibrium magnetization in the longitudinal direction is 90°. The flip angle α of the rf excitation pulse is determined by the area under the pulse, i.e. by the duration t_p and the amplitude B_1 of the magnetic rf field. For a rectangular pulse $\alpha = \gamma B_1 t_p$ (eqn. (1.1.3)). Either B_1 or t_p can be adjusted to find the values for the 90° pulse. The amplitude B_1 is determined by the current in the rf coil which is delivered by the rf amplifier of the spectrometer. The maximum power of the rf amplifier is attenuated by an amount determined in the pulse program where also the pulse duration is set. The *power attenuation* is measured in *deciBels* or dB. A deciBel is one tenth of the logarithm to base 10 of the ratio of the output to the input power. For example, −20 dB is a *power ratio* of $10^{-(0.1\times20)} = 0.01$. Because the rf power is proportional to B_1^2, −20 dB power attenuation specifies an *amplitude ratio* of 0.1.

Typically, the *transmitter power* and consequently the value of B_1 are fixed so that the flip angle is adjusted by incrementing the *pulse widths* from scan to scan while observing an FID or an echo. The corresponding parameters are defined in the `Calibration window`. Typical values are listed in Tab. 2.6.1. In the example shown in Fig. 2.6.1, the initial 90° pulse length is 8 μs. It is varied in 19 steps of 2 μs to find the sig-

Table 2.6.1. Default parameters for determining the 90° pulse width with the NMR-MOUSE® PM25

Parameter	**Value**
transmitter frequency ν_{rf}	13.8 MHz
transmitter attenuation for 90° pulse	−6 dB at 300 W
minimum rf pulse duration t_p	7 μs
maximum rf pulse duration t_p	45 μs
receiver gain	31 dB
dwell time Δt	0.5 μs
number n_{acq} of complex points	14–50
echo time t_E	50–120 μs
number n_E of echoes	32
recycle delay t_R	300 ms
number of scans n_s	8

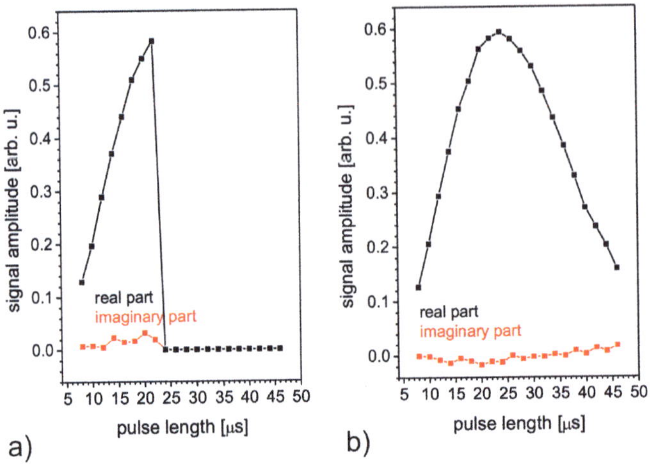

Figure 2.6.1. Pulse length calibration using a PM25 NMR-MOUSE with a copper sulfate solution. (a) Echo amplitudes as a function of the pulse length after eight increments of the pulse duration. (b) Echo amplitudes scanned with 20 incremental pulse durations. The optimum pulse length t_p for the 90° pulse is 24 μs.

nal maximum. For the NMR-MOUSE the signal is an echo from the test sample. When starting with too short an rf pulse, the signal amplitude passes through a maximum with increasing pulse length, and then decreases again. The pulse length at maximum signal amplitude is that of the 90° pulse.

The other, less frequently used way of setting the rf pulse flip angle is to decrease the transmitter attenuation at a fixed but too short duration t_p of the pulse and look for the transmitter power that gives maximum signal. The higher the rf power, the shorter the pulse duration for a given flip angle and the larger the *excitation bandwidth*. To excite a large sensitive volume in an inhomogeneous $\boldsymbol{B}_0$ field, high rf power is needed.

Given the duration t_p and the amplitude B_1 of the 90° pulse, the parameters of the 180° pulse are obtained either by doubling the pulse length and maintaining the amplitude or by reducing the *transmitter power* attenuation by 6 dB to double the amplitude while maintaining the pulse duration. Although both procedures lead to a 180° pulse, their effects differ when applied in strongly inhomogeneous field. In measurements with the NMR-MOUSE, echoes are excited with a 90° pulse followed by one or more 180° pulses. If the *flip angle* of the pulses is adjusted via the *pulse length*, both pulses will have different widths and will excite the spins in volumes of different size leading to characteristic interference effects, which are observed in the shapes of the echoes. These are less pronounced when the flip angle is adjusted via the pulse amplitude, so that the flip angle is typically varied via the pulse amplitude in experiments with the NMR-MOUSE. Radio frequency pulses of low amplitude and correspondingly long pulse length excite small sensitive volumes in inhomogeneous magnetic fields and are used when thin slices are to be excited in measurements with the Profile NMR-MOUSE.

2.7 Pulse sequences and parameters

In a homogeneous or weakly inhomogeneous magnetic field, the NMR parameters are set with a *single-pulse program* that measures the *free induction decay* (FID, Fig. 2.7.1a). In inhomogeneous fields, an echo (*Hahn echo*) or multi-echo program (*CPMG*, Fig. 2.7.1b) are used to acquire a signal.

Setting the experimental parameters

The *experimental parameters* depend on the sample and the pulse sequence as well as on the hardware (NMR-MOUSE, type of magnet). The most important parameters to be adjusted are summarized in Tab. 2.7.1 for the single-pulse experiment and the CPMG experiment.

The resonance frequency and often a user-adjustable offset to the resonance frequency can be defined. The sum of both specifies the *transmitter frequency* ν_{rf}. When setting up an experiment, the transmitter offset is set to zero in a first scan, and a large *spectral width* is chosen with a short *dwell time* Δt to find the NMR signal. The dwell time is the time interval between successively sampled data points and the sampling rate $1/\Delta t$ sets the spectral widths. For example, $\Delta t = 1\,\mu s$ sets a spectral width of

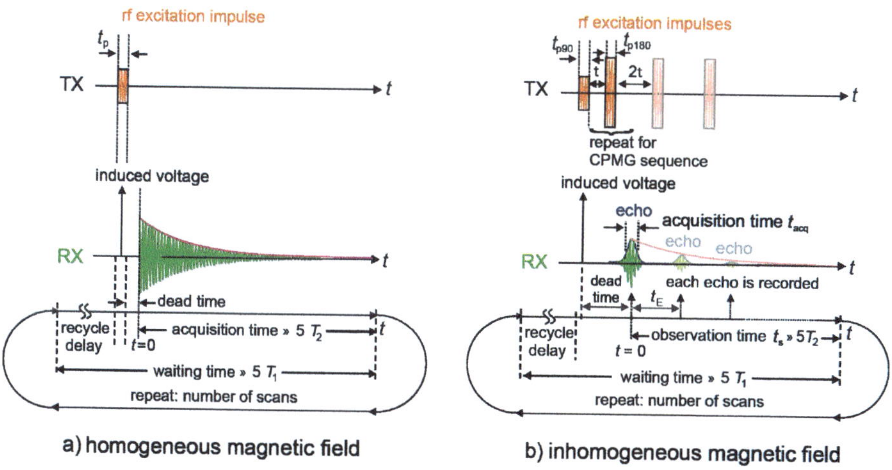

Figure 2.7.1. Basic pulse sequences in NMR. (a) Single-pulse excitation for measurement of the free induction decay. It is preferably used in homogeneous fields. Time zero of the observation time is the beginning of the recorded signal, although the generated signal approximately starts in the center of the first rf pulse. The time shift between the zero points of the generated and the recorded signals is the dead time of the experiment. Note: recycle delay + acquisition time > 5 T_1. (b) Carr–Purcell–Meiboom–Gill (CPMG) sequence for measurement of an echo train in inhomogeneous fields. If only one echo is measured, the sequence is called the Hahn echo sequence. Note: recycle delay + observation time > 5 T_1.

±1000 kHz which is a good value for collecting echoes with the NMR-MOUSE. For ^{1}H spectroscopy at 40 MHz an FID would be sampled with 10 times the spectral widths of 12 ppm, that is, 5000 Hz. Given that positive and negative frequencies can be discriminated when sampling both magnetization components M_x and M_y, the spectral width is set to ±2.5 kHz corresponding to $\Delta t = 400$ µs. The transmitter offset is then adjusted to move the transmitter frequency to the center of the NMR spectrum, and subsequently the spectral width can be reduced to the frequency range of the spectrum, for example, to 500 Hz for ^{1}H NMR spectroscopy at 40 MHz.

The amplitude of the rf pulse determines the value of the rf magnetic field B_1. It is adjusted by attenuating the maximum transmitter power. For example, the proper 90° pulse may demand a transmitter power attenuation of −8 db. But because in the first measurement neither its proper pulse length t_p nor its amplitude are known, a short pulse width like 5 µs is used. Further setup-dependent parameters to be adjusted are the receiver gain and the receiver reference phase φ_{RX}. They are discussed next. Subsequently the parameters for acquisition of a CPMG pulse sequence are discussed. They are the *number* n_s *of scans*, the repetition time or *recycle delay* t_R, the *number* n_E *of echoes*, the *echo time* t_E, the *acquisition time* t_{acq}, and the *dwell time* Δt. Typical acquisition parameters applicable to measuring an FID at 42 MHz for ^{1}H NMR spectroscopy in homogeneous fields (Fig. 2.7.1a) and CPMG echo trains with the NMR-MOUSE in inhomogeneous fields (Fig. 2.7.1b) are summarized in Tab. 2.7.1. The CPMG parameters are given for two different NMR-MOUSE sensors, one with a depth range of 5 mm measuring *skin cream* and one with a depth range of 25 mm measuring water-saturated *rock* or *soil*.

Table 2.7.1. Default acquisition parameters for an FID and a CPMG echo train

Parameter	FID	CPMG	CPMG
Parameter	FID	CPMG	CPMG
device	spectroscopy magnet	NMR-MOUSE PM5: skin cream	NMR-MOUSE PM25: rock, soil
transmitter frequency ν_{rf}	42 MHz	17.1 MHz	13.8 MHz
transmitter attenuation of 90° pulse	−6 dB at 10 W	−8 dB at 300 W	−6 dB at 300 W
duration t_{p90} of 90° pulse and $t_{p180} = 2t_{p90}$	500 µs	2–5 µs (spacer dependent)*	7–36 µs (spacer dependent)*
receiver gain	31 dB	31 dB	31 dB
number n_E of echoes	–	256–600	128–1024
echo time $t_E \approx 2\tau$	–	32–60 µs	50–120 µs
dwell time Δt	4 ms	0.5 µs	0.5 µs
number n_{acq} of complex points	1024	16	32
acquisition time t_{acq}	4 s	8 µs	16 µs
observation time	4 s	$n_E \times t_E$	$n_E \times t_E$
number n_s of scans	4–256	4–256	4–256
recycle delay t_R	6 s	0.6 s	1 s
resolution	0.25 Hz	10 µm	210 µm

* A spacer increases the distance between NMR-MOUSE magnet and rf coil

Receiver gain

The *receiver gain* determines the amplification of the NMR signal to be digitized. The optimum value is the one for which the received signal extends through nearly the whole input range of the *analog-to-digital converter* (ADC). This range is called the acquisition window of the ADC. If the receiver gain is set too high, the signal maxima are clipped, and the received signal cannot properly be processed. If the gain is too low, the *dynamic range* of the instrument is not fully utilized, and very small signals may be lost in the discretization gaps of the analog-to-digital converter.

Receiver phase

Neglecting relaxation, the precessing nuclear magnetization induces a voltage with a resultant current in the rf coil proportional to $\cos\{\omega_0 t + \varphi_{RX}\}$ which oscillates at the *Larmor frequency* $\nu_0 = \omega_0/2\pi$ and possesses a phase φ_{RX} which accumulates while the signal passes through the receiver electronics. The radio frequency ω_0 is typically converted to audio frequency by mixing the induced signal in the receiver with a cosine and a sine wave from the rf frequency synthesizer with frequency $\nu_{rf} = \omega_{rf}/2\pi$ and phase φ_{TX}. The signals produced by the *mixer* at the difference frequency are digitized and stored for further processing. They are referred to as $u(t) = M_0 \cos\{(\omega_{rf} - \omega_0)t + \varphi_{TX} - \varphi_{RX}\}$ and $v(t) = M_0 \sin\{(\omega_{rf} - \omega_0)t + \varphi_{TX} - \varphi_{RX}\}$. Together they form the complex magnetization $M(t) = u(t) + \mathrm{i}\, v(t) = M_0 \exp\{\mathrm{i}\Omega t + \mathrm{i}\phi\}$, where $\Omega = \omega_{rf} - \omega_0$ is the audio frequency and $\phi = \varphi_{TX} - \varphi_{RX}$ is the phase offset.

To record clean echoes and echo trains, the *receiver phase* φ_{RX} needs to be adjusted from the spectrometer panel by the operator in such a way that $\phi = \varphi_{TX} - \varphi_{RX} = 0$ and all echoes appear in only one channel of the receiver. This adjustment helps to avoid a *phase correction* in relaxation measurements after data acquisition. The whole procedure may be bypassed when recording *magnitude data* in the time domain. But magnitude data should be avoided whenever possible since, for example, when analyzing experimental relaxation curves which in the presence of measurement noise leads to a systematic error in the fit parameters from magnitude data. In practice the receiver phase φ_{RX} is adjusted in such a way that the imaginary part of the recorded signal is zero while the real part has maximum intensity and is positive (Fig. 2.7.2).

Phase cycling

NMR data are usually measured multiple times, and the data from each scan are added to improve the *signal-to-noise ratio* (SNR). From scan to scan, the transmitter and receiver phases φ_{TX} and φ_{RX}, respectively, are incremented cyclically in steps, which, for convenience are multiples of 90°, so that the *number n_s of scans* should be integral multiples of four. These *phase increments* and their variations from scan to scan are determined by the operator in addition to the receiver reference phase φ_{RX}. This procedure is called *phase cycling*. It is routinely used in most NMR experiments to reduce

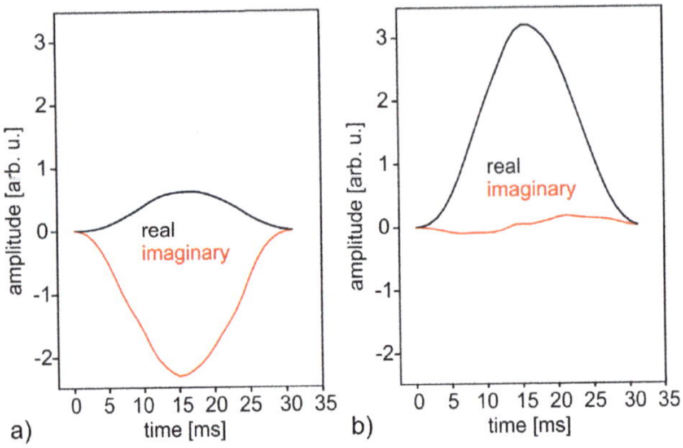

Figure 2.7.2. Adjusting the receiver phase offset by example of the echo sum from a CPMG echo train. (a) Without phase adjustment, the real component $u(t)$ (black, positive) and the imaginary component $v(t)$ (red, negative) are both non-zero. (b) When the receiver phase offset φ_{RX} is adjusted to $-\varphi_{TX}$, the imaginary part scatters around zero while the real part has maximum positive amplitude.

spectrometer artifacts. Phase cycles are also employed to select particular magnetization components from the measured response by cancelling unwanted ones in the signal sum. The rf *pulse phases* and the *receiver phases* are specified in the pulse program together with appropriate phase cycling schemes that indicate how the phases change from scan to scan.

There are different notations for the rf phases. The four quadrature phases x, y, $-x$, $-y$ corresponding to phase increments of $\varphi = 0°, 90°, 180°, 270°$ are written in shorthand as $n \times 90°$ with $n = 0, 1, 2, 3$. The phases for *single-pulse excitation* (Fig. 2.7.1a) and a *CPMG sequence* (Fig. 2.7.1b) are listed in Fig. 2.7.3. When running a single-pulse experiment, for example, the rf pulses are applied with the transmitter phase $\varphi_{TX} + \varphi_{90}$ and the signal is acquired with the receiver phase set to $\varphi_{RX} + \varphi_{acq}$ stepping cyclically from scan to scan through the values of φ_{90} and φ_{acq} (Fig. 2.7.3a). The *phase cycle* for the single-pulse excitation is known as *CYCLOPS* for CYClically Ordered Phase Sequence. For relaxation measurements and in multidimensional NMR,

a) Single-pulse exc.

φ_{90}	φ_{acq}
x	$-y$
y	x
$-x$	y
$-y$	$-x$

b) CPMG sequence

φ_1	φ_2	φ_{acq}
x	y	$-y$
x	$-y$	$-y$
$-x$	y	y
$-x$	$-y$	y

Figure 2.7.3. Cycles for the rf phase increments of the excitation pulses and the receiver. a) Single-pulse excitation. b) CPMG sequence. The phase increments are denoted by $x, y, -x, -y$ are $0°$, $90°$, $180°$, and $270°$, respectively.

more than one excitation pulse are applied, so that the number of steps in a phase cycle is larger, and a minimum of three phases needs to be set for the CPMG sequence (Fig. 2.7.3b).

Number of scans

The *number n_s of scans* defines the number of times an echo train or an FID is measured. The data acquired in each scan are added to improve the *signal-to-noise ratio*. When the *signal-to-noise ratio* is good, the minimum number of scans is determined by the *phase cycle*, i.e. 4 for FIDs and CPMG trains given the phase cycles listed in Fig. 2.7.3. If the signal-to-noise ratio is bad, then it is improved by adding the signals of many scans and n_s needs to be high. Usually the signal is contaminated by random noise so that the signal-to-noise ratio improves with the square root of the number of scans. Thus doubling the signal-to-noise ratio requires multiplying the number of scans and with it the measuring time by a factor of four.

Recycle delay

The recycle delay is the time between the end of the data acquisition period where the FID, an echo or an echo train are measured, and the start of the next scan. The time between two scans is the sum of the observation time and the *recycle delay t_R* (Fig. 2.7.1). It should be of the order five times the *longitudinal relaxation time T_1* to recover 99% of the thermodynamic equilibrium magnetization. Usually the recycle delay is much larger than the observation time, so that the recycle delay t_R is a good approximation of the time between to scans. Before determining T_1, t_R may be set to an arbitrary value, for example 1 s.

Once T_1 is known, the recycle delay can be shortened if the flip angle is chosen smaller than 90°, so that more scans can be acquired in a given time. Richard Ernst has calculated the excitation flip angle α_E which gives the optimum *signal-to-noise ratio* for a given delay $t_{acq} + t_R$ between the excitation pulses in a single-pulse experiment. This angle is known as the *Ernst angle*, and it follows from the equation

$$\cos \alpha_E = \exp\{-(t_{acq} + t_R)/T_1\}. \quad (2.7.1)$$

For short t_R, the signal-to-noise ratio can be improved with *signal averaging* by a factor up to $2^{1/2}$ with the Ernst angle as the flip angle of the excitation pulse by acquiring more scans in a given time.

In some cases, the recycle delay may be chosen longer than needed to recover the thermodynamic equilibrium magnetization. This situation arises when measuring trains of many echoes so that the rf power dissipated in the coil raises the temperature of the probe. Then long recycle delays are needed to cool down the equipment.

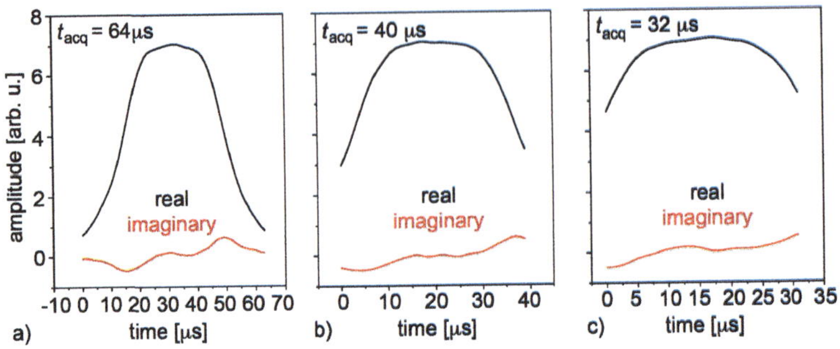

Figure 2.7.4. Effect of the acquisition time t_{acq} in an echo experiment on sampling the echo signal. The echo is centered in the acquisition window. The shorter t_{acq} is, the less of the echo is recorded. (a) For long t_{acq} the complete echo is recorded but also noise without signal. (b) For short t_{acq} only signal is recorded along with noise, but some signal is wasted. The optimum value for t_{acq} is between the settings for (a) and (b).

Echo time

The *echo time* t_E must be longer than or at least equal to the sum of the pulse length t_{p180}, twice the dead time of the receiver electronics, and the acquisition time t_{acq} of the echo. Taking as an example a train of 10 μs long pulses, an acquisition time of 20 μs, and a dead time of 30 μs, $t_E \geq 90$ μs. The acquisition window is centered to the middle of the interval between two 180° pulses of a CPMG sequence (Fig. 2.7.1b), and the optimum length of the *acquisition time* t_{acq} is in-between the *pulse length* t_p and $1.5 \times t_p$ for the NMR-MOUSE (Fig. 2.7.4). The instrument *dead time* can be determined by initially assigning a value to t_E which is equal to the length of a pulse plus the acquisition time of the echo and then subsequently increasing it in small increments until the echo is no longer distorted.

Number of echoes

To determine the *number* n_E *of echoes*, n_E is first set to a low value like 8 to 20, and a CPMG train is acquired with a small number n_s of scans. From the recorded signal the final number of echoes needed to measure the desired signal is estimated. If only total spin density needs to be determined, it is sufficient to record the initial part of the echo train with a small number of echoes. For a relaxation analysis, the complete echo train needs to be recorded. Depending on the type of sample, the number of echoes may be very low, like 30 for hard matter, or very high like 3000 for soft matter.

Acquisition time and dwell time

The maximum *resolution* obtainable in an NMR *spectrum* depends on the inverse of the observation time, which is equal to the acquisition time when measuring the impulse response (Fig. 2.7.1a). As an example, to obtain 1 Hz resolution it is necessary to acquire the signal for 1 sec. With digital equipment the signal is sampled stroboscopically, and neighboring data points are separated by the *dwell time* Δt. The *sampling rate* $1/\Delta t$ defines the *spectral width* or the frequency range within which peaks in the spectrum can unambiguously be assigned to frequencies or chemical shifts. If the spectral width is 500 Hz, then the maximum dwell time is 4 ms given that the transmitter frequency ν_{rf} is centered in the middle of the spectrum, and that real and imaginary parts of the transverse magnetization are recorded to distinguish positive and negative frequencies.

The dwell time Δt and the *acquisition time* t_{acq} together define the number $t_{acq}/\Delta t$ of points that are recorded. Setting these parameters properly is especially important not only when measuring *spectra* in homogeneous magnetic field, but also when measuring depth profiles with the NMR-MOUSE using CPMG sequences to generate echo trains in the inhomogeneous stray field of the sensor (Fig. 2.7.1b). In echo experiments the acquisition time t_{acq} denotes the time for acquisition of one echo. The longer it is, the narrower the slice from which the signal is acquired with the NMR-MOUSE, that is, the higher the spatial resolution. Note that the maximum acquisition time is limited by the echo time and the duration for which the echo signal dominates the noise signal.

2.8 Data processing

The instrument manufacturer usually supplies a *data processing software* package along with the *operating software*. But third party software or self-written routines can also be employed for processing the measured data. The `Prospa` program saves the acquired data in an ASCII file to a location specified by the working directory and the experiment name. This data file can be imported to Origin® or a similar program for data processing, or one can use the fit routines of `Prospa` for processing.

The basic data processing operations are described in Chapter 3 in the context of the different types of measurements. The two basic data acquisition schemes addressed in this chapter are the measurement of the *free induction decay* or FID by *single-pulse excitation* in a homogeneous magnetic field (Fig. 2.7.1a) and the measurement of a CPMG echo train in an inhomogeneous magnetic field (Fig. 2.7.1b). FIDs are processed by *Fourier transformation* into NMR spectra (Fig. 1.3.1a), which reveal the chemical structure of molecules or provide NMR images of an object. *Echo trains* are processed by inverse *Laplace transformation* into distributions of relaxation times (Fig. 1.3.1b).

A simpler way to analyze echo trains is to fit the CPMG decay with a model function e.g. the sum of two exponentially decaying functions (Fig. 2.8.1). This approach works well and with good accuracy for CPMG data acquired from soft matter with the NMR-MOUSE. The fit parameters are the amplitudes of the two exponential functions and their decay time constants T_{2short} and T_{2long}. These constants are the transverse NMR relaxation times, which identify two components in the sample whose concentrations are given by the amplitudes of the two exponential functions. Large data sets of spectra or echo trains are often analyzed with *data mining* software, which works out similarities and differences between the data records in the set based on statistical principles.

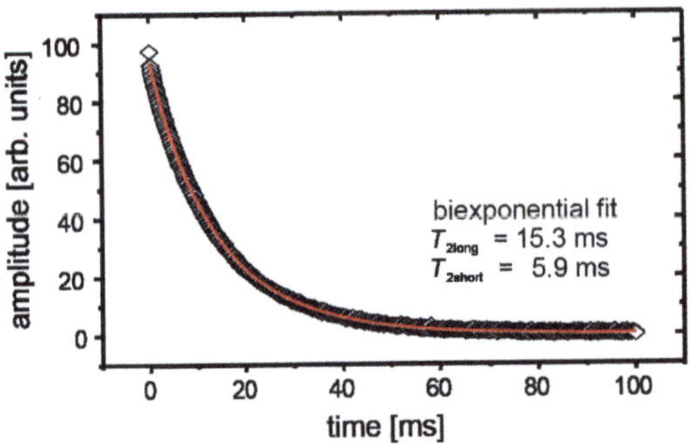

Figure 2.8.1. Bi-exponential fit of a CPMG echo train of a collagen sample measured with a Bruker Minispec mq 50 spectrometer. The two different T_2 values identify two main components with different molecular mobility.

2.9 Summary

- The main hardware components to connect when setting up the NMR equipment are the magnet with the radio frequency probe, the console, and the computer
- Open sensors like the NMR-MOUSE are sensitive to external noise
- Tuning and matching is recommended when measuring with simple Halbach magnets. It is rarely required for single-sided devices but should be checked regularly
- The rf pulse flip angle is preferably calibrated by variation of the rf power when measuring with the NMR-MOUSE. Alternatively it can be adjusted by changing the pulse length
- The most important acquisition parameters for CPMG measurements are the echo time, the number of echoes, the repetition time, the dwell time, the acquisition time, and the number of scans

2.10 Further reading

Levitt M. Spin Dynamics. Hoboken: Wiley; 2007. Chapter 4.

Keeler J. Understanding NMR Spectroscopy. Chichester: Wiley & Sons; 2005.

Berger S, Braun S. 200 and More NMR Experiments. Weinheim: Wiley-VCH; 2004.

Fukushima E, Roeder SBW. Experimental Pulse NMR: A Nuts and Bolts Approach. Reading: Addison Wesley; 1981.

The spectrometer manual of the instrument manufacturer.

3 Types of measurements

In NMR measurements, many different parameters can be determined. The three basic ones are obvious from the *impulse response* corresponding to the free induction decay signal (Fig. 1.1.2c). They are the (1) initial signal amplitude, (2) the decay time constants, and (3) the frequencies of the impulse response. These quantities refer to (1) the *spin density* corresponding to the number of nuclei within the sensitive volume, (2) the *relaxation times* and *self-diffusion coefficients*, which provide information about rotational and translational molecular motion, and (3) the *resonance frequencies*, which report about the identity of the magnetic nuclei and their chemical environment. Typically these parameters assume many values in a given sample or object. One says that they are distributed. Then not just one particular spin density, relaxation time or diffusion coefficient, and frequency need to be determined, but distributions of these parameters. Distributions of spin densities are distributions of concentrations of nuclei with particular properties. Spatial distributions of this kind are known as *images* and are determined by *magnetic resonance imaging* (*MRI*) or *magnetic resonance tomography* (*MRT*). The particular properties attributed to the nuclei define the image *contrast* in addition to mere concentration differences. Important properties are the relaxation times and the self-diffusion coefficients. Their distributions are typically analyzed when characterizing fluids in porous media, suspensions and emulsions. Another important property is the NMR frequency. A distribution of NMR frequencies is familiar to every chemist as an NMR *spectrum*, which is measured most frequently to identify the chemical structure and conformation of molecules in solution.

Each type of parameter is determined in a suitable measurement procedure, which is represented by an rf *pulse sequence* in NMR. There are literally hundreds of different types of such procedures. The general features of the basic procedures are discussed in the following together with the default instrument settings for measurements and with the procedures to extract these parameters from the experimental data. This chapter addresses each parameter type introduced above referring to amplitude, decay, and frequency of the impulse response in separate sections, however, with one modification: Due to the outstanding importance of imaging, distributions of concentrations are not discussed in the context of signal amplitudes but in a separate section on imaging.

3.1 Spin density

3.1.1 Introduction

Spin density denotes the number of spins in the sensitive volume, i.e. the spin concentration. It is directly proportional to the total magnetization of that volume and defines the *amplitude* of the impulse response. Spin density differs from gravimetric density because the number of spins in a given volume depends on the material chemistry and the isotope distribution. In conventional NMR where the sample rests inside a glass tube centered in the rf coil, the signal amplitude needs to be normalized to the sample volume in the coil which varies with the diameter, the positioning, and the filling height of the glass tube. In single-sided NMR, the sensitive volume typically rests completely inside the sample, and normalization is not required in these cases.

3.1.2 Objective

Measurements of the spin density M_0 can assist detecting the presence of an undesirable component in the sample, serve to compare different samples, or map variations in one sample of one type based on the density of the nucleus measured. The spin density depends on the sample chemistry. For example, water (H_2O) has a high proton (^1H) density and teflon (polytetrafluoroethylene, PTFE: $-[CF_2-CF_2-]_n$) has zero proton density but a high fluorine density. Because the rf pulse sequence and the state of the spectrometer electronics can affect the signal amplitude, relative values are used to quantify the spin density and spin-density contrast,

$$\text{relative spin density:} \quad M_0/M_{0\text{ref}} = s(0)/s_{\text{ref}}(0), \tag{3.1.1}$$

$$\text{spin-density contrast:} \quad C_M = (M_0 - M_{0\text{ref}})/M_{0\text{ref}} = [s(0) - s_{\text{ref}}(0)]/s_{\text{ref}}(0), \tag{3.1.2}$$

where the spin density M_0 is proportional to the measured impulse response $s(0)$ at time $t = 0$, and $s_{\text{ref}}(0)$ is the *impulse response* of a reference sample, such as water, determined with the same parameter and spectrometer settings. These definitions equally apply to conventional NMR and single-sided NMR.

3.1.3 Further reading

Blümich B. NMR Imaging of Materials. Oxford: Clarendon Press; 2000.

Coates GR, Xiao L, Prammer MG. NMR Logging. Houston: Halliburton Energy Services; 1999.

Kimmich R. NMR Tomography. Diffusometry, Relaxometry. Berlin: Springer; 1997.

Callaghan PT. Principles of Nuclear Magnetic Resonance Microscopy. Oxford: Clarendon Press, 1991.

Mansfield P, Morris PG. NMR Imaging in Biomedicine. Adv Magn Reson. Suppl. 2. New York: Academic Press; 1982.

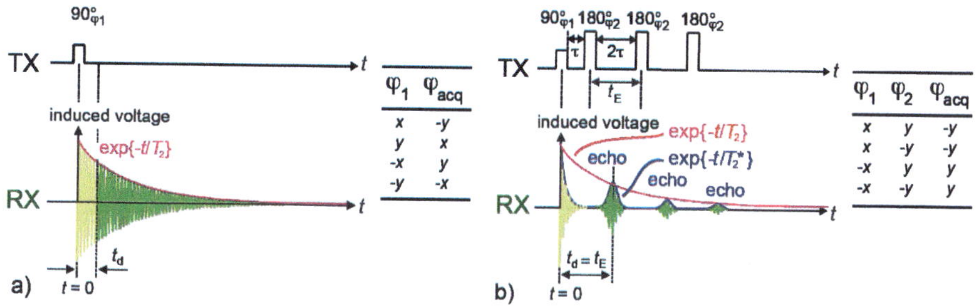

Figure 3.1.1. Pulse sequences and phase cycles for estimation of the spin density corresponding to the signal amplitude at $t = 0$ from signal measurements for $t > 0$. Note that $t = 0$ is located in the center of the rf pulse. TX denotes transmitter and RX receiver. (a) Measurement in a homogeneous field. The impulse response or FID (dark) can be measured only for times t larger than the dead time t_d. The signal for $0 \leq t < t_d$ (light) cannot be accessed. (b) Measurement in an inhomogeneous field. The envelope of a train of CPMG echoes cannot be determined for times shorter than the echo time t_E. Therefore the envelope needs to be extrapolated from the experimental data to $t = 0$. Note that the delay 2τ between two pulses is smaller than the echo time t_E. Often a two-step phase cycle is used instead of the full one, which employs the phases in lines 1 and 3 only.

3.1.4 Theory

The spin density M_0 is proportional to the amplitude $s(0)$ of the signal following a single excitation pulse (Fig. 3.1.1a) or to the amplitude of a CPMG echo train (Fig. 3.1.1b). In NMR the *impulse response* $s(t)$ is referred to as the *free induction decay* or *FID*. Because the rf circuit takes a dead time t_d to recover after the excitation pulse, the FID cannot be measured at $t = 0$, but only for times $t \geq t_d$. For a CPMG train measured in an inhomogeneous field, the minimum time delay t_d from the center of the first pulse to the first sampled data point is the echo time t_E. If t_d is much smaller than the transverse relaxation time, $s(t_d)$ is a good approximation of the signal amplitude $s(0)$. This condition is often violated for hard matter and for fluids in small pores. Then $s(t)$ significantly decays during t_d, and the signal amplitude needs to be estimated by extrapolation of the measured signal $s(t > t_d)$ to $t = 0$. A simple way to estimate the signal amplitude from CPMG echo trains is to determine the amplitude of the first echo at short echo time provided the shortest T_2 is long enough (Fig. 3.1.1b). This condition usually applied to soft matter and fluids in large pores of porous media. Moreover, integrating the initial part of the echo envelope is a convenient way to enhance *signal-to-noise ratio*.

3.1.5 Hardware

Spin density can be measured with any type of NMR instrument because such measurements require the least demanding hardware. The minimum hardware needed consists of a spectrometer, a computer, and a probe with a magnet, which does not

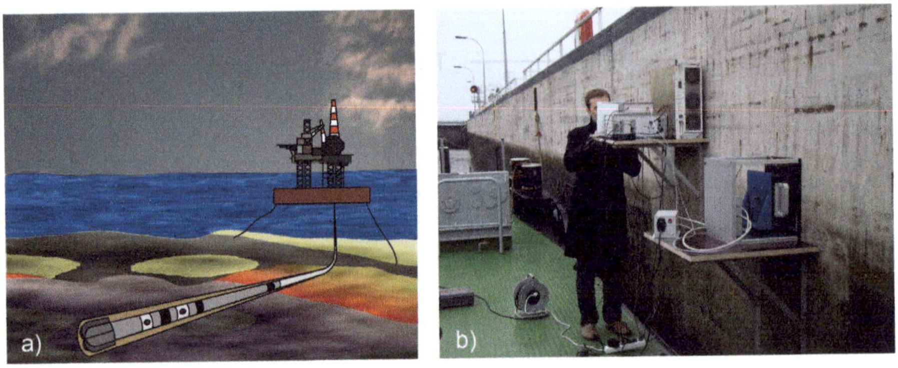

Figure 3.1.2. Special NMR sensors capable of measuring spin density. (a) NMR tool for logging-while-drilling. The magnet and the spectrometers electronics are part of a drill string advancing a hole in an oil well. The NMR signal from fluids in the borehole wall is measured while drilling. (b) Setup to measure spin density depth profiles in the concrete walls of a lock. The NMR-MOUSE is mounted on a lift turned sideways (foreground) to adjust the distance to the wall. The sensor and the spectrometer with a separate high-power amplifier (background) are placed on boards bolted to the wall of the lock.

need to produce a highly homogeneous field (Fig. 1.4.1). This is why spin density can already be measured with the inhomogeneous stray fields of a magnet and an rf coil. Stray-field NMR is applied in well-logging NMR and logging-while-drilling NMR where the magnet resides inside the hole of a well (Fig. 3.1.2a) to measure the signal from fluids in the surrounding rock formation. It is also applied with unilateral NMR sensors, like the NMR-MOUSE (Fig. 3.1.2b) and related devices.

3.1.6 Pulse sequences and parameters

The *single-pulse excitation* is used in homogeneous fields to measure the impulse response or FID (Figs. 3.1.1a and 2.7.1a), and a *Hahn echo* or a multi-echo *CPMG sequence* (Figs. 3.1.1b and 2.7.1b) are used to acquire data in inhomogeneous fields. The recycle delay, the pulse width or amplitude, the acquisition time, in the case of the CPMG sequence also the *inter-echo time* t_E or alternatively the *inter-pulse spacing* 2τ and the number n_E of echoes, as well as the *number* n_s *of scans* are parameters defined by the operator (Tab. 2.7.1). The number of scans defines the number of times the experiment is repeated. The data acquired in each scan are added to improve the signal-to-noise ratio.

Because the signal amplitude needs to scale with all spins of one kind in the sample, the *recycle delay* between successive scans is chosen large enough to allow all magnetization components formed by these spins to recover their thermodynamic equilibrium values between scans. The longest *longitudinal relaxation time* $T_{1\text{long}}$ has to be estimated or determined (Section 3.2.6) to set the recycle delay t_R to 5 $T_{1\text{long}}$. The *observation time* should cover the complete signal decay of duration 5 T_2, where T_2 is the *transverse relaxation time*. T_2 is estimated from the decay of the *impulse response* in a homogeneous field or the decay of a *CPMG echo train* in an inhomogeneous field. A reasonable value for ^{1}H in liquids is 0.8 s giving an observation time of 4 s. When a *free induction decay FID* is sampled in a homogeneous field, the observation time is identical to the *acquisition time* t_{acq}, and the sampling rate is equal to or larger than half the widths of the spectrum given that a complex signal $s(t)$ is sampled so that signal from positive and negative frequencies can be discriminated and that the receiver reference frequency is in the center of the spectral acquisition window. With a 12 ppm *chemical shift range* for protons, the *spectral width* at 40 MHz is 480 Hz. Therefore, when setting the transmitter frequency to the middle of the spectrum, the minimum sampling rate is $1/\Delta t = 1/250$ Hz so that the dwell time Δt is 4 ms. Then, in a 4 s observation time, 1000 complex data points are sampled for an FID. Note that this calculation estimates minimum requirements. Today most spectrometers sample the signal at much higher rates to improve the *signal-to-noise ratio* and maintain the phase fidelity by *digital filtering*.

When acquiring an echo train in the strongly inhomogeneous field of the NMR-MOUSE, each echo is sampled in an acquisition window of width t_{acq}, which covers the duration of the *echo* and is centered at each echo maximum (Fig. 2.7.4). Typically, a short dwell time Δt of 1 µs is chosen and t_{acq} is about 10 to 20 µs. The echo train should last for the complete decay of the *transverse magnetization* so that the observation time is 5 T_2. With the echo time t_E, the number n_E of echoes follows as $n_E = 5T_2/t_E$. Given an *echo time* of $t_E = 1$ ms and $T_2 = 0.8$ s, 4000 echoes are observed in 4 s. This value applies to liquid samples. For solids, T_2 is much shorter, e.g. 1 ms for *rubber*. With $t_E = 100$ µs, $n_E = 50$ echoes are acquired. It is important to limit the *number of echoes*, because the rf power dissipated in the coil can heat up the coil and eventually the sample near it or even burn the coil. If many echoes are acquired, longer *recycle delays* may be employed to dissipate the heat.

For the acquisition of an FID or a CPMG train, the *number* n_s *of scans* needs to be set sufficiently high to obtain a good signal-to-noise ratio. Typically n_s is an integral multiple of the number of phases to be cycled in the pulse program. This is why n_s is often a multiple of four or eight. For fluid samples, $n_s = 4$ may be a good value to start with, for solid samples, $n_s = 100$ or higher should be chosen.

Depending on prior use of the spectrometer, the correct setting of the *transmitter frequency* ν_{rf} and the 90° *pulse width* t_p need to be assured. Both are determined in the general setup procedure of the instrument. The width of the 180° pulse in the CPMG sequence is set to twice the 90° pulse width t_p or the *transmitter power* is attenuated by 6 dB less for the 180° pulse than for the 90° pulse. Default parameters to acquire an *FID* of liquids in a homogeneous field and a *CPMG echo train* from soft solid matter in a strongly inhomogeneous field are summarized in Tab. 2.7.1. Common issues are listed in Tab. 3.1.1.

Table 3.1.1. Common issues encountered when measuring CPMG echo trains

- The spectrometer frequency ν_{rf} is wrong
- The probe is not tuned and matched to the spectrometer: a conductive or highly dielectric sample may change the settings for tuning and matching
- The 90° pulse width has not been set correctly
- The 180° pulse width or amplitude have not been set correctly
- The observation time is set wrong and part of the signal is cut off when set too short or noise is acquired at long times when set too long
- The recycle delay is too short and signal contributions with long T_1 are suppressed
- The number of scans is insufficient
- The number of echoes is set too high so that the rf circuit of the probe heats up
- The receiver phase φ_{RX} is not adjusted, and the second receiver channel shows non-zero signal

3.1.7 Beginner's level measurement

Spin density is determined from the initial amplitude of an FID or a CPMG echo train $s(t)$ at $t = 0$. To a first approximation, receiver *dead time* t_d (Fig. 3.1.1) and rf *pulse width* are neglected in determining the *spin density*. When storing the experimental data for subsequent analysis, the dead time for the FID and the *echo time* for a CPMG train need to be stored along with them so that the first data point of the recorded signal is stored as $s(t_d)$ at $t = t_d$ and not at $t = 0$. Proper consideration of the dead time is particularly important for signals with rapidly decaying components such as the signals from *polyethylene* and grey *cement* (Fig. 3.1.3). Polyethylene is a semi-crystalline polymer with amorphous domains, giving a slowly relaxing signal, and with crystalline domains, giving a rapidly relaxing signal. A short dead time t_d amounts to short echo time $t_E = t_d$ in a CPMG sequence (Fig. 3.1.1b). By shortening the echo time, the recorded signal starts earlier (Fig. 3.1.3a) and more data points can be acquired for the signal from the crystalline domains so that *crystallinity* can be determined with higher accuracy for example, from a bi-exponential fit to CPMG decays acquired with the NMR-MOUSE. Grey cement is the most popular building material. For example, *concrete* is a mixture of grey cement and stones. But grey cement has paramagnetic impurities,

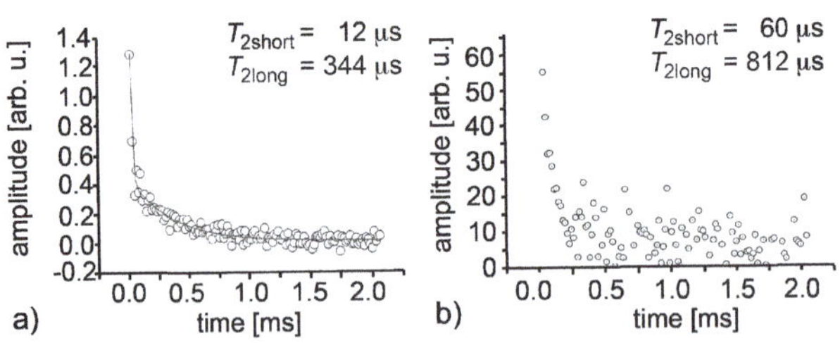

Figure 3.1.3. CPMG decays measured with the NMR-MOUSE at an echo time of $t_E = 16\,\mu s$. (a) Low-density polyethylene (LDPE). (b) Mortar in grey concrete. The initial rapid decay can only be sampled with a short echo time.

which severely shortens the initial decay of the water signal. With a short echo time of 16 μs, this decay can be observed much better (Fig. 3.1.3b).

3.1.8 Advanced level measurements

The *spin density* is the signal amplitude at time zero of the FID or the CPMG echo train. Because the signal at time zero is hidden underneath the first pulse and blanked by the instrument *dead time*, it needs to be extrapolated from the experimental data (Fig. 3.1.1). The signal at time zero is estimated in practice by recording *solid echoes* for solid materials and *Hahn echoes* for liquids at different echo times and extrapolating the echo envelope to zero echo time. The solid echo (Fig. 5.1.5b) compensates the evolution of the magnetization under the impact of the *dipole-dipole interaction*, which is the dominating interaction of the protons in solids, while the Hahn echo (Figs. 2.7.1b, 3.2.3b and 5.1.5a) compensates the evolution of the magnetization due to differences in precession frequency which in liquids comes from magnetic field inhomogeneity and chemical shift differences. In weakly inhomogeneous fields, the *FID* can be measured instead and the initial part extrapolated to the center of the excitation pulse given the pulse width and the dead time. At longer times, the effects of field inhomogeneity contribute more and more to the signal decay, so that Hahn or CPMG echoes need to be measured that are void of effects of field inhomogeneity in weakly inhomogeneous fields (Fig. 3.1.4).

When acquiring echo trains in a strongly inhomogeneous $\boldsymbol{B}_0$ field, a few peculiarities arise because the rf pulse always rotates the magnetization around the axis of the *effective field* in the coordinate frame, which rotates around the $\boldsymbol{B}_0$-field direction with

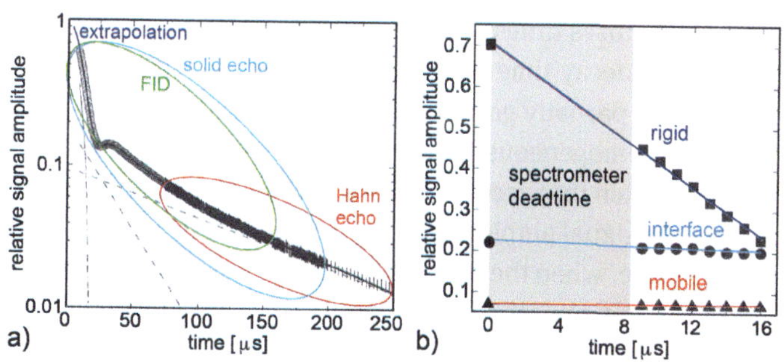

Figure 3.1.4. Extrapolation of the signal to zero time by example of a polyethylene sample in a weakly inhomogeneous field (adapted from [1] with permission). (a) Composition of the time domain signal from data measured with solid echoes and Hahn echoes. The solid-echo data agree with the FID. The missing data near the time origin have been extrapolated by fitting the experimental data with the sum of an Abragam function and two exponentials. (b) Change of the component amplitudes in the fit function with varying solid-echo time. A significant change is observed for the rigid component represented by the Abragam function.

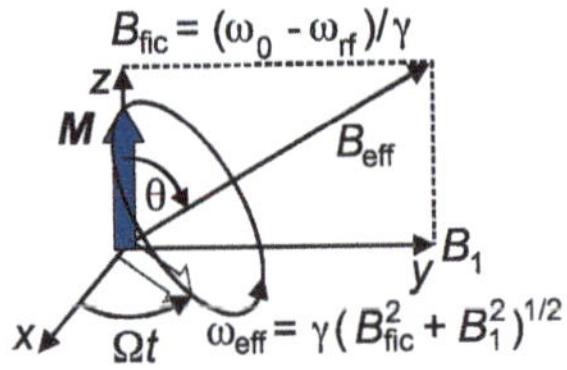

Figure 3.1.5. The magnetization $\boldsymbol{M}$ always precesses around the effective field $\boldsymbol{B}_{\rm eff}$, which is the vector sum of the fictive off-resonance field $\boldsymbol{B}_{\rm fic}$ in the frame rotating with the rf frequency $\omega_{\rm rf}$ and the $\boldsymbol{B}_1$ field, which is static in this rotating frame. The effective field $\boldsymbol{B}_{\rm eff}$ is inclined from the off-resonance magnetic field $\boldsymbol{B}_{\rm fic}$ towards the rf magnetic field $\boldsymbol{B}_1$ by the angle $\theta = \arctan\{B_1/B_{\rm fic}\}$. The precession frequency of the magnetization during the rf pulse is $\omega_{\rm eff}$. The rf pulse flip angle $\alpha = \omega_{\rm eff} t_{\rm p}$ is specified on resonance where $\omega_0 = \omega_{\rm rf}$ so that $\boldsymbol{B}_{\rm fic} = 0$. This is why a 90° pulse will not rotate an off-resonance magnetization component from the z-axis fully into the transverse plane.

the transmitter frequency $\omega_{\rm rf} = 2\pi\nu_{\rm rf}$ (Fig. 3.1.5). The effective field is the vector sum of two fields. One is the *offset field* or *fictive field* $\boldsymbol{B}_{\rm fic} = (\omega_0 - \omega_{\rm rf})/\gamma\, \boldsymbol{z}$, which points along the direction $\boldsymbol{z} = (0, 0, 1)^{\rm t}$ of the applied static magnetic field $\boldsymbol{B}_0$ and whose magnitude is determined by the offset of the resonance frequency ω_0 from transmitter frequency $\omega_{\rm rf}$. The other field is the rf field $\boldsymbol{B}_1$ along an axis in the transverse plane, for example, along $\boldsymbol{y} = (0, 1, 0)^{\rm t}$. Then the effective field is given by $\boldsymbol{B}_{\rm eff} = B_1\boldsymbol{y} + (\omega_0 - \omega_{\rm rf})/\gamma\, \boldsymbol{z}$. The angle θ by which the direction of this field is inclined to the z axis is given by $\tan\{\theta\} = (\gamma B_1)/(\omega_0 - \omega_{\rm rf})$. The larger the *frequency offset* $\omega_0 - \omega_{\rm rf}$, the more the pulse rotation axis deviates from the y axis. As a consequence, the refocusing pulses in a CPMG train are less and less able to invert the magnetization with increasing resonance offset. Then all but the first echo are sums of *Hahn echoes* and *stimulated echoes* (Fig. 3.2.3), because stimulated echoes are produced in general by three and more pulses when the flip angle of none of the pulses is 180° and *Hahn echoes* are produced in general with any two pulses unless the first pulse is a 180° pulse. Furthermore, the echo train decays with a decay time constant $T_{\rm 2eff}$, which is longer than T_2, because the effective relaxation is partially governed by T_1, and the first echo in a CPMG train measured in strongly inhomogeneous fields is smaller than the second echo. The latter consequence has to be taken into account when extrapolating the echo decay curve to zero time to estimate the signal amplitude corresponding to the spin density. The first consequence is less severe, when the first pulse and all subsequent pulses generating the echo train have the same width. This is why the 180° refocusing pulses in CPMG echo trains are preferably set by not doubling the width of the initial 90° excitation pulse but by doubling the amplitude.

Conventionally the time delays in a *CPMG sequence* are set in an approximation valid for *echo times* long compared to the *pulse widths*. In this limit the time $t_{\rm E}$ between two echoes is approximately equal to the time 2τ between two 180° pulses, and the delay between the first two pulses in a CPMG sequence is τ (Fig. 3.1.1b). But this is not exact, because the pulse widths are neglected in this approximation. For short echo times, the pulse durations have to be taken into account to maximize the echo amplitudes. Then, when setting the flip angle of the 180° pulse by doubling the width t_{90} of the 90° pulse, the delay between the first two pulses has to be shorter than τ by $2t_{90}/\pi$ [2]. Note that the spectrometer software may assume the validity of the long echo time limit and approximate $t_E \approx 2\tau$. This is why for critical applications the time lag between echoes should be checked on an *oscilloscope* and compared to the settings of the delays in the *pulse program*.

3.1.9 Data processing

The NMR data are obtained as complex numbers $s(t) = u(t) + iv(t)$. If properly adjusted, only the real channel $u(t)$ carries signal from a CPMG echo train, and apart from noise $v(t) = 0$. If signal is found in both channels, $u(t)$ and $v(t)$, $s(t)$ needs to be *phase corrected* by multiplication with $\exp\{-i\phi\}$ to rotate the complete signal into the real channel prior to further analysis. Alternatively, the magnitude of $s(t)$ can be evaluated with a constant offset which accounts for the magnitude of the noise signal. This procedure, however, should be used with caution because a systematic error is introduced into the analysis if the signal-to-noise ratio is low.

Correct signal amplitudes are obtained by fitting the experimental data with model functions (Tab. 3.1.2) so that the signal amplitude at zero time can be extrapolated (Fig. 3.1.4). In simple cases and often for soft matter, the *fit function* is a *mono-exponential function* or a *bi-exponential function*, the sum of two exponential functions (Fig. 3.1.6a). For rubber and rigid solids like amorphous polymers with a complex network of magnetic *dipole-dipole interactions* between the spins, a *Gauss function* is used to fit the observed signal at short observation time. For crystalline powders the *Abragam function* gives a good fit. Accurate decay envelopes at short time are observed with an FID following single-pulse excitation in an inhomogeneous field and with *Hahn echoes* or *solid echoes* (Fig. 3.1.4) in inhomogeneous fields. *Spin-lock effects* and *off-resonance effects* may lead to an apparent non-exponential decay in multi-echo trains like the CPMG echo train. The echo envelopes of such trains can often nicely be fitted with a *stretched exponential function*, which also applies to decays of signal in matter with a *distribution of relaxation times*. Stretching the

Table 3.1.2. Fit functions

Name	Expression	Shape
Gauss	$s(0)\exp\{-1/2(t/T_2)^2\}$	Fig. 3.1.6b for $b = 2$
Abragam	$s(0)\exp\{-1/2(t/T_2)^2\} \times \sin\{at\}/(at)$	
Exponential	$s(0)\exp\{-t/T_2\}$	Fig. 3.1.6b for $b = 1$
Stretched exponential	$s(0)\exp\{-(t/T_2)^b\}$	
Kohlrausch	$s(0)\exp\{-(1/b)(t/T_2)^b\}$	Fig. 3.1.6b for arbitrary b

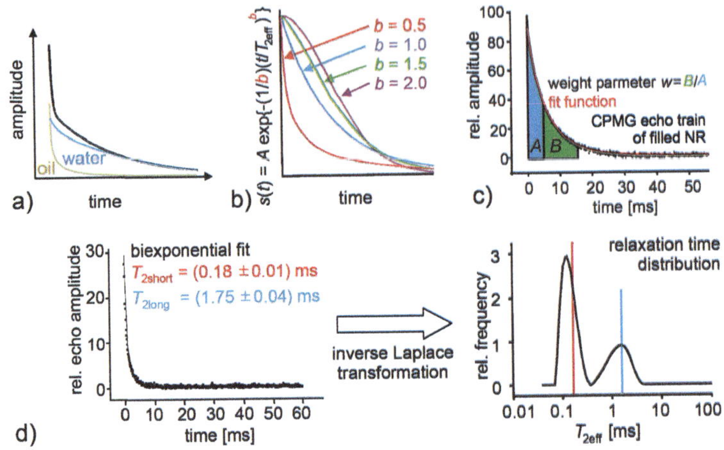

Figure 3.1.6. Analysis of transverse relaxation decays. (a) Decomposition of the relaxation signal of water and oil in terms of a sum of two fit functions. (b) Kohlrausch function for different exponential parameters b. $b = 1$ produces the exponential decay function and $b = 2$ the Gauss function. (c) Definition of the weight parameter w as the ratio of two definite integrals of the echo envelope. (d) A bi-exponential relaxation decay produces a bimodal distribution of relaxation times with the peak positions centered at the two relaxation times and the peak widths determined by the inversion algorithm. The inverse Laplace transform of the echo envelope is typically displayed on a logarithmic scale to give the distribution of relaxation times.

time axis is avoided by dividing the exponent by its power (Fig. 3.1.6b). This function was derived by Rudolf Kohlrausch to describe an accelerated rate process. His son Friedrich simplified this *Kohlrausch function* into the stretched exponential function. Both functions in many cases provide the same fit quality as a bi-exponential function but with one less fit parameter. The relaxation time T_2 extracted from a fit with the Kohlrausch function is close to the long relaxation time in a bi-exponential fit.

The transverse proton magnetization decay of multi-component systems like *semi-crystalline polymers* is well described by a sum of an Abragam function and two exponentials corresponding to the crystalline/amorphous interface and the amorphous domains (Fig. 3.1.4a). The relative amplitudes of the constituent functions determine the relative proton densities of the system components (Fig. 3.1.4b). To convert them to gravimetric material densities, the different densities of the morphological domains need to be known.

In comparative studies, relative signal amplitudes are needed with respect to the amplitude from a reference sample (eqn. (3.1.1)). Although the spectrometer electronics may be stable over months, it is advisable to scale signal amplitudes relative to the amplitude of a readily available *reference material*, for example, water. In many studies, *relaxation-weighted spin densities* suffice for comparison of material properties in place of absolute spin densities. Then extrapolation of the signal to zero time and fitting a model function to the experimental data may not be necessary.

Also other parameters can be extracted from the signal. For example, the total signal may be normalized to its initial value and integrated. When approximating the signal by a sum of exponential functions

$$s(t) = \sum_i s_i(0)\exp\{-t/T_{2\mathrm{eff},i}\} \tag{3.1.3}$$

the sum of all echoes from a multi-echo train is proportional to the integral of the envelope of the FID, which corresponds to a spin-density weighted *average relaxation time* $\langle T_{2\mathrm{eff}}\rangle$. After normalization of the signal $s(t)$,

$$\int_0^\infty s(t)/s(0)\mathrm{d}t = \sum_i x_i T_{2\mathrm{eff},i} = \langle T_{2\mathrm{eff}}\rangle, \tag{3.1.4}$$

where $x_i = s_i(0)/\sum_i s_i(0)$ is the spin-density fraction of nuclei of component i.

Alternatively, ratios and sums of partial integrals can be defined to produce suitable parameters that provide adequate contrast to differentiate variations in material properties. For example, the *weight parameter* w (Fig. 3.1.6c)

$$w = t_1/(t_2 - t_1)\int_{t_1}^{t_2} s(t)\mathrm{d}t \Big/ \int_0^{t_1} s(t)\mathrm{d}t \tag{3.1.5}$$

has been used to characterize morphology variations in semi-crystalline polymers. The integration limits in the definition of w need to be adjusted for maximum *contrast* $C_w = (w - w_{\mathrm{ref}})/w_{\mathrm{ref}}$ (eqn. (3.1.2)).

While an evaluation by a fit function relies on a proper choice of this function, an integration or integration by parts does not. The most general fit function is given by eqn. (3.1.3), which can be written in integral form as

$$s(t) = \int_0^\infty S(p)\exp\{-tp\}\mathrm{d}p, \tag{3.1.6}$$

identifying the signal $s(t)$ as the Laplace transform of the distribution $S(p)$ of *relaxation rates* $p = 1/T_{2\mathrm{eff}}$, so that $S(p)$ can be obtained from $s(t)$ by *inverse Laplace transformation*. Because the relaxation time is usually plotted on a logarithmic scale (Fig. 3.1.6d), the difference between $\ln\{1/T_{2\mathrm{eff}}\}$ and $\ln\{T_{2\mathrm{eff}}\}$ is just a change in sign. In the presence of measurement noise in the signal to be transformed, the inverse Laplace transformation needs to be regularized to avoid instabilities in the algorithm. This in some way implies that a fit function with a limited number of exponentials is being transformed.

The resultant *distribution of relaxation times* often shows two or more maxima. This distribution can be fitted with peak functions or integrated by parts to obtain concentrations of components that differ markedly in their relaxation times. In a way this procedure is equivalent to determining component amplitudes from fitting the experimental data with bi- or tri-exponential functions. But while different types of fit functions can easily be used, the original Laplace transform method only applies to simple *exponential relaxation* and needs to be generalized to the use of other basis functions for the analysis of solids and other matter containing components with inherent non-exponential relaxation. Independent of the nature of the relaxation process, the total integral of the Laplace transform equals the total signal amplitude. Sometimes signal amplitudes are calculated from integrals of Laplace transforms to make use of the noise filtering property, which is inherent to the regularization procedure of the particular algorithm. Nevertheless, accurate total amplitudes are also obtained only when extrapolating the measured signal decay to zero time. Otherwise the distribution of relaxation times is cut off at short relaxation times.

3.1.10 References

[1] Hedesiu C, Demco DE, Kleppinger R, Adams-Buda A, Blümich B, Remerie K, Litvinov VM. The effect of temperature and annealing on the phase composition, molecular mobility and the thickness of domains in high-density polyethylene. Polymer. 2007; 48: 763–777.

[2] Hürlimann MD. Optimization of timing in the Carr–Purcell–Meiboom–Gill sequence. Magn Reson Imag. 2001; 19: 375–378.

3.2 Relaxation and diffusion

3.2.1 Introduction

Relaxation denotes the return of a system from a non-equilibrium state to equilibrium. The characteristic time for this process is the *relaxation time*. NMR relaxation is an important source of information about *molecular motion*. This motion is slow in solids and fast in liquids, and consequently relaxation times depend on the timescale of the molecular motion, which modulates nuclear spin interactions like the magnetic *dipole-dipole coupling* of nuclear spins to electron spins, the dipole-dipole coupling of nuclear spins with each other, and the shielding of the spins from the applied field by the electrons orbiting in the molecule. In addition to the timescale, the relaxation times depend on the geometry of the molecular motion.

One distinguishes two basic relaxation types in NMR: the *longitudinal relaxation* or *spin-lattice relaxation* described by the relaxation time T_1 and the *transverse relaxation* or *spin-spin relaxation* described by the relaxation time T_2. Longitudinal relaxation denotes the return of magnetization components along the direction of the main magnetic field $\boldsymbol{B}_0$ (the z-axis) to their thermodynamic equilibrium values (Fig. 1.1.1). By convention, this direction is the z-direction. Transverse relaxation denotes the decay of magnetization orthogonal to the direction of the magnetic field. This magnetization is directly detected in an NMR experiment. Its decay in inhomogeneous fields is governed by different mechanisms, in particular rotational motion and *translational diffusion*, which can be separated with the help of Hahn echoes.

3.2.2 Objective

An important objective of NMR in materials science is the establishment of relationships between *NMR parameters* and *material properties*. Because most material properties change with the molecular dynamics, i.e. with the free volume on the molecular scale, NMR relaxation times are important quantities for materials characterization. It is difficult to establish accurate theoretical relationships between NMR relaxation times and material properties except in special cases, so that the NMR *relaxation times* T_1 and T_2 including the *diffusion coefficient* D and material properties are often correlated by means of data from reference samples. The most common pulse sequences for the determination of T_1, T_2 and D are treated below. T_1 is treated first because it needs to be known to set the waiting time between successive scans to $5\,T_1$. Relaxation data and diffusion data are evaluated by extracting relaxation times or diffusion coefficients and component amplitudes from fits of model functions and by inverse Laplace transformation.

3.2.3 Further reading

Kimmich R. Principles of Soft-Matter Dynamics. Dordrecht: Springer; 2012.

Callaghan PT. Translational Dynamics & Magnetic Resonance. Oxford: Oxford University Press; 2011.

Casanova F, Perlo J, Blümich B, editors. Single-Sided NMR. Berlin: Springer; 2011.

Price W. NMR Studies of Translational Motion: Principles and Applications. Cambridge: Cambridge University Press; 2009.

Song YQ. Novel Two-Dimensional NMR of Diffusion and Relaxation for Material Characterization. In: Stapf S, Han SI, editors. NMR Imaging in Chemical Engineering. Weinheim: Wiley-VCH; 2006.

Blümich B. NMR Imaging of Materials. Oxford: Clarendon Press; 2000.

Kimmich R. NMR Tomography. Diffusometry, Relaxometry. Berlin: Springer; 1997.

Callaghan PT. Principles of Nuclear Magnetic Resonance Microscopy. Oxford: Clarendon Press; 1991.

Slichter C. Principles of Magnetic Resonance. 3rd ed. Berlin: Springer; 1990.

Freeman R. A Handbook of Nuclear Magnetic Resonance. Harlow: Longman Scientific Technical; 1987.

3.2.4 Theory

Longitudinal relaxation refers to the build-up of *thermodynamic equilibrium magnetization* along the direction of the magnetic field. This process changes the *energy* of the spins making up the nuclear magnetization and involves an energy transfer between the spins and the surrounding electrons, which are called the lattice. The characteristic time for this energy transfer is the *spin-lattice relaxation time* or *longitudinal relaxation time* T_1. *Transverse magnetization* does not change the energy but the *entropy*, i.e. the order of the spins. In a perfectly homogeneous magnetic field, transverse magnetization follows the envelope of the FID. In their thermodynamic equilibrium before and directly after excitation with a single rf pulse, all magnetization components are aligned in the same direction and start to precess in the transverse *xy* plane, inducing a voltage in the NMR receiver coil (Fig. 1.1.2b). Yet, each component experiences different randomly fluctuating local fields, permanently changing their precession frequencies in different ways, so that the vector sum $\boldsymbol{M}$ of magnetization components decreases in magnitude and the coherence of precession between the spins is lost (Fig. 1.1.2c). The time constant of this decay is the *transverse relaxation time* or *spin-spin relaxation time* T_2.

Both relaxation times T_1 and T_2 are defined in their simplest form in the *Bloch equation*

$$\frac{\mathrm{d}\boldsymbol{M}(t)}{\mathrm{d}t} = \gamma \boldsymbol{M}(t) \times \boldsymbol{B} - \mathbf{R}[\boldsymbol{M}(t) - \boldsymbol{M}_0] \tag{3.2.1}$$

with the relaxation matrix

$$\mathbf{R} = \begin{bmatrix} 1/T_2 & 0 & 0 \\ 0 & 1/T_2 & 0 \\ 0 & 0 & 1/T_1 \end{bmatrix}. \tag{3.2.2}$$

Transverse and longitudinal relaxation processes occur simultaneously. They refer to orthogonal magnetization components. In bulk liquids T_2 is approximately equal to T_1 whereas in solids or porous media T_2 can become much smaller than T_1. For water in natural soil systems a T_1/T_2 ratio of 2 up to 5 is often found, in solid polymers this ratio can be two or three orders of magnitude.

In inhomogeneous fields, the destructive interference of magnetization components is enhanced and the impulse response decays faster with $T_2^* < T_2$. However, the impact of field inhomogeneity can be removed by forming *echoes* spaced apart by the *echo time* t_E (Fig. 1.1.2d). These echoes are attenuated only by T_2 relaxation unless the molecules can change position between places with different field strengths within the echo time. The molecules of fluids can indeed move around by translational Brownian motion, which is characterized by the *self-diffusion coefficient* D. If they do so in the pores of a porous medium, the transverse magnetization is further relaxed by collisions with the pore wall.

For a fluid in a pore with *surface-to-volume ratio* S/V the transverse relaxation rate measured with a CPMG sequence is given by

$$\frac{1}{T_2} = \frac{1}{T_{2\text{bulk}}} + \rho_2 \left(\frac{S}{V}\right)_{\text{pore}} + \frac{D(\gamma G t_{\text{E}})^2}{12}, \tag{3.2.3}$$

where $T_{2\text{bulk}}$ is the relaxation of the bulk fluid, ρ_2 the *surface relaxivity* of the pore wall, and G the gradient of the magnetic field $\boldsymbol{B}_0$ inside the pore. This equation is valid in the limit of fast diffusion where the molecules can diffuse across a pore many times within the echo time t_{E}. The gradient G inside the pore is caused either by distortions of the applied magnetic field $\boldsymbol{B}_0$ due to differences in magnetic susceptibility inside and outside the pore or by the inhomogeneity of the applied field. The equation demonstrates that the *relaxation time distribution* scales with the distribution of surface-to-volume ratios and maps the *pore size distribution* probed by fluids saturating porous media. The equation also demonstrates that the transverse relaxation rate is enhanced by translational *diffusion* of molecules in the gradient of the magnetic field. NMR with time-invariant or *constant field gradients* and with *pulsed field gradients* is a highly appreciated method for measuring translational *diffusion*.

The *longitudinal relaxation time* T_1 does not vary much for the protons in different chemical groups of liquid molecules. Its value is often in the range of a few seconds. In contrast to T_2, it is not affected by molecular diffusion in inhomogeneous magnetic fields, so that T_1 relaxometry is a standard tool for investigating such systems, and in the fast diffusion limit one finds in analogy to eqn. (3.2.3)

$$\frac{1}{T_1} = \frac{1}{T_{1\text{bulk}}} + \rho_1 \left(\frac{S}{V}\right)_{\text{pore}}. \tag{3.2.4}$$

3.2.5 Hardware

The hardware for measuring relaxation is the same as that for measuring spin density (Section 3.1.5). Because there are no stringent requirements to field homogeneity, conceptually simple *closed magnets* as well as *open magnets* can be employed (Fig. 1.2.1). For example, a low-field desktop spectrometer like the Bruker Minispec can be used, which in its standard configuration is connected to a permanent magnet that operates at a fixed, elevated temperature to eliminate magnetic field drift from temperature variations (Fig. 3.2.1a). For diffusion measurements, the hardware should include a unit that generates *pulsed gradient fields* unless the measurement is conducted in a magnetic stray field with an inherent gradient. Portable stray-field sensors like the NMR-MOUSE are convenient tools to measure relaxation and diffusion (Fig. 3.2.1b). When positioned on a step-motor-operated sled, *depth profiles* of relaxation and diffusion parameters can be measured in automatic mode. The NMR-MOUSE is not temperature stabilized. Due to the high gradient of the sensor, a temperature change of the magnet results in a slight shift of the position of the sensitive volume. Because both relaxation

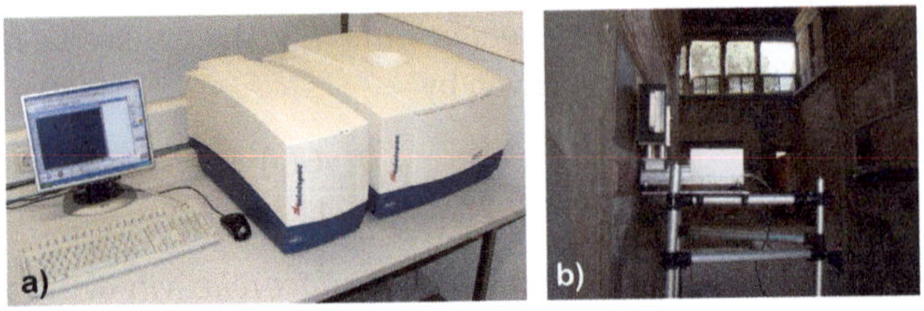

Figure 3.2.1. Low-field NMR equipment for measuring relaxation and diffusion. (a) Bruker Minispec with a temperature stabilized 1 T magnet (right). (b) NMR-MOUSE set up on a positioning sled in Herculaneum to measure a depth profile in a painted wall.

times and diffusion coefficients depend on the temperature of the object, the sample temperature should be recorded along with the NMR data. This applies in particular to relaxation times of soft materials like *rubber* and diffusion coefficients of fluids.

3.2.6 Longitudinal relaxation

Theory

Longitudinal magnetization is also called *polarization*. Longitudinal relaxation is the restoration of the *thermodynamic equilibrium* state after, for example, changing the magnitude or the direction of the magnetic field, the magnitude of the nuclear magnetic polarization, or the direction of the magnetization. The magnetic field, which the sample or the object experiences, starts to change the nuclear polarization as soon as the sample is placed inside the magnet or the NMR-MOUSE is placed near the object. The polarization can be attenuated by *saturation pulses* or enhanced by expert methods of *hyperpolarization*, which are not discussed in this book. The direction of the magnetization is altered from its equilibrium direction along the axis of the magnetic field by rf pulses. The value of the thermodynamic equilibrium polarization follows the *Boltzmann distribution* (1.1.1). At room temperature, the high temperature approximation is valid, and the polarization is directly proportional to the magnitude of the static magnetic field $\boldsymbol{B}_0$. This is one reason why high-field NMR spectroscopy is employed to analyze solutions at low concentration.

The functional dependence of the longitudinal magnetization $M_z(t)$ as it approaches its equilibrium value M_0 can be obtained by solving the *Bloch equation* (3.2.1) starting with the particular initial conditions defined by the experiment. There are two standard experiments to measure longitudinal relaxation. In the *saturation recovery* experiment, magnetization recovery starts from zero (Fig. 3.2.2a), and in the *inversion recovery* experiment, magnetization recovery starts from negative equilibrium magnetization (Fig. 3.2.2b). Both experiments lead to a relaxation function of the form

$$M_z(t_0) = M_0\,(1 - f\,\exp\{-t_0/T_1\}) \tag{3.2.5}$$

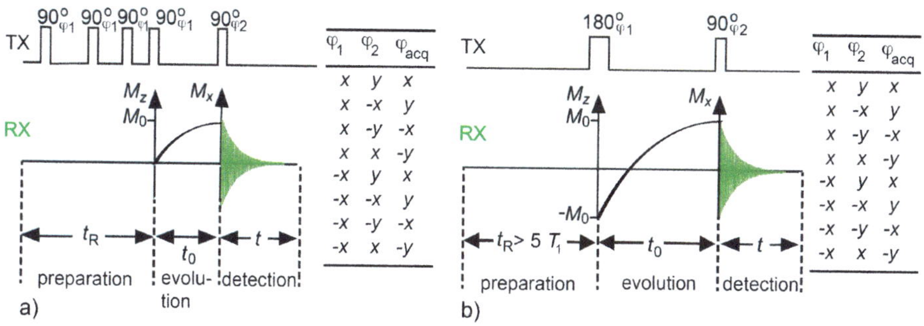

Figure 3.2.2. Pulse sequences for measuring longitudinal relaxation curves and phase cycles. TX denotes transmitter and RX receiver. (a) Saturation recovery sequence. (b) Inversion recovery sequence.

for a simple fluid like water. The factor f depends on the type of experiment performed. For the saturation recovery experiment $f = 1$, and for the inversion recovery experiment $f = 2$. The value of T_1 is governed by interactions of the nuclear spins among each other and with the spins of the surrounding electrons in the molecules. These interactions are modulated by molecular motion, that is, by the translational and rotational motions of the molecules. Due to the interaction between nuclear and electron spins, T_1 relaxation is accelerated by the presence of paramagnetic centers, e.g. dissolved ions with unpaired electrons like Cu^{2+} and paramagnetic centers at interfaces in porous media. This is why *copper sulfate* is added to water to reduce T_1 and why *relaxation agents* are used to generate contrast in medical magnetic resonance imaging.

Pulse sequences and parameters

T_1 relaxation concerns the change of *longitudinal magnetization* M_z. Because M_z cannot be measured directly by NMR, it must be converted to observable *transverse magnetization* M_x or M_y with a 90° pulse. The initial amplitude of the transverse magnetization then corresponds to the value of the longitudinal magnetization immediately before application of the pulse. The transverse magnetization is observed either by recording an *FID* in a homogeneous magnetic field or by recording a *CPMG echo train* in an inhomogeneous field.

Any experiment to measure longitudinal relaxation starts by moving the longitudinal magnetization away from its equilibrium value. Two obvious non-equilibrium states are (a) zero longitudinal magnetization corresponding to $M_z(0) = 0$ and (b) negative equilibrium magnetization corresponding to $M_z(0) = -M_0$. Consequently, there are two standard methods to measure longitudinal relaxation. With the saturation recovery method (Fig. 3.2.2a), the thermodynamic equilibrium magnetization is first destroyed by a series of about five 90° pulses with logarithmically decreasing time delays between them, and with the inversion recovery method (Fig. 3.2.2b), the

Table 3.2.1. Default acquisition parameters for measuring T_1 relaxation with the PM25 NMR-MOUSE for water saturated rock using a 10 mm spacer

Parameter	Saturation recovery	Inversion recovery
transmitter frequency ν_{rf}	13.8 MHz	13.8 MHz
transmitter attenuation for 90° pulse	−6 dB at 300 W	−6 dB at 300 W
duration t_p of 90° pulse	13 μs	13 μs
dwell time Δt	0.5 μs	0.5 μs
acquisition time t_{acq}	8 μs	8 μs
recovery time t_0	0.02–0.8 s	0.02–0.8 s
echo time t_E	80 μs	80 μs
number n_E of echoes	10	10
recycle delay t_R	0 s	1.5 s
number n_s of scans	16	16

thermodynamic equilibrium magnetization is first inverted by a 180° pulse. Either experiment is repeated for different recovery times t_0 through a range $0 < t_0 < 5\,T_1$, and the initial amplitude of the recorded signal is plotted as a function of t_0 to obtain the longitudinal relaxation curve (Fig. 3.2.2). With the *saturation recovery* method the relaxation curve scales only half the dynamic range of that measured with the *inversion recovery* method, but the measurement is faster and simpler, because no recycle delays are needed between subsequent scans. On the other hand, inversion recovery has twice the dynamic range, but takes more time, because complete recovery of the longitudinal magnetization between scans needs to be assured with an estimated recycle delay t_R sufficiently long to allow the recovery of all magnetization components, the T_1 of which is unknown at the time or the measurement.

Parameters suitable for measuring the relaxation of water magnetization in *porous rock* with an NMR-MOUSE are summarized in Tab. 3.2.1. When measuring soft matter like *rubber* or *skin cream*, the parameters for detection of an FID or a CPMG train (Fig. 3.1.1) are collected in Tab. 2.7.1 and can be used in the detection periods of the saturation recovery and inversion recovery pulse sequences.

Measurements

Closed magnets typically have a magnetic field sufficiently homogeneous that the entire sample volume in the rf coil can be excited with the rf pulses. Then it is sufficient to detect the *FID* in the inversion recovery or saturation recovery experiments as depicted in Fig. 3.2.2. If the detection field $\boldsymbol{B}_0$ is highly inhomogeneous such as in stray field experiments, a *Hahn echo* or *CPMG sequence* are required for detection (Fig. 3.1.1b). Because the saturation recovery method is less prone to errors, its use is preferred over that of the inversion recovery sequence. To minimize systematic errors in the determination of relaxation parameters, the data need to be acquired in phase-sensitive mode. Common issues encountered when measuring *longitudinal relaxation* curves are those also encountered when measuring *spin density*. They are summarized in Tab. 3.1.1 above.

Data processing

In simple fluids and solids, T_1 relaxation curves are usually mono-exponential, and according to eqn. (3.2.5) $1/T_1$ is the slope of the line obtained when plotting $\ln\{f/(1 - M_z(t_0)/M_0)\}$ versus t_0. Alternatively, T_1 can be obtained by fitting eqn. (3.2.5) to the experimental data. In either case, the data need to be available in phase-sensitive mode. If signal is observed in the real and the imaginary data channels, the experimental data set $u(t) + \mathrm{i}\,v(t)$ needs to be phase corrected by multiplication with $\exp\{-\mathrm{i}\,\phi\}$ with the angle ϕ suitably chosen to rotate the complete signal into the real channel. If the data were acquired with the inversion recovery method, T_1 can be estimated from the time t_{zero} where the recovery curve crosses the zero line: $T_1 = t_{\mathrm{zero}}/\ln\{2\}$. For heterogeneous samples, the relaxation curves are often multi-exponential. Instead of fitting a weighted sum of relaxation terms (3.2.5) in this case, distributions of relaxation times can be derived from the recorded signal by inverse *Laplace transformation* using the inversion kernel defined in eqn. (3.2.5).

3.2.7 Transverse relaxation

Theory

Transverse relaxation denotes the decay of the transverse magnetization, for example, $M_x(t)$. According to the *Bloch equation* (3.2.1) this decay is mono-exponential,

$$M_x(t) = M_x(0)\exp\{-t/T_2\}. \qquad (3.2.6)$$

Here T_2 is the *transverse relaxation time*. It is the characteristic decay time of the impulse response (Fig. 3.1.1a) and specifies the *line width* $\Delta\nu_{1/2}$ in NMR spectra of liquids according to

$$\Delta\nu_{1/2} = 1/(\pi T_2). \qquad (3.2.7)$$

The longer the FID lasts, the longer T_2 is and the narrower the line in the NMR *spectrum*.

The decay of the transverse magnetization arises from destructive interference of magnetization components, which precess with different frequencies around the magnetic field (Fig. 1.1.2b). These different frequencies result from different chemical environments, which give rise to the chemical shift liquids, from an inhomogeneous magnetic field, which varies across the sample volume, and from random translation and rotation of molecules and chemical groups, which modulate anisotropic spin interactions. The destructive interference of transverse magnetization components due to random motion is irreversible while the destructive interference caused by the other two mechanisms is reversible. This interference is undone in the center of a *Hahn echo* (Fig. 3.2.3a) and in the centers of *CPMG echoes* (Fig. 3.1.1b). The decay of the echo envelope, therefore, is caused only by the irreversible interference from random molecular motion. The associated *relaxation rate* is $1/T_2$. The relaxation rate associated with the signal decay due to both reversible and irreversible decay is $1/T_2^*$ where $T_2 > T_2^*$ (Fig. 3.1.1b). The goal of relaxation measurements is to inquire about the motions that lead to the irreversible decay, so that relaxation measurements are done with Hahn echoes and with CPMG echo trains because in the echo maxima the reversible decay is undone.

The destructive interference of magnetization components precessing with constant frequency is reversed to constructive interference by a 180° pulse. The constructive interference builds up an *echo*, which was discovered by Erwin Hahn [1]. In the echo maximum, constructive interference changes to destructive interference, and the echo decays in the same manner as the FID. The formation of echoes can be repeated many times as discovered by Carr and Purcell [2]. Meiboom and Gill found that the echo train is less affected by pulse flip angle imperfections if the initial 90° pulse and the subsequent 180° pulses are 90° out of phase [3]. The resultant *CPMG sequence* is the standard sequence for rapid measurement of transverse relaxation decays (Fig. 3.1.1b).

The formation of the *Hahn echo* is explained in Fig. 3.2.3. Following the initial 90° pulse, the magnetization components constituting the total nuclear magnetization precess with different frequencies, so that the vector sum of all magnetization components eventually approaches zero (Fig. 3.2.3a,b). The 180° pulse of the Hahn echo flips the transverse magnetization components around the axis of the rf magnetic field $\boldsymbol{B}_1$. This introduces a 180° phase jump in the transverse magnetization components orthogonal to $\boldsymbol{B}_1$ and puts the fast magnetization components behind the slow ones (Fig. 3.2.3b). After the same time $t_{E/2}$ it took the magnetization components to move apart, they refocus in the echo maximum by constructive interference.

Such an echo can also be generated when dividing the 180° pulse into two 90° pulses separated by a delay t_m (Fig. 3.2.3c). The second 90° pulse converts half of the *transverse magnetization* into *longitudinal magnetization* which is then converted back to transverse magnetization by the third 90° pulse. A so-called *stimulated echo* or *indirect echo* is then observed after the third pulse and a *direct echo* of *Hahn echo* after the second pulse. Both echoes ideally have half the amplitude of the maximum Hahn echo generated by the 180° pulse as the second pulse (Fig. 3.2.3a). During the time t_m between the second and the third pulses of the stimulated echo sequence, half of the magnetization is stored as longitudinal magnetization and relaxes with T_1, which is often appreciably longer than T_2, so that stimulated echoes can be observed after much longer times than Hahn echoes.

The stimulated echo finds many applications in NMR, for example, for measuring diffusion and analyzing rate processes by two-dimensional Fourier and Laplace exchange NMR. In these scenarios the time t_m serves to extend the timescale to allow longitudinal magnetization components originating from transverse magnetization components with particular resonance frequencies or with particular transverse relaxation times to exchange by various dynamic processes to become associated with transverse magnetization components with other resonance frequencies or other relaxation times. This is why t_m is called the *mixing time* or *mixing period*.

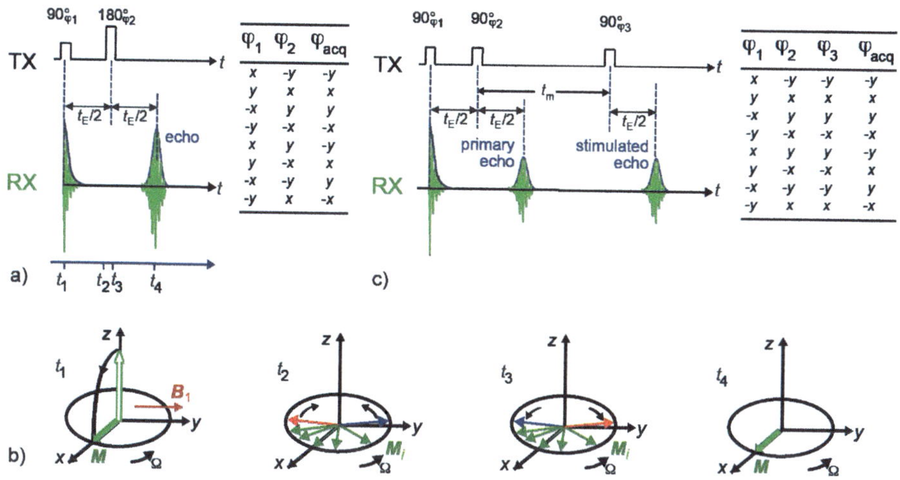

Figure 3.2.3. Basic echo pulse sequences. TX denotes transmitter and RX denotes receiver. (a) Hahn echo pulse sequence and complete phase cycle. (b) Precession of magnetization components relative to the transmitter frequency ω_{rf}. Schematic snapshots at the times t_1 to t_4 indicated in (a). A 90°_y pulse rotates the magnetization from the magnetic field direction z along the x direction. Each magnetization component M_i precesses at a different frequency Ω_i. A 180°_x pulse rotates all magnetization components around the x-axis. The magnetization components exchange positions and maintain their sense of precession. All components align again along the x-axis at time $t_4 = t_E$ and form the Hahn echo. (c) Stimulated echo pulse sequence and phase cycle. During the mixing time t_m there is non-equilibrium longitudinal magnetization, which relaxes with T_1.

When a *CPMG echo train* is generated in the strongly inhomogeneous fields of stray-field sensors, the infinite range of resonance offsets $\Omega = \omega_0 - \omega_{rf}$ leads to a wide distribution of effective rf pulse flip angles (Fig. 3.1.5). The proper flip angles of 90° and 180° are only fulfilled at exact resonance for $\omega_0 = \omega_{rf}$. At some *resonance offset* Ω the nominal 180° pulse becomes an effective 90° pulse so that the first three pulses of a CPMG sequence generate a *stimulated echo*. In consequence, the echoes observed with a CPMG sequence in a strongly inhomogeneous magnetic field are sums of direct Hahn echoes and indirect stimulated echoes, and the echo train decays with an effective relaxation time $T_{2\text{eff}} > T_2$ [4, 5]. Furthermore, the first echo is always a direct, Hahn echo while the second one is the sum of a direct and an indirect, stimulated echo, so that the first echo is smaller than the second, a fact that needs to be taken into account when evaluating CPMG echo trains by fitting model functions [6].

Pulse sequences and parameters

The standard pulse sequences for measuring transverse magnetization decays are the *Hahn echo sequence* (Fig. 3.2.3a) and the *CPMG sequence* (Fig. 3.1.1b). Several echoes need to be measured to sample the complete magnetization decay. Because the sig-

Table 3.2.2. Default echo acquisition parameters for the PM25 NMR-MOUSE for water saturated rock

Parameter	Hahn echo	CPMG sequence
transmitter frequency ν_{rf}	13.8 MHz	13.8 MHz
transmitter attenuation for 90° pulse	−6 dB at 300 W	−6 dB at 300 W
duration t_p of the 90° pulse	13 μs	13 μs
dwell time Δt	1 μs	1 μs
echo time t_E	120 μs–50 ms	120 μs
number n_E of echoes	1	2000
number n_s of scans	128	128
recycle delay $t_R = 5\,T_1$	2.5 s	2.5 s

nal initially decays quickly and then more slowly at later times, the echoes can be spaced on a logarithmic timescale, i.e. the echo time can be doubled from one echo to the next. This can readily be achieved with the Hahn echo sequence, when independent measurements are done for each echo. With the CPMG sequence, all echoes are measured in a single shot, but at constant intervals t_E. This is why the CPMG sequence is usually preferred over the Hahn echo sequence when measuring transverse relaxation decays. Moreover, the impact of translational *diffusion* can be ameliorated when choosing short echo times in the CPMG sequence. This is particularly important when observing relaxation in heterogeneous media such as fluids in porous *rock*, suspensions and emulsions where the applied magnetic field is locally distorted by differences in magnetic susceptibility. These give rise to inhomogeneous fields in the pores, which are approximately described by internal field gradients G. In combination with *diffusion*, these contribute to the attenuation of the transverse magnetization (eqn. (3.2.3)).

When measuring rapidly relaxing magnetization, short echo times need to be employed. Whereas the *echo time* t_E is the separation between echoes, the pulse program often calls the time delay 2τ between the 180° pulses the echo time, neglecting the pulse width (Fig. 3.1.1b). Note that correctly $t_E = 2\tau + t_{180}$, where t_{180} is the duration of the 180° pulse. Typical starting values for measuring transverse relaxation in water-saturated rock are listed in Tab. 3.2.2.

Beginner's level measurements

To begin, the general instrument parameters such as transmitter frequency, frequency offset, receiver gain, receiver phase, and 90° pulse length are set as described in Chapter 2. Then the appropriate pulse sequence program must be loaded: CPMG or Hahn echo. The further settings of the spectrometer are the same as for measuring *spin density* (Tab. 2.7.1). But in contrast to spin-density measurements, the complete T_2 decay must be recorded by choosing a sufficiently high *number of echoes*. This number can be a few thousand for liquid samples, and the high number of rf pulses passing through the rf coil of the probe may lead to significant heating of the coil. One remedy is to increase the echo time t_E, but this may increase the influence of diffusion on T_2 when liquids are measured, and one should compare the results from different parameter settings. The other remedy is to increase the recycle delay.

The *recycle delay* t_R should be set to about 5 T_1. T_1 can be long for samples containing liquids, e.g. $T_1 \approx 3$ s or more for pure water. T_1 should be determined first (Section 3.2.6) if not known. A fast alternative to determining T_1 is to measure a single Hahn echo with a few scans only, for example, $n_s = 4$, and repeat the measurement with increasing values of the recycle delay. When t_R is sufficiently long the echo intensity approaches a constant value. Common issues encountered in relaxation measurements are summarized in Tab. 3.1.1.

Advanced level measurements

For decades, NMR relaxation curves were measured with the intention to analyze them in terms of relaxation models or simply to extract relaxation times and component amplitudes (Section 3.1.9). Driven by the needs of the oil industry, relaxation curves are being analyzed more and more in terms of *relaxation time distributions*, which are formally obtained by inverse *Laplace transformation* of the relaxation curve (eqn. (3.1.6)). The Laplace transformation expresses the signal in terms of a sum of exponential functions in a way similar to the *Fourier transformation*, which expresses the signal in terms of a sum of harmonic functions. In contrast to harmonic functions, different exponential functions are not orthogonal to each other, so that the inverse Laplace transformation is inherently unstable and not unique. Nevertheless, the algorithms produce useful results, which provide novel insights, in particular into the pore space of fluid-filled solids and soft matter like *rock*, *food* and *biological tissue*.

In analogy to *multi dimensional Fourier NMR* spectroscopy methods, which are of outstanding importance in modern molecular analysis, *multi-dimensional Laplace NMR* methods were developed once an efficient inverse *2D Laplace transformation* algorithm had become available [7]. The interest in these schemes derives from the fact that *Laplace NMR* can be performed even in strongly inhomogeneous fields [8]. From the great variety of 2D Laplace methods known by now, the two most important ones which address relaxation are the *T_1-T_2 correlation experiment* and the *T_2-T_2 exchange experiment*.

In the T_1-T_2 correlation experiment transverse relaxation is detected with a CPMG sequence applied to partially relaxed longitudinal magnetization. According to the two ways of preparing partially relaxed longitudinal relaxation (Fig. 3.2.2), inversion recovery and saturation recovery, there are two versions of the T_1-T_2 correlation experiment. The saturation recovery version (Fig. 3.2.4a) is easier to set up. The pulse sequence is repeated for all values of the longitudinal magnetization recovery period, which is now called the *evolution period* t_1, and the associated CPMG echo decays are measured as a function of the *detection period* t_2 and stored row by row in matrix form.

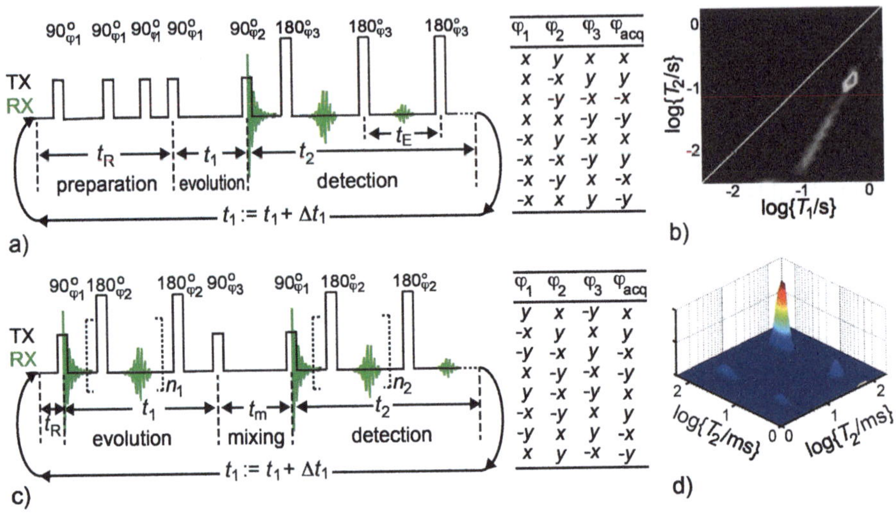

Figure 3.2.4. 2D Laplace NMR of distributions of relaxation times. The experiments are cycled through a range of t_1 values. (a) T_1-T_2 correlation pulse sequence and phase cycle. T_1 is interrogated with a saturation recovery sequence and detected with a CPMG sequence. (b) T_1-T_2 correlation map of water saturated Allermoehe sandstone with a permeability of 1.85 mD and a porosity = 6%, measured with an echo time of $t_E = 0.15$ ms in a 0.22 T Halbach magnet (adapted from [9] with permission). (c) Pulse sequence and phase cycle for T_2-T_2 exchange NMR. T_2 is probed with CPMG sequences in the evolution and detection periods by varying n_1 and n_2 independently. Both periods are separated by a mixing period t_m during which the transverse magnetization encoded in the evolution period t_1 is stored as longitudinal magnetization. (d) T_2-T_2 exchange map for water-saturated, spherical silica particles [10]. Cross peaks arise from protons diffusing back and forth between two different relaxation environments.

This matrix is then inverted by an inverse 2D Laplace transformation into a *2D distribution of relaxation times*. To invert the data the correct kernels of the Laplace transformation applicable to inversion or saturation recovery and transverse magnetization decay need to be chosen. The resultant 2D relaxation time distribution shows a ridge in the 2D plane of relaxation times T_1 and T_2 (Fig. 3.2.4b).

For water-saturated sandstone one often finds ridges parallel to the diagonal, which indicate a constant T_1/T_2 ratio for all *pore sizes*, suggesting that the walls of the small and large pores are of the same kind with the same *surface relaxivities* ρ_1 and ρ_2 defined in eqns. (3.2.3) and (3.2.4). For an Allermöhe sandstone sample with low porosity and low permeability this ridge is not parallel to the diagonal (Fig. 3.2.4b) when measured with an echo time of 150 μs, and the smaller T_2 is, the larger the T_1/T_2 ratio. Because small relaxation times refer to small pores, this observation suggests that smaller pores give rise to larger field inhomogeneity and consequently to larger gradients due to differences in magnetic susceptibility of pore water and rock matrix, and that the faster signal decay for small pores is the result of diffusion during the echo time t_E of the CPMG sequence. The validity of this explanation was proven by measuring the same sample again but with a shorter echo time of 60 μs and observing a ridge parallel to the diagonal. Thus T_1-T_2 correlation experiments are suitable tools to probe *local field gradients* in fluid-filled porous media [9].

The *relaxation exchange* experiment (Fig. 3.2.4c) probes the pore space geometry by observing the occurrence of cross peaks between relaxation peaks on the diagonal at different relaxation time coordinates along the two axes of the 2D map (Fig. 3.2.4d). The sequence consists of CPMG sequences for both the *evolution period* t_1 and the *detection period* t_2, which are separated by a *mixing period* t_m during which longitudinal magnetization components can exchange. In fluid-filled porous media, this exchange is accomplished by the migration of protons from one relaxation center to another. Unless the fluid experiences a pressure gradient this migration usually proceeds by translational *diffusion*. The T_2-T_2 exchange experiment therefore is capable of probing translational diffusion in a homogeneous magnetic field, as long as the exchange kinetics that give rise to the cross peaks can be analyzed and the distances between relaxation centers are known. Vice versa, if the *diffusion coefficient* is known, distance constraints between relaxation centers can be derived from T_2-T_2 exchange NMR experiments. The mixing time t_m sets the time for the molecules to exchange sites. It is limited by T_1. But in liquids, T_1 may not be much larger than T_2, so that the maximum durations of the evolution period t_1 and the detection period t_2 are of the same order as the duration of the mixing period t_m. In consequence, the site exchange already proceeds during the evolution and detection periods, and quantitative kinetic parameters can be derived from such exchange maps only by modeling the experimental exchange maps.

Data processing

Transverse relaxation decays are processed in the way described in Section 3.1.9. Typically model functions are fitted to extract relevant parameters like component *amplitudes* and *relaxation times* from the relaxation signals of heterogeneous samples. An alternative to fitting the data with a *model function* is to compute the inverse Laplace transform to arrive at a *distribution of relaxation times*. The peak areas in such a distribution can be integrated to obtain the concentrations of different components that contribute to the recorded NMR signal with different relaxation times.

Two-dimensional distributions of relaxation times and in particular relaxation exchange maps need to be modeled for quantitative analysis, because dynamic phenomena such as the exchange of magnetization between relaxation sites i and j with *exchange rate* $k_{ij} = 1/\tau_{ij}$ can impact the apparent relaxation rate $1/T_2$. By collecting the longitudinal or the transverse magnetization components from different relaxation sites in a vector $\boldsymbol{M}$, the magnetization vector at time t can be calculated with the relaxation matrix $\mathbf{R}$ and the matrix $\mathbf{K}$ of exchange rates given the magnetization vector at time t_0 [10],

$$\boldsymbol{M}(t) - \boldsymbol{M}_0 = \exp\{-(\mathbf{R} + \mathbf{K})(t - t_0)\}[\boldsymbol{M}(t_0) - \boldsymbol{M}_0]. \qquad (3.2.8)$$

With the help of this formula, the magnetization can be calculated at any moment in time for a relaxation experiment involving n relaxation sites, where n is the dimension of the vector $\boldsymbol{M}$ and of the square matrices $\mathbf{R}$ and $\mathbf{K}$, and the evolution of magnetization giving rise to the relaxation exchange map of Fig. 3.2.4d can be modeled.

3.2.8 Translational diffusion

Theory

The thermally activated random movement of the molecules from one point in space to another under the impact of intermolecular collisions is called *translational diffusion*. It was discovered by Robert Brown who studied the random movement of pollen in a water drop under a microscope; hence it is also called *Brownian motion*. The *self-diffusion coefficient* D is the mathematical constant to quantify Brownian motion at thermal and chemical equilibrium. Einstein and Smoluchowski showed that the motion of one particle can be describe as a statistical fluctuation with mean square displacement

$$\langle R^2(\Delta)\rangle = 2D\Delta \tag{3.2.9}$$

for motion along one spatial dimensional during time Δ, where D is the coefficient of translational diffusion. For random motion in three dimensions, $\langle R^2(\Delta)\rangle = 6D\Delta$. According to Stokes and Einstein, the diffusion coefficient D relates to viscosity η following

$$D = k_B T/(6\pi\eta R), \tag{3.2.10}$$

where is k_B the Boltzmann constant and T the temperature.

Diffusion in a concentration gradient is called *inter-diffusion*; diffusion solely due to Brownian motion is called *self-diffusion*. NMR and light scattering are common methods to study self-diffusion, whereby NMR can also be applied to optically opaque media. In bulk fluids and gases the diffusion is free, and the diffusion coefficient is independent of the observation time Δ in eqn. (3.2.9). If, however, the *diffusion length* $l_d = (\langle R^2(\Delta)\rangle)^{1/2}$ is restricted by the size of a pore bearing the fluid or gas, then an effective diffusion coefficient $D_{eff}(\Delta)$ is observed which depends on the diffusion time Δ. This situation typically applies to fluids in porous media, and in the short diffusion time limit, the effective diffusion coefficient depends on the inverse pore radius or the surface-to-volume ratio S/V,

$$\lim_{\Delta\to 0} \frac{D_{eff}(\Delta)}{D_0} = 1 - \frac{4\sqrt{D_0\Delta}}{9\sqrt{\pi}}\frac{S}{V}. \tag{3.2.11}$$

The displacement of spins by random diffusion but also by coherent flow is probed in NMR with inhomogeneous magnetic fields. The most common approach parallels the one followed in *magnetic resonance imaging* (Section 3.3), where the magnetic field usually depends linearly on space so that it can be described as the sum of a constant, homogeneous field B_0, which by convention points along the z-direction, and a *gradient field* ***Gr*** which points in the same direction but varies linearly in space in different directions (Fig. 3.3.1, eqn. (3.3.1)), for example along the x-direction, so that the magnetic field $G_x x$ is added to the homogeneous field, where G_x is the *field gradient* and x is the space coordinate which specifies the position of the spin. With the NMR frequency being proportional to the field strength (eqn. (1.1.2)) spins at different positions x experience different resonance frequencies, and a spin that diffuses along the gradient direction changes its resonance frequency as it moves from one position $x = r$ to the next. The precession angle (Fig. 3.3.2a) associated with this change in resonance frequency is analyzed to derive the self-diffusion coefficient.

Pulse sequences and parameters

The principle of measuring self-diffusion by NMR is to measure the mean square displacement $\langle R^2 \rangle$ that magnetization components experience in a given diffusion time Δ. To this end, initial and final positions of the spins are identified following the principles of magnetic resonance imaging (Section 3.3) by measuring their initial and final resonance frequencies ω_i and ω_f, respectively, in a magnetic gradient field with field gradient $G = G_x, G_y$, or G_z. Suitable pulse sequences, therefore, must contain at least two time intervals, one where the initial positions of the spins are tagged and one where the final positions are read out. Given that the measurement requires inhomogeneous magnetic fields, these pulse sequences are echo sequences, in particular the *Hahn echo* (Fig. 3.2.3a) with the *CPMG echo train* as its extension and the *stimulated echo* (Fig. 3.2.3c).

The gradients G are active in the first and the second half of the echo time t_E in the Hahn echo (Fig. 3.2.5a) and the stimulated echo (Fig. 3.2.5b). The 180° pulse separating both halves has the effect of inverting the sign of the gradient field before the pulse, so that the effective gradient G_{eff} changes sign. Consequently initial and final positions are marked with different signs, and the position difference, that is, the diffusion length covered during the diffusion time $\Delta = t_m$ determines the signal decay. Depending on how the field gradient is generated, whether in a time-invariant fashion with the NMR-MOUSE (Fig. 3.2.5c,d) or in a pulsed fashion (Fig. 3.2.5e,f) and depending on the pulse sequence used (Fig. 3.2.5a,b), the measured signal depends differently on the diffusion coefficient D and on the other pulse sequence parameters (Fig. 3.2.5). Irrespective of the pulse sequence, diffusion is a random process so that different spins experience different diffusion lengths. For free diffusion, the average precession angles $(\omega_f - \omega_i)\Delta$ show a Gaussian distribution with zero mean, so that the NMR signal is attenuated by diffusion similar to signal attenuation by transverse relaxation. When confined to pores of regular size, the Gaussian approximation does not apply and the magnetization shows diffraction effects, which relate to the uniform pore size.

Experiments with constant gradients are easy to perform with *stray-field NMR*, but the sensitivity is low, because the rf pulses are applied in the presence of the gradient

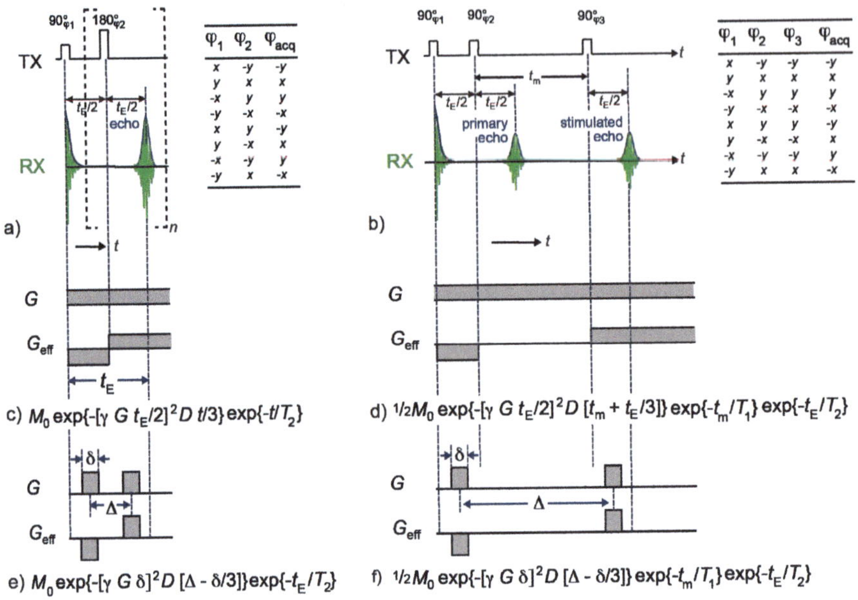

Figure 3.2.5. Pulse sequences, phase cycles and signal attenuation factors for measuring molecular translational diffusion. (a) Hahn echo sequence ($n = 1$) and CPMG sequence ($n > 1$). (b) Stimulated echo sequence. Each can be used together with time invariant field gradients (c, d) and with pulsed field gradients (e, f). The 180° pulse has the effect of inverting the sign of all applied gradients G prior to the pulse, so that the effective gradients G_{eff} form anti-phase pairs of equal area but opposite sign. In either case the echo amplitude is recorded. Note that (b) refers to both the Hahn echo where the signal is recorded as a function of the echo time t_{E} and the CPMG echo train where the signal is recorded as a function of the echo envelope decay time $t = n_{\text{E}}t_{\text{E}}$. In the stimulated echo experiment with time-invariant field gradients t_{m} corresponds to the diffusion time Δ.

so that only a slice is excited from the whole sample, while with *pulsed field gradients* in a homogeneous field the entire sample volume is excited and the *signal-to-noise ratio* is higher. Neglecting relaxation the echo amplitude of the stimulated echo is half the amplitude of the direct Hahn echo, so that the Hahn echo sequence is used when the signal-to-noise ratio is low. On the other hand, the diffusion time of the Hahn echo is limited by the *echo time* t_{E}, while it can be extended by the *mixing time* t_{m} with the stimulated echo, so that slower diffusion is probed with the stimulated echo sequence. Typical parameters applicable to the stimulated echo sequence and an NMR-MOUSE with a time invariant field gradient (Fig. 3.2.5a,c) are summarized in Tab. 3.2.3. The diffusion-encoded echo is detected with a CPMG train at short echo time $t_{\text{E}} = 50\,\mu\text{s}$ in an inhomogeneous field or with an FID in a homogeneous field. Parameters for diffusion measurement with pulsed field gradients in homogeneous field using the stimulated echo sequence (Fig. 3.2.5b,f) are listed in Tab. 4.1.2 in the context of *DOSY NMR*. In a diffusion experiment, either the time $t_{\text{E}}/2$ during which the gradients affect the transverse magnetization in a constant gradient experiment, or the gradient ampli-

Table 3.2.3. Parameters for measuring diffusion of water with an NMR-MOUSE

Parameter	Value
magnet	NMR-MOUSE PM25
transmitter frequency ν_{rf}	13.8 MHz
transmitter attenuation for 90° pulse	−6 dB at 300 W
duration t_p of 90° pulse	7 μs
dwell time Δt	0.5 μs
recycle delay t_R	2 s
number n_s of scans	16
observation time t_{acq}	10 μs
number n_{acq} of data points	20
diffusion time $t_m = \Delta$	20 ms
diffusion encoding time $t_E/2$	0.02 ms–1 ms
number n_G of diffusion encoding steps	16
gradient	7 T/m

tude δ in a pulsed gradient experiment are varied on a logarithmic scale from scan to scan in something like 10 steps.

Beginner's level measurements

Diffusion coefficients of pure solvents are conveniently measured with the NMR-MOUSE using the *stimulated echo* sequence (Fig. 3.2.5d) by varying the encoding time $t_E/2$ in suitable increments either linearly or preferably on a logarithmic scale. Because of the high gradient G of the NMR-MOUSE, signal attenuation by relaxation can often be neglected compared to signal attenuation by diffusion. Then the natural logarithm of the relative signal amplitude is proportional to a cubic time variable according to

$$\ln\{s(t_E)/s(0)\} = -\gamma^2 G^2 (t_E/2)^2 (t_m + 1/3\, t_E) D\,, \tag{3.2.12}$$

and the diffusion coefficient can be calculated from the slope of the linear relationship with the known value of the gradient G (Fig. 3.2.6). Although the measurement is simple, attention should be paid to the issues listed in Tab. 3.2.4.

Table 3.2.4. Common issues when measuring diffusion with the NMR-MOUSE

- The diffusion encoding time t_E is set too high and the signal vanishes before detection
- The diffusion encoding time t_E or the diffusion time $t_m = \Delta$ are too low so that the signal attenuation by diffusion is minor
- The sample is porous and surface relaxation leads to significant signal attenuation in addition to diffusion
- The fluid content in the sample is so low that diffusion cannot be measured
- The receiver phase is misadjusted and the signal phase needs to be adjusted after data acquisition

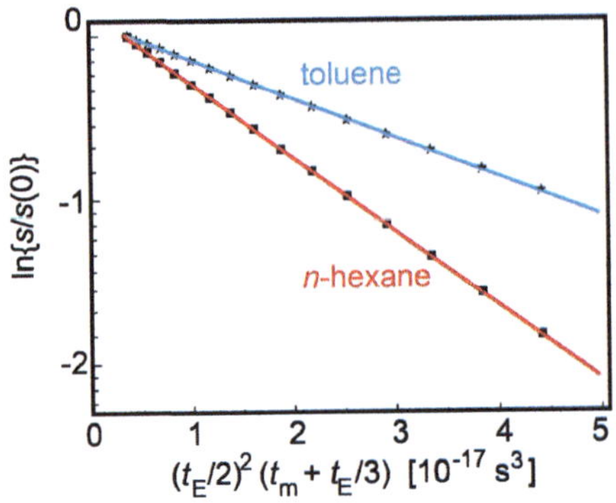

Figure 3.2.6. Plot of the natural logarithm of the relative signal attenuation of pure *n*-hexane and pure toluene at room temperature measured with a CPMG detected stimulated echo sequence by varying the echo time t_E. The diffusion coefficients are calculated from the slopes of the lines as $4.53 \cdot 10^{-9}$ m^2/s and $2.37 \cdot 10^{-9}$ m^2/s for *n*-hexane and toluene, respectively.

Advanced level measurements

Translational diffusion is a molecular transport phenomenon and as such important to a wide variety of phenomena in chemical engineering and the life sciences. Similar to relaxation, diffusion provides a means of discriminating different types of molecules in complex solution and of probing the geometry of confinements of fluids on the micrometer scale. In this regard, 2D *T_2-D correlation* NMR provides a means of characterizing complex solutions by two separation variables, relaxation and diffusion. It is particularly useful in characterizing *emulsions* and *suspensions* (Fig. 4.3.3) and multi-phase fluids like oil and water in porous media (Fig. 7.1.6). The standard pulse sequence consists of a *stimulated echo* sequence with echo time t_{E1} to encode diffusion followed by a *CPMG echo train* with echo time t_{E2} to detect the transverse magnetization decay for a set of different echo times t_{E1} (Fig. 3.2.7a). In consideration of the flip angle distribution resulting from resonance offset in inhomogeneous fields (Fig. 3.1.5), the transverse relaxation time is an effective relaxation time $T_{2\text{eff}}$, and a suitable phase cycle is required to select the direct echoes in the CPMG echo train and discard the stimulated echoes [8]. If the gradient G is sufficiently strong, then signal attenuation by T_1 and T_2 relaxation can be neglected during the evolution period for diffusion encoding. Furthermore, if t_{E2} is sufficiently short, diffusion can be neglected in the attenuation of the CPMG echo train during the detection period. In this approximation, the NMR signal detected from a single pore then decays according to

$$s(t)/s(0) = \exp\{-[\gamma G t_{E1}/2]^2 D (\Delta + 1/3\, t_{E1})\} \exp\{-n\, t_{E2}/T_{2\text{eff}}\}. \tag{3.2.13}$$

For a porous medium, this expression has to be summed over all pores with their own effective diffusion coefficients D and relaxation times $T_{2\text{eff}}$.

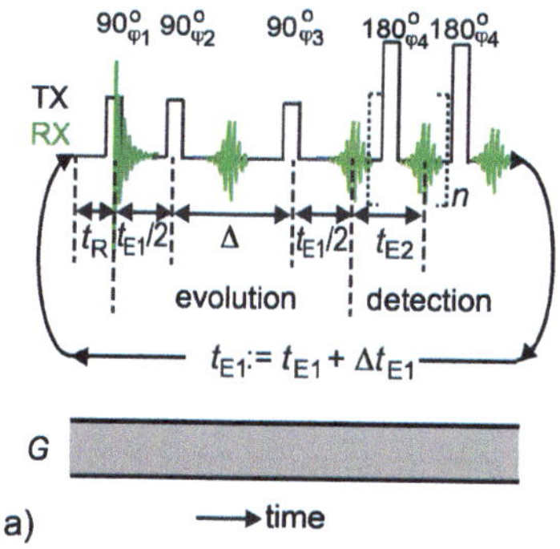

a)

φ_1	φ_2	φ_3	φ_4	φ_{acq}
x	x	x	y	-x
-x	x	x	y	x
x	-x	x	y	x
-x	-x	x	y	-x
x	x	-x	y	x
-x	x	-x	y	-x
x	-x	-x	y	-x
-x	-x	-x	y	x
y	x	x	x	y
-y	x	x	x	-y
y	-x	x	x	-y
-y	-x	x	x	y
y	x	-x	x	-y
-y	x	-x	x	y
y	-x	-x	x	y
-y	-x	-x	x	-y

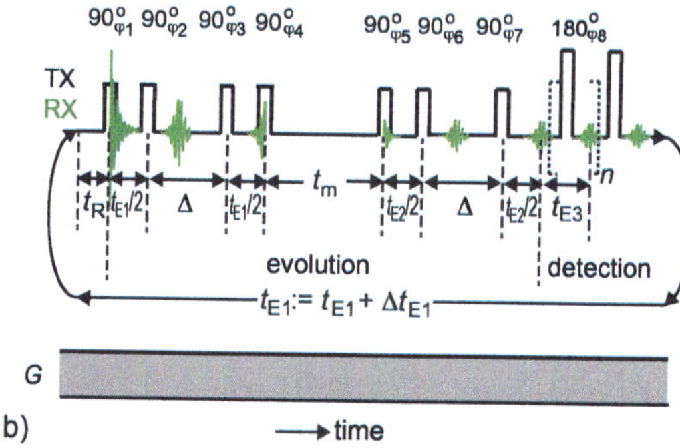

b)

pulse	initial phase	repeats
φ_1	x	×2
φ_2	x	×2
φ_3	x	×1
φ_4	x	×2
φ_5	x	×4
φ_6	x	×1
φ_7	x	×4
φ_8	y	×1
φ_{acq}	-x	

Figure 3.2.7. Pulse sequences (left) and phase cycles (right) for 2D Laplace NMR with distributions of diffusion coefficients for stray-field NMR devices with a time-invariant and uniform gradient G. Diffusion is encoded with the stimulated echo sequence. (a) Correlation of the distributions of diffusion coefficients and transverse relaxation times [8]. (b) Diffusion-diffusion exchange experiment [11].

Another interesting 2D diffusion experiment is the *D-D exchange* experiment (Fig. 3.2.7b). It correlates the *distribution of diffusion coefficients* at an initial time with that observed a time t_m later. It is reminiscent of the *T_2-T_2 exchange* experiment (Fig. 3.2.4c), where two relaxation time distributions are correlated by translational diffusion during a mixing time t_m, but different from the T_2-T_2 exchange experiment the diffusion encoding is fast compared to relaxation, so that the resulting 2D Laplace maps are better approximations of the probability that a molecule with an initial diffusion coefficient D_i has a final diffusion coefficient D_f some time t_m later. This *joint probability density* must be derived from T_2-T_2 exchange maps by computer simulation because the relaxation encoding and detection times are of similar order of magnitude as the mixing time. On the other hand, both dimensions of a D-D exchange map are measured indirectly point-by-point, while the final relaxation dimension in a T_2-T_2 exchange map is scanned directly with a CPMG sequence, so that the measurement time for a D-D exchange map is considerably longer than that for a T_2-T_2 exchange map.

The pulse sequence in Fig. 3.2.7b employs stimulated echoes separated by the mixing time t_m. The echo times t_{E1} and t_{E2} are varied independently for diffusion encoding, and for each pair of echo times the signal is detected with a CPMG sequence so that the echoes can be added for signal-to-noise improvement. The phase cycle of this sequence has 128 steps. For each of the eight rf pulses in the sequence, the phase is repeated the number of times indicated in the phase table in Fig. 3.2.7b independent of the other phases in order to select the correct transformations of longitudinal into transverse magnetization and vice versa under the impact of the rf pulses.

In an n-hexane-filled bed of 2 µm diameter zeolites with 0.8 nm pores, two pore sizes are observed, the nanometer pores in the grains and micrometer pores between the grains. Within each pore, the diffusion of n-hexane molecules is restricted leading to a small effective diffusion coefficient in the small pores and a large one in the large pores. The pore sizes and the pore connectivity are probed by the *D-D exchange* maps (Fig. 3.2.8), where cross peaks at the coordinates of different diffusion coefficients identify molecules moving from one pore type to the other during the mixing time t_m. With increasing mixing time, the cross-peak integrals grow, and the growth rate approximates the exchange rate with which molecules migrate between pores.

A detailed analysis accounts for diffusion and relaxation, while for short mixing times relaxation can be neglected, so that a peak in a D-D exchange map can be approximated by

$$s(t_{E1}/2, t_m, t_{E2}/2)/s(0, t_m, 0) = a_{ij}(t_m)\exp\{-b_1(t_{E1}/2)D\}\exp\{-b_2(t_{E2}/2)D\}, \qquad (3.2.14)$$

where $b = (\gamma G t_E/2)^2(\Delta + t_E/3)$ in the notation of Fig. 3.2.7b for the diffusion time $t_m = \Delta$. For $i = j$, the coefficient a_{ij} is the integral of a peak on the diagonal corresponding to spin density in a particular size of pore, and for $i \neq j$ it is a cross-peak intensity which builds up with the exchange time t_m.

D-T_2 maps and D-D exchange maps are important tools for advanced studies of porous media which provide 2D maps that correlate a distribution of diffusion coef-

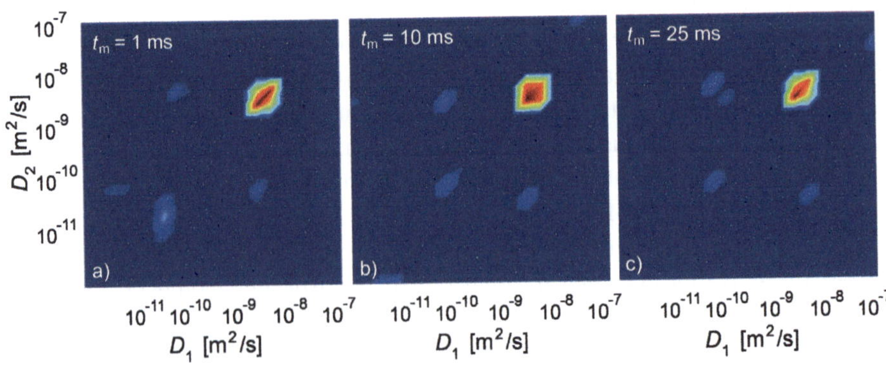

Figure 3.2.8. D-D exchange maps of n-hexane in a pack of nanoporous zeolite particles for different exchange times. (a) $t_m = 1$ ms. (b) $t_m = 10$ ms. (c) $t_m = 25$ ms (adapted from [11] with permission). Experimental parameters: Profile NMR-MOUSE, $\nu_{rf} = 11.7$ MHz, $G = 11.5$ T/m, $5.5\,\mu s \leq t_E/2 \leq 1.6$ ms, $\Delta = 1.6$ ms, $1\,\text{ms} \leq t_m \leq 90$ ms.

ficients with a distribution of relaxation times or two distributions of diffusion coefficients measured at different times. Because such studies do not demand chemical shift resolution, they can be executed with simple portable magnets.

Data processing

The experimental data are processed by 2D inverse Laplace transformation to obtain 2D maps with distributions of relaxation times and diffusion coefficients. Depending on whether the experimental data yield a D-T_2 map or a D-D map the proper kernel has to be used in the 2D Laplace inversion routine. These kernels are the expressions eqns. (3.2.13) and (3.2.14), respectively, and they are provided with the Laplace inversion software. D-D exchange maps are a better approximation of a joint probability density of a molecule being at an initial position at $t_m = 0$ and a final position a time $t_m > 0$ later than T_2-T_2 exchange maps, because the times to encode diffusion can be made short compared to the mixing time contrary to the times to encode relaxation. Nevertheless, a quantitative analysis of D-D exchange maps demands the assistance of computer simulations.

3.2.9 References

[1] Hahn EL. Spin echoes. Physical Review. 1980; 80:580–594.

[2] Carr HY, Purcell EM. Effects of diffusion on free precession in nuclear magnetic resonance experiments. Phys Rev. 1954; 94:630–638.

[3] Meiboom S, Gill D. Modified spin-echo method for measuring nuclear relaxation times. Rev Sci Instrum. 1958; 29:688–691.

[4] Balibanu F, Hailu K, Eymael R, Demco DE, Blümich B. Nuclear magnetic resonance in inhomogeneous magnetic fields. J Magn Reson. 2000; 145:246–258.

[5] Hürlimann MD, Griffin DD. Spin dynamics of Carr–Purcell–Meiboom–Gill-like sequences in grossly inhomogeneous B_0 and B_1 fields and application to NMR well logging. J Magn Reson. 2000; 143:120–135.

[6] Hürlimann MD. Optimization of timing in the Carr–Purcell–Meiboom–Gill sequence. Magn Reson Imaging. 2001; 19:375–378.

[7] Song YQ, Venkataramanan L, Hürlimann MD, Flaum M, Frulla P, Straley C. T_1-T_2 correlation spectra obtained using a fast two-dimensional Laplace inversion. J Magn Reson. 2002; 154:261–268.

[8] Hürlimann MD, Venkataramanan L. Quantiative measurement of two-dimensional distribution functions of diffusion and relaxation in grossly inhomogeneous fields. J Magn Reson. 2002; 157:31–42.

[9] Anferova S, Anferov V, Arnold J, Talnishnikh E, Voda MA, Kupferschläger K, et al. Improved Halbach sensor for NMR scanning of drill cores. Magn Reson Imag. 2007; 25:474–480.

[10] van Landegham M, Haber A, d'Espinose de Lacaillerie JB, Blümich B. Analysis of multisite relaxation exchange NMR. Conc Magn Reson. 2010; 36A:153–169.

[11] Neudert O, Stapf S, Mattea C. Diffusion exchange NMR spectroscopy in inhomogeneous magnetic fields. J Magn Reson. 2011; 208:256–261.

3.3 Imaging

3.3.1 Introduction

Magnetic resonance imaging (*MRI*), also known as *magnetic resonance tomography* (*MRT*), is a well-established technique in medical diagnostics. It is appreciated for its wide variety of contrast and its sensitivity to soft matter unlike *X-ray tomography* or *computer tomography* (*CT*). MRI also finds applications in materials science, biology and chemical engineering, for example in studying elastomer products, plants and fluid transport. With the recent availability of desktop instruments, MRI can now be used also outside the laboratory, for example, to study the growth of plants in the greenhouse and the field and to optimize production processes by non-invasive inline inspection.

In the context of compact NMR, MRI is often understood to imply the use of desktop instruments with closed magnets. But images can also be acquired in the inhomogeneous *stray field* of magnets. In fact, the Profile NMR-MOUSE is designed to measure 1D profiles with high spatial resolution. In a general context, MRI refers to measuring 1D, 2D, or 3D spatially resolved NMR signals from an object. Here, 1D refers to a linear trace through an object or a 1D *projection* of an object, 2D refers to a particular *slice* through the object or a projection of the 3D object onto a 2D plane. Note that projection means integration of the signal along the space coordinates, which are not resolved in the image.

In MRI, the spatial information is imprinted in the measured NMR signal by the NMR frequency, which is made dependent on the position by making the magnetic field change with position. A linear dependence of the NMR frequency on position is achieved with constant field gradients. The image is then obtained as the Fourier transform of the measured data. It is also possible to scan images point by point by moving a sensitive slice through the object. This procedure is used with the Profile NMR-MOUSE to scan 1D *depth profiles*. The amplitude of such a depth profile is defined by the *contrast* parameter chosen, for example, the *spin density*, a *relaxation* parameter, the *diffusion coefficient*, or the w-parameter (eqn. (3.1.5)) as described in the preceding sections. In the following, only imaging schemes are introduced which acquire the image information in *Fourier space* and not point by point in real space.

3.3.2 Objective

MRI is used with the intention to create images of NMR properties or functions of NMR properties of an object. Important NMR properties that can be imaged are the *spin density*, spatially-resolved *relaxation times*, the *diffusion coefficient*, and velocity components. Also functions of one or more of these parameters can be imaged. In this case one refers, for example, to *relaxation-weighted spin density* images. To measure images of NMR parameters, the frequencies and phases of the NMR signal from a sample must be modulated in space by exploring magnetic field gradients in combination with rf pulses and suitable data acquisition schemes. The decision on the particular choice of the imaging sequence and the parameter settings depend on the object and the features one wants to visualize. As a general rule, the parameters are set to maximize the image *contrast* in order to differentiate different structures in the image. The contrast is defined as a relative difference of pixel intensities (eqn. (3.1.2)). Typically, the pixel intensity as such does not map an object property such as the modulus of *cross-link density* in an absolute way but rather in a relative way by comparison with the intensity of another pixel.

3.3.3 Further reading

Casanova F, Perlo J, Blümich B, editors. Single-Sided NMR. Berlin: Springer; 2011.

Stapf S, Han SI. editors. NMR Imaging in Chemical Engineering. Weinheim: Wiley-VCH; 2006.

Kose K. Compact MRI for Chemical Engineering. In: Stapf S, Han SI, editors. NMR Imaging in Chemical Engineering. Weinheim: Wiley-VCH; 2006.

Blümich B. Essential NMR. Berlin: Springer; 2005.

Blümich B. NMR Imaging of Materials. Oxford: Clarendon Press; 2000.

Kimmich R. NMR Tomography. Diffusometry, Relaxometry. Berlin: Springer; 1997.

Callaghan PT. Principles of Nuclear Magnetic Resonance Microscopy. Oxford: Clarendon Press, 1991.

3.3.4 Theory

Gradient fields and field gradients

Imaging explores magnetic fields that vary in space. The spatial variation is expressed by a Taylor series expansion of the way the magnetic field changes in space. The first term of this expansion relevant for imaging is the *linear field* term with a *constant field gradient*. Apart from *stray-field magnets*, MRI magnets typically have relatively homogeneous fields $\boldsymbol{B}_0$, and magnetic *gradient fields* are added to this by means of a set of three *gradient coils* inside the magnet, which generate magnetic fields in the same direction as the magnet, but with amplitudes varying linearly in space (Fig. 3.3.1).

Depending on the direction in which the field amplitude varies, the gradient points in the x, y, or z direction of the coordinate system defined by the three gradient coils. These gradient fields are typically turned on and off when needed in the pulse

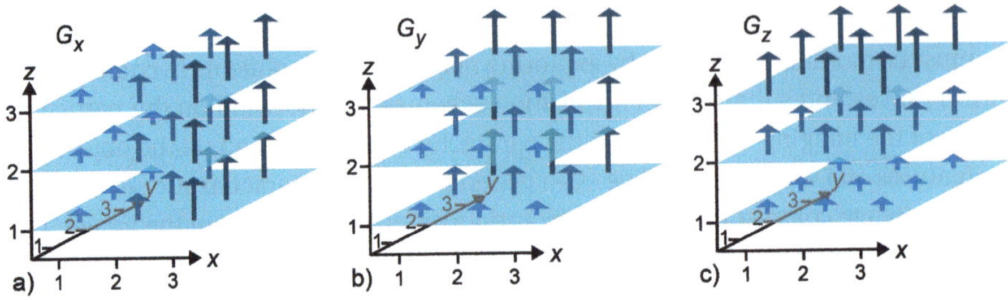

Figure 3.3.1. Illustration of gradient fields. The arrows in the planes symbolize the magnitude and the direction of the magnetic field. In each drawing the field changes linearly in one space direction illustrating a constant x-gradient G_x (a), a constant y-gradient G_y (b) and a constant z-gradient G_z (c).

sequence. Although the coils produce magnetic fields, it is the gradient of these fields that enters into the description of MRI, so that one conventionally speaks of *pulsed field gradient* NMR. It is the custom to collect the three gradients $G_i = \mathrm{d}B_z/\mathrm{d}x_i$, $i = x, y, z$ in a *gradient vector* $\boldsymbol{G} = (G_x, G_y, G_z)^\mathrm{t}$, because the vector description of MRI is valid in most cases of practical interest, where the strength of the maximum gradient field is much lower than the strength of $\boldsymbol{B}_0$. It is important to note that the directions of the *field gradient vector* and the *gradient field vector* are not the same. Depending on which coil is powered up, different gradient fields add to the constant field $\boldsymbol{B}_0$ of the magnet resulting in a sum field, the magnitude of which is well approximated by

$$B_z = B_0 + \boldsymbol{Gr}, \tag{3.3.1}$$

where $\boldsymbol{r} = (x, y, z)^\mathrm{t}$ is the position vector and $\boldsymbol{Gr}$ is the *gradient field*.

Fourier space

The acquisition of *depth profiles* with the NMR-MOUSE may be conceived as a form of one-dimensional *imaging*. In this case the information of one pixel after another is acquired at different depth. Magnetic resonance imaging is generally understood to deliver two- or even three-dimensional images.

Typically the image information is acquired not in real space as in depth profiling with the NMR-MOUSE, but in *Fourier space* or ***k**-space*, where $\boldsymbol{k}$ is the *wave vector*. The image is then obtained by *Fourier transformation* of the signal acquired for different values of $\boldsymbol{k}$. Each element of the wave vector is a wave number corresponding to the inverse period of a magnetization wave in the object (Fig. 3.3.2c). This wave is produced by the transverse magnetization rotating in a magnetic field with a constant *gradient* G. The gradient vector $\boldsymbol{G}$ with components G_x, G_y, G_z specifies the change of the applied magnetic field $\boldsymbol{B}$ in the x, y, and z directions, where the direction of $\boldsymbol{B}$ by convention identifies the z-direction of the reference coordinate system. If the gradient is applied in z-direction (Fig. 3.3.2b), $G_x = G_y = 0$ and $G_z = \partial B_z/\partial z$, so that the

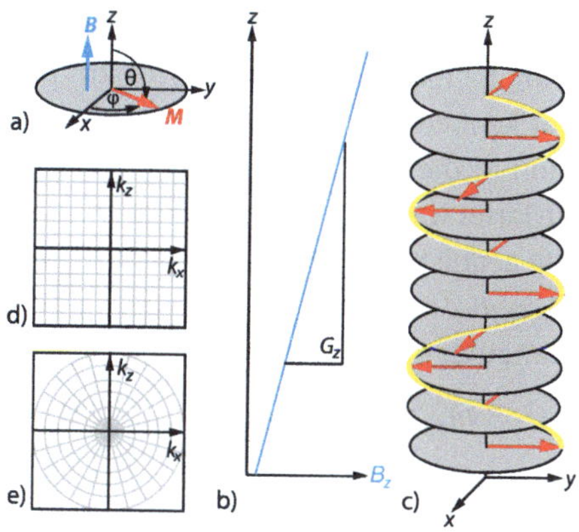

Figure 3.3.2. Concepts of magnetic resonance imaging. (a) The phase of the transverse magnetization is the angle φ, which the component of the magnetization vector $\boldsymbol{M}$ encloses with the x-axis of the laboratory coordinate system. (b) The magnetic field B_z varies linearly along the space coordinate z of the image. The field has a constant gradient G_z. (c) The transverse magnetization from voxels at different positions z, along which the gradient G_z of the magnetic field $\boldsymbol{B}$ is constant, forms a spiral wave, because in a given time the magnetization precesses faster in higher than in lower field. (d) In Fourier imaging, the NMR data are sampled in $\boldsymbol{k}$-space on a rectangular grid. (e) In back-projection imaging, the NMR data are sampled in $\boldsymbol{k}$-space on a grid defined in cylindrical or spherical coordinates.

NMR frequency depends linearly on z,

$$\omega(z) = \gamma B_z(z) = \gamma(B_0 + G_z z). \tag{3.3.2}$$

Because in NMR the magnetization vector is measured in the transverse plane, the *precession angle* or the *phase* φ of the NMR signal is measured (Fig. 3.3.2a). With the gradient $\boldsymbol{G}$ being controlled in magnitude and direction during the measurement by modulating currents through the gradient coils surrounding the object, the real and imaginary parts of the measured transverse magnetization are conveniently described by the magnetization *amplitude* and phase. For time-varying gradients, the phase is given by the time integral of the *NMR frequency* ω,

$$\varphi(z,t) = \int_0^t \omega(z,t')\mathrm{d}t' = \gamma \int_0^t B_z(z,t')\mathrm{d}t' = \gamma B_0 t + \gamma \int_0^t G_z(t')\mathrm{d}t' z = \omega_0 t + k_z z, \tag{3.3.3}$$

where k_z is the *wave number* of the transverse magnetization at position z. In general, the *wave vector* is given by the time integral of the *gradient vector*,

$$\boldsymbol{k}(t) = \gamma \int_0^t \boldsymbol{G}(t')\mathrm{d}t'. \tag{3.3.4}$$

To acquire imaging data this gradient integral needs to be varied in accordance with the required *spatial resolution* and image dimensions or *field of view*.

Spatial resolution and field of view

To modulate a component of the $\boldsymbol{k}$ vector, for example, $k_z = \gamma \int_0^t G_z(t')\mathrm{d}t'$ the gradient G_z may be varied in increments ΔG_z or kept constant and time t is stepped in increments Δt. Encoding spatial information in the NMR signal in the first way is called *phase encoding*, and encoding it in the latter way is called *frequency encoding* with Δt being the *sampling interval* during data acquisition. Choosing n_p values for the gradient amplitude by phase encoding, the *spatial resolution* $1/\Delta z$ and the *field of view* z_max are given for the commonly used, rectangular gradient pulses of duration t by

$$\text{Phase encoding, resolution:}\quad 1/\Delta z = \gamma n_{\mathrm{p,max}} \Delta G_z t/(2\pi) = k_{z,\mathrm{max}}/(2\pi), \tag{3.3.5a}$$

$$\text{field of view:}\quad z_\mathrm{max} = 2\pi/(\gamma \Delta G_z t), \tag{3.3.5b}$$

For frequency encoding, the spatial resolution is limited by the line width $\Delta\omega_{1/2} = 2/T_2^*$ and not the maximum gradient strength $G_{x,\mathrm{max}} = n_{\mathrm{f,max}} \Delta G_x$ as in the case of phase encoding,

$$\text{Frequency encoding, resolution:}\quad 1/\Delta x = \gamma G_x T_2^*/2, \tag{3.3.5c}$$

$$\text{field of view:}\quad x_\mathrm{max} = 2\pi/(\gamma G_x \Delta t). \tag{3.3.5d}$$

Because frequency encoding is used during signal detection, the *frequency encoding gradient* is also called *read gradient*.

Construction of the image

The image information is measured by acquiring NMR signal for a range of points in two- or three-dimensional $\boldsymbol{k}$-*space* from which the image can be constructed. A key difference among imaging methods is the way in which the magnetization is traced through $\boldsymbol{k}$-space. If the traces are on a rectangular grid one speaks of *Fourier imaging* (Fig. 3.3.2d), if the traces follow polar coordinates one speaks of *back-projection imaging* (Fig. 3.3.2e). Other traces are also conceivable. In either case, the image $M_z(\boldsymbol{r})$ is essentially constructed from the complex NMR signal

$$s(\boldsymbol{k}) = M_x(\boldsymbol{k}) + \mathrm{i}M_y(\boldsymbol{k}) = \int M_z(\boldsymbol{r})\mathrm{e}^{\mathrm{i}\boldsymbol{kr}}\mathrm{d}\boldsymbol{r} \tag{3.3.6}$$

by inverse *Fourier transformation*. This equation is valid under the assumption that the longitudinal magnetization $M_z(\boldsymbol{r})$ is converted into transverse magnetization by a 90° pulse, and that transverse relaxation and chemical shift can be neglected.

Image contrast

Although the effects of *relaxation* and *chemical shift* are neglected in eqn. (3.3.6), these quantities are a highly important asset to MRI, because they can be used to generate *contrast* in the image, by which different features not recognizable by other imaging methods can be identified. The simplest type of contrast is *relaxation contrast*, which introduces a *relaxation weight* to the *spin density* $M_z(\boldsymbol{r})$ due to the *relaxation times*

$T_1(\boldsymbol{r})$ and $T_2(\boldsymbol{r})$ being different for different voxels, i.e., when the relaxation times are functions of the voxel position $\boldsymbol{r}$.

T_1 contrast is introduced by shortening the *recycle delay* t_R between scans, so that magnetization components with long T_1 cannot fully recover. T_2 *contrast* is introduced by incorporating a *spin echo* in the imaging pulse sequence and adjusting the *echo time* t_E so that magnetization components with short T_2 are lost and those with long T_2 are maintained to form the image. Including the relaxation weights, in the echo maximum the NMR signal (3.3.6) becomes

$$s(\boldsymbol{k}) = M_x(\boldsymbol{k}) + \mathrm{i}M_y(\boldsymbol{k}) \\ = \int [1 - \exp\{-t_R/T_1(\boldsymbol{r})\}] \exp\{-t_E/T_2(\boldsymbol{r})\} M_z(\boldsymbol{r}) \exp\{\mathrm{i}\boldsymbol{k}\,\boldsymbol{r}\}\mathrm{d}\boldsymbol{r}. \quad (3.3.7)$$

If several images are measured for different values of the *recycle delay* t_R and the *echo time* t_E, the *relaxation times* T_1 and T_2 can be extracted at each position $\boldsymbol{r}$ to obtain relaxation *parameter images* $T_1(\boldsymbol{r})$ and $T_2(\boldsymbol{r})$.

3.3.5 Hardware

The hardware for NMR *imaging* is basically the same as that for NMR *relaxometry* or NMR *spectroscopy*, however, with the additional capability of rapidly turning linear magnetic *gradient fields* on and off. These gradient fields are generated by electric currents through the gradient coils and add to the homogeneous magnetic field $\boldsymbol{B}_0$ provided by the permanent magnet. The timing and the strength of these currents are controlled by the imaging pulse sequence and are indicated in the pulse sequence diagrams by traces specifying the gradient strength G. The magnetic field $\boldsymbol{B}_0$ is not required to be as homogeneous for imaging as it is for spectroscopy, so that more simple magnets can be employed, unless one wants to resolve an NMR spectrum in each voxel, a method called *spectroscopic imaging*. Figure 3.3.3a depicts a 0.5 T desktop *Halbach magnet*, which can be employed to image objects of small animals and soft matter with diameters up to 4 cm. The field is sufficiently homogeneous to resolve the line splitting from ethanol in the ^{1}H NMR spectrum of a 1 cm^3 size sample. Three *gradient coils*, one for each space direction, are hidden inside the magnet. *Gradient modulators* and *gradient amplifiers* are needed to drive the currents in the gradient coils to generate pulsed magnetic gradient fields.

Although gradient coils for pulsing gradient fields have been implemented on stray-field sensors for imaging (Fig. 1.3.3b), the signal-to-noise ratio in 2D imaging applications with the *NMR-MOUSE* is insufficient for practical applications due to the strong permanent gradient in the direction away from the surface of the magnet. For this reason, only the time-invariant gradient of the NMR-MOUSE is used in practice for 1D depth profiling. Compact imaging devices in practical use are therefore desktop imagers (Fig. 3.3.3a) and the Profile NMR-MOUSE (Fig. 3.3.3b).

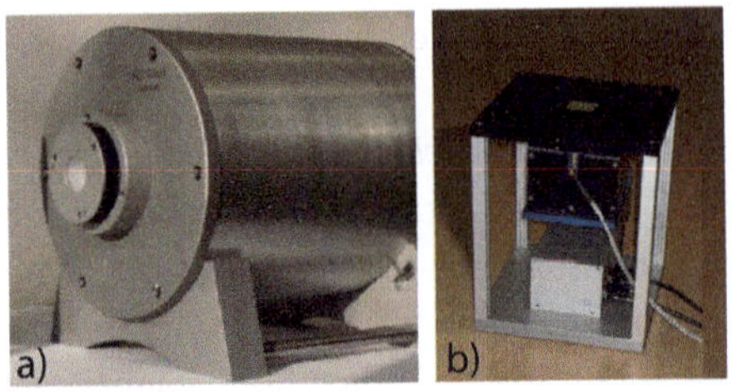

Figure 3.3.3. Compact magnets for ^{1}H NMR imaging. (a) 0.5 T magnet with a 4 cm diameter open access hole. (b) Profile NMR-MOUSE PM5 mounted on a lift for depth profiling of samples placed on the top plate of the device.

3.3.6 Pulse sequences and parameters

To image a 3D object, the gradient fields have to be applied with the field gradients in all three directions of space and stepped over a range of values. Although there is a large number of different imaging schemes, most of them provide images of 2D slices through the object and employ echoes (Fig. 3.3.4). All slice-selective 2D *Fourier imaging* methods follow a general pattern with three successive steps,

- slice selection,
- phase encoding of one space axis,
- frequency encoding of the other space axis.

Slice selection

When imaging a slice parallel to the *xy* plane, a *slice selection* gradient G_z is applied in the z-direction while an rf excitation pulse of duration t_p is applied. Due to the spread of resonance frequencies induced by the gradient field, only the magnetization components within the frequency bandwidth $\Delta\omega \approx 2\pi/t_p$ of the rf pulse are excited, i.e., rotated away from the z-axis. The center frequency of the pulse corresponds to a particular position z_0 in the object and the bandwidth $\Delta\omega = \gamma G_z \Delta z$ to a particular slice width Δz. By varying the *pulse length* for a given value of G_z the *slice thickness* can be adjusted. The quality of slice selection not only depends on the pulse length but also on the pulse shape. A rectangular pulse possesses a sinc-shaped frequency distribution and excites an ill-defined slice profile with many side lobes. Vice versa a sinc-shaped pulse possesses a rectangular frequency distribution and excites the magnetization in a rectangular slice.

Spin-echo imaging

Imaging based on a *Hahn echo* is called *Hahn echo imaging* or *spin-echo imaging* (Fig. 3.3.4a). Several echoes need to be generated to measure a complete image. In the interval following the slice selective pulse, the transverse magnetization precesses

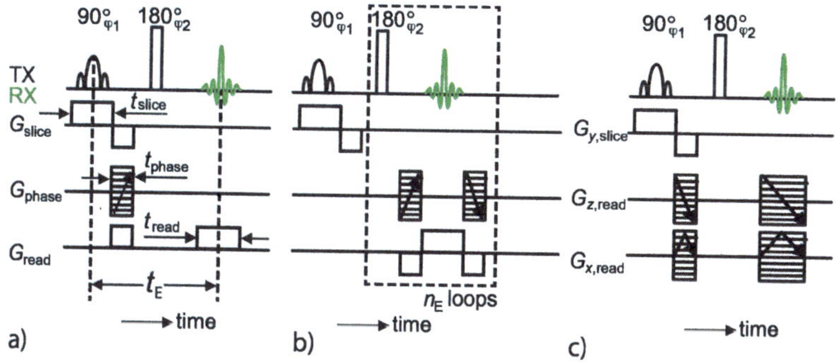

φ_1	φ_2	φ_{acq}
x	-y	-y
y	x	x
-x	y	y
-y	-x	-x
x	y	-y
y	-x	x
-x	-y	y
-y	x	-x

Figure 3.3.4. Pulse sequences and phase cycle for imaging based on Hahn echoes. (a) Hahn echo or spin-echo imaging. The phase-encoding gradient is applied in the first half of the echo time t_E. (b) RARE (rapid imaging with T_1 enhancement). The phase-encoding gradient is applied in the second half of the echo time t_E before the echo is formed and rewound afterwards so that it can be set to a new value in the next echo time. This is why the minimum echo time is longer for RARE than for Hahn echo imaging. (c) Back-projection imaging. The read gradients are varied from scan to scan in angle increments $\Delta\varphi$ with $G_{z,\text{read}} = G_{z,\max} \cos\varphi$ and $G_{x,\text{read}} = G_{x,\max} \sin\varphi$.

freely, and a gradient pulse is stepped in increments of ΔG_{phase} from echo to echo through a range of positive and negative gradient values G_{phase} for phase encoding of the space information in the indirectly detected dimension of the $\boldsymbol{k}$-space data matrix. While the gradient pulse lasts, the components of the transverse magnetization lose phase coherence and the signal amplitude decreases. Magnetic field inhomogeneity from the magnet and from susceptibility differences within the object leads to further signal loss. But this loss can be regained by forming a *Hahn echo* using a 180° pulse. Following this pulse, the echo is generated in the presence of a *frequency-encoding* gradient for direct detection of the image information in the second $\boldsymbol{k}$-space dimension. This Hahn echo coincides with a read-gradient echo, maximizing the echo amplitude. Such a *gradient echo* is generated by two gradient pulses with identical time integrals, which are applied with opposite signs unless the two gradient pulses are separated by a 180° rf pulse. The gradient echo arises when the time integral of the complete gradient modulation function vanishes. Upon completion of the imaging experiment, the 2D data matrix is Fourier transformed from $\boldsymbol{k}$*-space* into the image in real space.

Multi-echo imaging

The time to acquire a complete image is shortened if several echoes of a multi-echo train are measured for different values of the phase-encoding gradient following the pulse sequence for *RARE imaging* (Rapid Acquisition with Relaxation Enhancement, Fig. 3.3.4b). With this scheme, multiple traces through $\boldsymbol{k}$-space can be acquired in one shot. RARE imaging requires a somewhat larger *echo time* than *spin-echo imaging* be-

Table 3.3.1. Acquisition parameters for ^{1}H MRI of rubber

Parameter	Spin-echo imaging	RARE imaging
transmitter frequency [MHz]	9.045 MHz	9.045 MHz
transmitter attenuation of 90° pulse	12 dB at 100 W	12 dB at 100 W
duration t_p of selective 90° pulse	50 µs	50 µs
shape of the elective pulse	Gauss	Gauss
slice thickness	5 mm	5 mm
transmitter attenuation of 180° pulse	−6 dB at 100 W	−6 dB at 100 W
duration t_p of the 180° pulse	50 µs	50 µs
phase-encoding field of view	35 mm	35 mm
number n_p of phase-encoding steps	64	6
number n_E of echoes	1	6
frequency encoding field of view	35 mm	35 mm
number n_{data} of data points in read direction	64	64
dwell time Δt	10 µs	10 µs
echo time t_E	1000 µs	1400 µs
number n_s of scans	32	32
recycle delay t_R	70 ms	70 ms

cause the space-encoding phase of the transverse magnetization needs to be reset by a gradient echo in each echo time so that a new trace through ***k**-space* can be followed with the subsequent echo. Default acquisition parameters for spin-echo imaging and RARE imaging of natural *rubber* are listed in Tab. 3.3.1. They refer to an image with 64 × 64 pixels, whereby 11 RARE traces are required to collect the data for one image. When the shaped, selective initial rf pulse of the sequence is replaced by a short, non-selective pulse, the slice thickness is extended across the length of the sensitive volume, and a 2D projection is obtained instead of an image. Note that spin-echo imaging and RARE imaging scan ***k***-space in Cartesian coordinates (Fig. 3.3.2c). The image is constructed from the data by straightforward 2D Fourier transformation. This is why both schemes belong to the class of *Fourier imaging* methods.

Back-projection imaging

The simplest imaging technique is *back-projection imaging*. Free induction decays or spin echoes are measured in the presence of time-invariant gradients. The Fourier transform of such a signal is a *projection*. It is the signal integral of the spin density in the space directions perpendicular to the gradient direction. An image can be reconstructed from many such projections acquired with the gradient pointing in different directions. Instead of rotating the gradient direction from one acquisition to the next, the sample can be rotated also. Two-dimensional back-projection imaging (Fig. 3.3.4c) requires a slice-selective pulse followed by a read-gradient echo which coincides with the spin echo. There is no phase-encoding gradient, and ***k***-space is scanned radially in cylindrical coordinates by *frequency encoding* (Fig. 3.3.2d). The image is obtained from the measured data by extrapolation of the data set from *cylinder coordinates* to *Cartesian coordinates* and subsequent 2D Fourier transformation.

3.3.7 Beginner's level measurements

If the shape of an object like a rubber gasket or fitting varies in only two dimensions, the third dimension does not need to be resolved spatially and all rf pulses of the imaging sequence can be short, non-selective pulses. The images of the rubber profiles depicted in Fig. 3.3.5 were acquired with *spin-echo imaging* using the parameters listed in Tab. 3.3.1 and the figure caption. Visually only the outer surface but neither the inner surfaces nor differences in the state of curing or the formulation of the carbon black filled rubber material within the profile cross section can be identified without touching or even destroying the profile. Simple MRI not only identifies the inner surfaces but also the *material density* and the *state of cure* from differences in signal amplitude and relaxation times in different voxels. In spin-echo images, the image contrast is a blend of *spin density* or proton density, and *relaxation weights* according to eqn. (3.3.7), which can only be disentangled by measuring several images with different repetition times and echo times and subsequently fitting longitudinal and transverse relaxation curves to the amplitudes for each pixel in the set of images.

But even simple relaxation-weighted *spin-density images* acquired at low resolution in a minute or two already reveal the location of the inner surfaces and variation of the material properties by their pixel amplitude. For example, the bottom part of the fitting in Fig. 3.3.5a and the top parts of the fitting in Fig. 3.3.5b give higher signal from inside, indicating either higher proton density or higher network chain mobility due to the use of a different polymer or a different state of curing. Quite surprisingly, the fitting of Fig. 3.3.5b, despite it having an aluminum core, can still be imaged quite well at 0.2 T albeit at the expense of reduced *signal-to-noise ratio*.

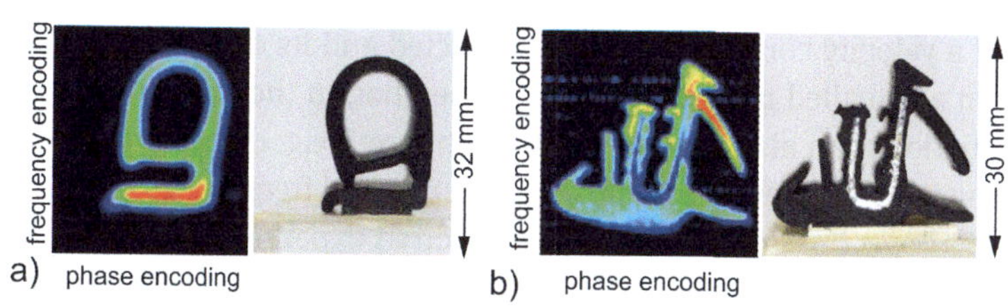

Figure 3.3.5. Spin-echo ^{1}H NMR images (left) of rubber fittings acquired from 5 mm thick slices at 0.2 T and photos (right). (a) Simple rubber fitting. 32 × 32 pixels, $t_E = 0.69$ ms, 32 scans, acquisition time 1 min 11 s. (b) Rubber fitting with inner aluminum liner. 64 × 64 pixels, $t_E = 1.011$ ms, 32 scans, acquisition time 2 min 43 s.

Table 3.3.2. Common issues in spin-echo imaging of elastomers

- The recycle delay is too short so that the object may be heated from deposited rf power or longitudinal magnetization cannot build up
- The recycle delay is too long and the acquisition time for the total image becomes to long
- The rf pulse flip angles are not adjusted and the signal amplitude is too low
- The maximum gradient strength is wrongly set given the line-width of the sample and the expected resolution
- The field of view set by the gradient increment does not match the size of the object
- The object is not centered in the magnet

Unless the software package of the instrument manufacturer provides a comfortable acquisition and processing environment, imaging experiments require training and experience. With reference to Hahn echo imaging, common issues are listed in Tab. 3.3.2.

3.3.8 Advanced level measurements

There is an abundance of imaging methods. They differ in the way the NMR signal is traced through $\boldsymbol{k}$-space, whether two or three dimensions of $\boldsymbol{k}$-space are imaged, and in the contrast by which different structures are discriminated. 2D *Fourier imaging* and *back-projection imaging* (Fig. 3.3.4) are by far the most popular classes of imaging methods. Fast-imaging methods avoid long recycle delays by using small rf pulse flip angles or echo trains. An example of the latter is *RARE imaging* (Fig. 3.3.4b).

Contrast is most often defined in a *magnetization filter* preceding the space-encoding pulse sequence (Fig. 3.3.6). By means of such filters a diffusion weight or a velocity weight may be imposed on the initial magnetization, the distribution of which is then interrogated by an NMR image. The procedure may be repeated several times with different contrast weights to obtain a series of images with varying contrast from which a contrast parameter like a relaxation time, a diffusion coefficient, and a velocity component can be extracted and its distribution in space then depicted in a so-called *parameter image*. Depending in the data processing, parameter distributions can also be assigned to each voxel, e.g. distributions of relaxation times, diffusion coefficients, velocity components, and frequencies. The latter case is referred to as *spectroscopic imaging*. The preferred experimental procedure deviates from the one described above in so far as the image is obtained by pure phase encoding, and the spectroscopic information is subsequently derived from an echo or a free induction decay acquired in a homogeneous field.

Even more advanced imaging methods integrate the contrast filter into the space encoding part of the imaging sequence to arrive at more compact imaging methods by which images can be acquired in short time to resolve time-dependent processes.

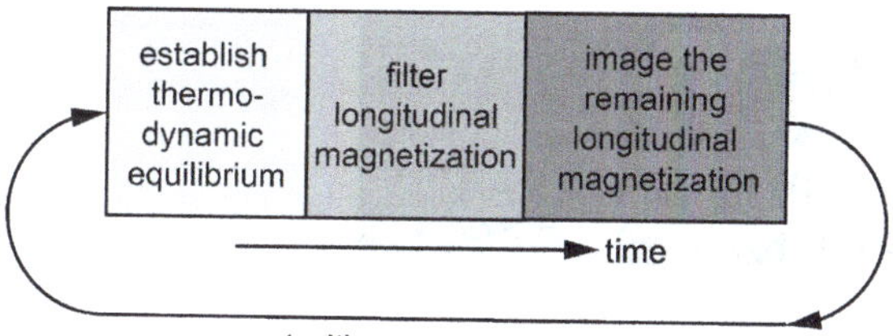

Figure 3.3.6. Parameter imaging. Starting from longitudinal magnetization in thermodynamic equilibrium, one part of the magnetization is retained in a filter, which generates image contrast. The other part passes the filter and its distribution in space is imaged. From a set of images acquired with different filter settings a parameter image can be calculated.

High time resolution is also needed when imaging is employed in process control, e.g. when monitoring the extrusion of rubber profiles.

3.3.9 Data processing

Data from NMR-imaging and NMR-spectroscopy experiments are processed in similar ways. Both demand a Fourier transformation as the basic processing operation. But prior to this operation the data are manipulated in different steps. In a first step, the complex decays of the transverse magnetization $s(\boldsymbol{k}) = M_x(\boldsymbol{k}) + iM_y(\boldsymbol{k})$ are truncated once they have decayed to the noise level. Any noise extending beyond the signal measured in $\boldsymbol{k}$-space only degrades the *signal-to-noise ratio* in the image. In a second step, the data are phase corrected by multiplication with a complex phase factor $\exp\{-i\phi\}$, so that the signal appears only in the real part of the recorded data and the imaginary part contains only noise,

$$s'(\boldsymbol{k}) = s(\boldsymbol{k})\exp\{-i\phi\}. \tag{3.3.8}$$

Then the apparent resolution is increased beyond the actual image resolution defined by the acquisition parameters in eqn. (3.3.5) by appending zeros to the measured signals. By this *zero filling*, the length of the data record from each FID is enlarged by a factor of at least 4 or more times its original value, so that the apparent resolution of the image is increased by this number. This procedure is done in all dimensions explored in $\boldsymbol{k}$-space. To further suppress noise, the FID is multiplied with its own exponential decay envelope. This procedure is called *apodization*. There are many apodization functions known. The one mentioned is known as the *matched filter*, because from a systems analytical point of view after Fourier transformation, the signal is then best filtered from the noise. The last elementary data processing step is the 2D or 3D *Fourier transformation*, which produces the NMR image from the $\boldsymbol{k}$-space data refined in the preceding steps.

All these steps are needed to produce good images. Depending on the application, further image processing steps may follow. For process control of rubber fittings,

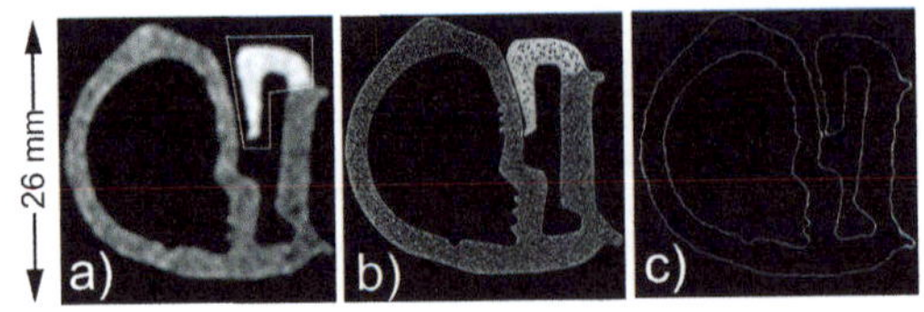

Figure 3.3.7. Image processing of an extruded elastomer foam fitting. (a) Spin-echo image acquired with two 100 μs long, slice selective echo pulses. The resultant slice thickness was 10 mm. The bright region is due to higher foam density. True resolution: 64 × 64 pixels corresponding to $(0.4\,\text{mm})^2$, echo time $t_E = 0.7$ ms, recycle time $t_R = 70$ ms, total image acquisition time 18 s. (b) CT scan with a pixel size of 29.7 μm and 40 min acquisition time. (c) Contours traced at half height of the pixel amplitude over the image background. Before Fourier transformation the experimental data were zero filled from 64 to 4092 points in each dimension. The accuracy of the contour position is by a factor of 10 higher than the true resolution of the image (adapted from [1] with permission).

e.g., the image contours need to be traced to locate inner surfaces which cannot be inspected by other techniques (Fig. 3.3.7). The positions of these surfaces can be specified after zero filling with an accuracy exceeding the pixel resolution set during the measurement. The accuracy is limited ultimately by the signal-to-noise ratio of the image and the apparent resolution determined by *zero filling*.

3.3.10 References

[1] Danieli E, Berdel K, Perlo J, Michaeli W, Masberg U, Blümich B, et al. Determining object boundaries from MR images with sub-pixel resolution: Towards in-line inspection with a mobile tomograph. J Magn Reson. 2010; 207:53–58.

3.4 Spectroscopy

3.4.1 Introduction

Spectrum is Latin and literally means image. When used in science, spectrum denotes the distribution of a characteristic quantity, which generally is energy or frequency. The spectrum of light is the light intensity as a function of color. In spectroscopy, a spectrum is a *distribution of frequencies*. Such distributions may be mono-modal, consisting of a single peak, or multi-modal, consisting of several peaks. If these peaks are broad the distribution may be called continuous, if they are narrow the spectrum is called discrete. Solid-state NMR spectra often resemble continuous distributions with particular line shapes unless special line-narrowing techniques are applied, while liquid-state spectra usually resemble discrete distributions. This is why they are also called *high-resolution spectra*. In high-resolution spectra, spin interactions which depend on the orientation of the molecule in the magnetic field are averaged out either by rapid isotropic, thermal motion of molecules in solution or by fast rotation of the sample around selected angles, for example, the *magic angle* of 54.7° for high-resolution ^{13}C solid-state NMR spectroscopy. The following discussion is limited to *liquid-state NMR spectroscopy*, where the magnetic *dipole-dipole interaction* between neighboring spins is averaged to zero and the orientation dependent *chemical shift* and *indirect coupling* are averaged to their mean values by the rapid tumbling of the solute molecules.

3.4.2 Objective

The objective is to measure the discrete distribution of NMR frequencies from nuclei of molecules in solution, where the orientation-dependent spin interactions vanish and narrow lines are observed in the spectra. These nuclei are primarily proton ^{1}H and carbon ^{13}C nuclei encountered in organic chemistry. The frequencies of these lines depend on the electronic environment of the observed nuclei, i.e. on the chemistry of the molecule. The beauty of NMR at reasonably high field of 0.5 T and more is that the frequencies in the NMR spectrum can conveniently be assigned to particular chemical groups, so that with some experience, the *chemical structure* can be read from the NMR *spectrum*. In this regard NMR spectra are easier to interpret than infrared spectra, where a normal coordinate analysis is often required to assign spectral features to the configuration and conformation of molecules. The primary goal of *high-resolution NMR spectroscopy* is indeed to determine *molecular structures*, and the different techniques of acquiring NMR spectra are there to refine the information needed for *chemical analysis*.

3.4.3 Further reading

Friebolin H. Basic One- and Two-Dimensional NMR Spectroscopy, 5th edition. Weinheim: Wiley-VCH; 2011.

Keeler J. Understanding NMR Spectroscopy, 2nd edition, Chichester: Wiley & Sons; 2010.

Levitt M. Spin Dynamics. Hoboken: Wiley; 2007.

Blümich B. Essential NMR. Berlin: Springer; 2005.

Berger S, Braun S. 200 and More NMR Experiments. Weinheim: Wiley-VCH; 2004.

3.4.4 Theory

Chemistry is the science of transforming and understanding chemical bonds. These are determined by the orbitals the electrons follow in an assembly of atoms that form or are part of a molecule. Because electrons possess an electric charge, their motion defines a current, which produces a magnetic field that adds to the applied magnetic field and affects the resonance frequency ν_0 of the observed nucleus. The primary effect at moderate field strength of 0.1 T and above is an observable shift of the resonance frequency. This shift is known as the *chemical shift*. It is measured as a relative difference δ of the observed frequency ν_0 from the resonance frequency ν_r of a reference compound, which is tetramethylsilane for ^{1}H and ^{13}C NMR spectroscopy,

$$\delta = (\nu_0 - \nu_r)/\nu_r. \tag{3.4.1}$$

The chemical shift is typically small for light nuclei. It ranges between 0 and 12 ppm for ^{1}H and 0 to 250 ppm for ^{13}C and as a relative quantity is independent of the strength of the applied magnetic field whereas the absolute frequency difference $\nu = \nu_0 - \nu_r$ is proportional to field strength (Fig. 3.4.1a). Different chemical groups in a molecule have different chemical shifts (Fig. 3.4.2b), and the chemical shifts are tabulated and can be predicted from the molecular structure (Fig. 3.4.2a) to facilitate the reconstruction of the chemical structure from the NMR spectrum.

In addition to the chemical shift, there is a second quantity that bears chemical information. This is the *indirect coupling* between nuclei that are connected by a path

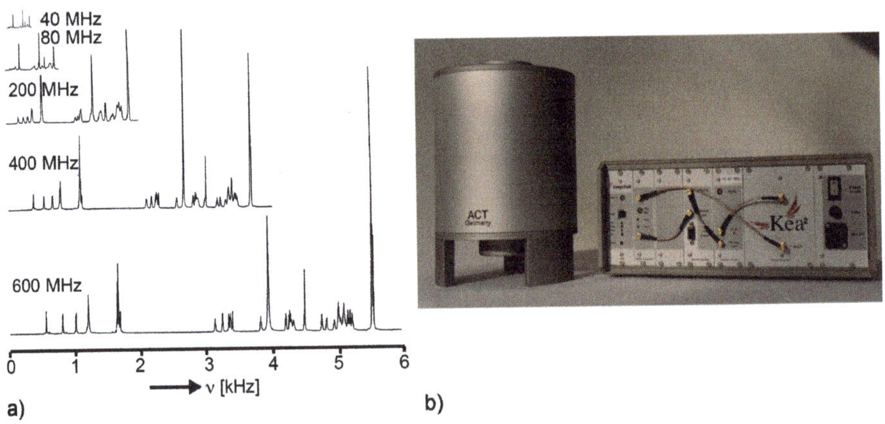

Figure 3.4.1. NMR spectroscopy. (a) ^{1}H NMR spectra measured at different Larmor frequencies $\nu_0 = \gamma B_0$ corresponding to different magnetic field strengths B_0. The spectra are plotted on the same frequency scale to illustrate the increase of chemical shift dispersion with increasing field strength. The vertical axis shows the signal amplitude due to thermodynamic equilibrium polarization, which, like the Larmor frequency, is also proportional to B_0. (b) Compact magnet for ^{1}H NMR spectroscopy at 42 MHz.

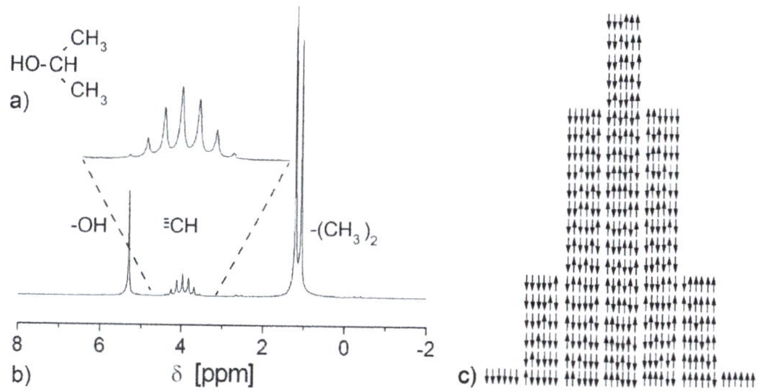

Figure 3.4.2. 42 MHz ^{1}H NMR spectroscopy of isopropanol. (a) The chemical formula identifies three magnetically distinct chemical groups: -OH, ≡CH, and $-CH_3$. (b) NMR spectrum of the undiluted compound. The ≡CH and $-CH_3$ resonances split by mutual J-coupling. The J-coupling to the -OH group has been suppressed by rapid exchange of the -OH group in the presence of a catalyst. (c) Possible orientations of the 6 methyl protons, which give rise to 7 distinct magnetic fields and are sensed by the ≡CH group via the indirect coupling. The phase-sensitive spectrum has been acquired in a single scan with 8 k data points and a 0.2 ms sampling interval. The data were processed by zero filling to 64 k data points, Fourier transformation and phase correction.

leading through one or more chemical bonds: The orientation of the magnetic moment of one spin affects the magnetic field produced by the electrons orbiting in the chemical bond of this nucleus, which in return is sensed by the magnetic moment of the nucleus at the other side of the chemical bond. The energy of this nuclear coupling mediated by the electrons of the chemical bond has been given the symbol J. This is why the indirect coupling of nuclei is also called *J-coupling*.

Note that there is also a *direct coupling* of magnetic moments through space that does not require electrons as messengers of the presence of the partner nucleus. The direct coupling is the *dipole-dipole interaction* (Fig. 5.1.1). Because it vanishes on average for molecules which tumble rapidly in all directions, such as the solute molecules in liquid solution, it is not observed in simple liquid-state NMR spectra. Yet it can be introduced, for example, by confining the solute molecules to anisotropic pores of a stretched gel, so that the dipole-dipole interaction is only partially averaged out, and the spectral changes from the *residual dipole-dipole interaction* are employed to assist a conformational analysis. In soft solid matter like *tendon* (Section 6.2) and *elastomers* (Section 5.1), the residual dipole-dipole interaction provides information about material properties via orientation-dependent *relaxation rates* (Figs. 5.1.3e and 6.2.3) and *multi-quantum build-up curves* (Fig. 5.1.6). Both the absolute values of dipole-dipole interaction and of the indirect coupling are independent of the magnetic field strength.

In liquid-state NMR spectra, the indirect and the direct couplings split the resonance lines from particular chemical groups. The splitting patterns depend on the nuclear angular momentum quantum number I of the spin and the number of spins at magnetically equivalent positions which couple to the observed resonance. For example, the three protons with spin quantum number $I = ½$ of a methyl ($-CH_3$) group give rise to a total spin 3/2. Because each of the three spins assumes either of two orientations ↑ and ↓ in the magnetic field, the three spins of the methyl group in an ensemble of identical molecules produce four different magnetic fields while two magnetically equivalent methyl groups give rise to total spin 3 and 7 different magnetic fields (Fig. 3.4.2c), which are sensed by the chemical groups coupled to the methyl groups. Vice versa, a methine (≡CH) group has total spin ½ and produces two different magnetic fields, which are sensed by its neighboring spins. In a solution there are typically to the order of 10^{20} or more solute molecules so that each combination of spins arises in the collection of all molecules. For example, the spectrum of isopropanol (Fig. 3.4.2a,b) gives rise to a septet for the methine group due to the combinations of magnetic fields from the two methyl groups (Fig. 3.4.2c), and the methyl group is observed as a doublet originating from the two different magnetic fields of the methine group. A further splitting from the indirect coupling to the OH group can also be observed but has been suppressed by the presence of a catalyst.

More complicated molecules such as proteins and mixtures of molecules have NMR spectra with many more lines. To aid in the analysis of the spectra, *multidimensional NMR spectra* can be acquired, which separate the lines crowding a one-dimensional spectrum into one or more further dimensions or correlate different lines to reveal connectivity between resonances in a 1D spectrum by cross peaks, which link different chemical shifts of coupled lines (Section 3.4.8).

3.4.5 Hardware

While 50 years ago field strengths near 1 T were good enough for measuring NMR spectra, today high-field strengths of 10 to 20 T are preferred, because the signal amplitude and the frequency spread or spectral dispersion are proportional to the field strength (Fig. 3.4.1a). But such high fields can only be obtained with costly and maintenance-intensive electromagnets made from superconducting wires. Magnets for spectroscopy at low field strength like 1 T can be fabricated as resistive electromagnets or as permanent magnets. But compared to high-field NMR spectra, the signal amplitude and the frequency range of the chemical shift of low-field spectra are compromised. Due to the latter fact, *low-field NMR spectroscopy* is sometimes referred to as *medium-resolution NMR spectroscopy*. Because permanent magnets are maintenance free they can be used when needed, and because modern magnets are small (Fig. 3.4.1b) they can be operated in the chemistry laboratory and under the *fume hood* for immediate control of reaction products and monitoring *chemical reactions* in real time. The molecules of interest are usually dissolved in a solvent. The solvent should be free of the observed nucleus in order to suppress the solvent signal in the solute spectrum. In ^{1}H NMR spectroscopy, deuterated solvents are commonly used.

At low magnetic field, good signal-to-noise ratio is achieved with large sample volume, high concentration of the solute, and *signal averaging*. The latter requires high temporal stability of the magnetic field, which may be challenged by varying temperature in the laboratory and the movement of metallic objects nearby the magnet. To stabilize the field against such impacts, a second NMR experiment is typically carried out which monitors the deuterium signal from the solvent or the fluorine signal of an external reference compound and locks this signal to a nominal reference frequency by adjusting the current in a coil inside the magnet, which shifts the B_0 field up or down. With the help of such a *field-frequency lock*, NMR spectra can be acquired over many hours without observable frequency shift of the resonance lines.

3.4.6 Pulse sequences and parameters

The standard pulse sequence to acquire NMR spectra consists of a single radio frequency excitation pulse, which generates the *free induction decay* (FID) in response to it (Figs. 3.1.1a and 3.4.3a). This decay is recorded and further processed into the NMR *spectrum*. Representative acquisition parameters for ^{1}H NMR spectra are summarized in Tab. 2.7.1. Typically, the excitation amplitude B_1 and the duration t_p of the excitation pulse are chosen to provide maximum signal amplitude, which is obtained for an excitation flip angle $\alpha = 90°$. But the pulse width specifies the *excitation bandwidth* ν_{max}. As a rule of thumb, $\nu_{max} = 1/t_p$. Given a *chemical shift range* $\delta_{max} = 12$ ppm for ^{1}H, the excitation bandwidth at 40 MHz is $\nu_{max} = 480$ Hz, so that one should choose $t_p \leq 4$ ms with the transmitter frequency centered in the middle of the spectrum.

The *FID* is sampled until the signal vanishes in the noise. At a noise level of 1%, this is for a time $t_{acq} = 5\,T_2^*$, because within that time the signal decays to less than 1% of its initial amplitude. Here T_2^* is the *transverse relaxation time* which includes signal attenuation from *magnetic field inhomogeneity*. Taking $T_2^* = 800$ ms, then $t_{acq} = 4$ s. The minimum rate $1/\Delta t$ at which the FID needs to be sampled is determined by the width of the spectrum. If the reference frequency of the rf transmitter is placed in the middle of the spectrum and the transverse magnetization is sampled as complex data points, then $1/\Delta t = \nu_{max}/2$ so that $\Delta t = 4$ ms, and 1000 data points are acquired in $t_{acq} = 4$ s with a *spectral resolution* of $\Delta\nu = 1/t_{acq} = 0.25$ Hz. This *resolution* can artificially be enhanced by *zero filling*, i.e. by artificially extending the data record with zeros before *Fourier transformation*. The recycle delay t_R follows the *acquisition period* t_{acq}. The sum $t_{acq} + t_R$ should be set to $5 \times T_1$ to recover 99% of the thermodynamic equilibrium magnetization. A reasonable estimate of T_1 for liquids is 2 s so that with $t_{acq} = 4$ s, the *recycle delay* is $t_R = 6$ s. Note that these considerations apply to NMR spectrometers with analogue filters that limit the signal bandwidth and filter the noise. Most NMR spectrometers today employ digital filters that process the signal sampled at a much higher rate in real time. Parameters representative for automatic acquisition of ^{1}H NMR spectra at low field are summarized in Tab. 3.4.1.

Table 3.4.1. Parameters for measuring ^{1}H NMR spectra of liquids with digital filtering

Parameter	Value
NMR frequency ν_0	42 MHz
power of rf amplifier	10 W
duration t_p of 90° pulse	10 µs
dwell time Δt	200 µs
recycle delay t_R	10 s
number n_s of scans	multiple of 4
observation time t_{acq}	3–6 s
number n_{data} of data points	16 k–32 k

The simplest *two-dimensional NMR* experiments are obtained by applying two or three excitation pulses and by varying the delay between two pulses before data acquisition (Fig. 3.4.3b,c). In this way, the thermodynamic equilibrium magnetization established after the *recycle delay* t_R is modulated by the pulses and time delays before the data are acquired during $t_{acq} = t_2$, i.e. the data *acquisition time* starts with non-equilibrium magnetization. From scan to scan, this non-equilibrium magnetization is changed by incrementing the *evolution period* t_1. For each value of t_1, the acquired magnetization is stored in a different row of a matrix. The resultant rows and columns of this matrix are labeled by the evolution period t_1 and the *detection period* t_2. By applying a *2D Fourier transformation* rows and columns are transformed from the time domain into the frequency domain, and a *2D spectrum* is obtained.

The two-pulse experiment is called *COSY* for *COrrelation SpectroscopY* (Fig. 3.4.3b). The three-pulse experiment is called *EXSY* for *EXchange SpectroscopY* if the delay of the first two pulses is varied or *multi-quantum* (MQ) experiment, if the delay between the second and the third pulses is varied (Fig. 3.4.3c). In either case new information is obtained in the resultant 2D spectra that is unavailable in conventional 1D spectra. A COSY spectrum identifies the lines that belong to the multiplets arising from the same mutual spin coupling. An EXSY spectrum reveals slow dynamic processes like conformation changes, which modulate the chemical shift. The same excitation scheme is used to identify resonances from protons at different chemical shifts, which relax in a concerted way, because they are physical neighbors, albeit not coupled by chemical bonds. One then speaks of *NOESY* or *Nuclear Overhauser Effect SpectroscopY*. This type of spectroscopy is the basic scheme for studying protein conformations in solution. Lastly, an MQ spectrum provides information similar to a COSY spectrum by identifying coupled spins, but not through the manipulation of one spin at a time, but of two coupled spins simultaneously.

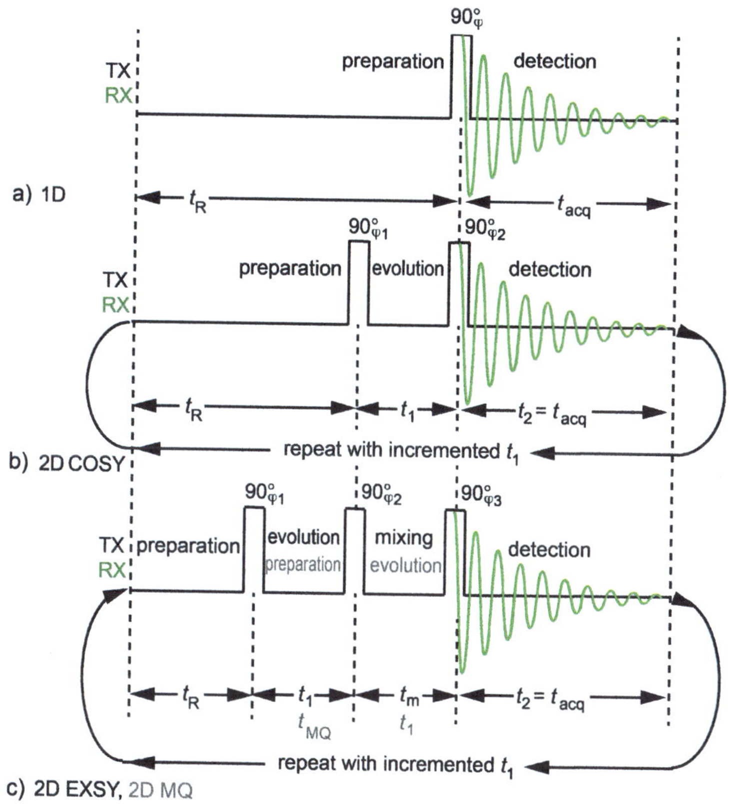

φ_1	φ_{acq}
x	*-y*
y	*x*
-x	*y*
-y	*-x*

φ_1	φ_2	φ_{acq}
x	*x*	*x*
x	*y*	*-x*
x	*-x*	*x*
x	*-y*	*-x*
-x	*x*	*-x*
-x	*y*	*x*
-x	*-x*	*-x*
-x	*-y*	*x*

repeat with *x* replaced by *y*

φ_1	φ_2	φ_3	φ_{acq}
x	*x*	*x*	*x*
-x	*x*	*x*	*-x*
x	*x*	*-x*	*-x*
-x	*x*	*-x*	*x*
x	*x*	*y*	*y*
-x	*x*	*y*	*-y*
x	*x*	*-y*	*-y*
-x	*x*	*-y*	*y*
x	*-x*	*x*	*-x*
-x	*-x*	*x*	*x*
x	*-x*	*-x*	*x*
-x	*-x*	*-x*	*-x*
x	*-x*	*y*	*-y*
-x	*-x*	*y*	*y*
x	*-x*	*-y*	*y*
-x	*-x*	*-y*	*-y*

Figure 3.4.3. Basic pulse sequences for NMR spectroscopy and phase cycles. (a) 1D NMR spectroscopy. Following the recycle delay t_R during which thermodynamic equilibrium magnetization is established, the free induction decay is acquired during the acquisition time t_{acq}. Note, that the diagram neglects the receiver dead time, as it is usually very short compared to the acquisition time. (b) 2D correlation spectroscopy (COSY). The initial magnetization before data acquisition during time t_2 is prepared in a non-equilibrium state with an rf pulse and an evolution time t_1, which is varied systematically from scan to scan. A 2D spectrum is obtained by Fourier transformation of the recorded magnetization over t_1 and t_2. (c) The two-pulse scheme can be extended to include more pulses. Depending on which pulse separation is varied in the three-pulse scheme, one obtains a 2D exchange spectrum (EXSY: t_m is fixed) or a 2D multi-quantum NMR spectrum (MQ: t_{MQ} is fixed).

3.4.7 Beginner's level measurements

The simplest experiment that gives an NMR spectrum is also the most often used experiment. It consists of applying an rf pulse, measuring the *impulse response* or the *free induction decay* (*FID*) and processing the data into a *spectrum*. The pulse sequence in Fig. 3.4.3a depicts the essentials of the experiment neglecting the receiver *dead time* (Fig. 3.1.1a). The excitation pulse is a 90° rf pulse for maximum amplitude of the FID. This pulse is applied after the *recycle delay* t_R during which the nuclear magnetization relaxes to thermal equilibrium. The FID is sampled for the duration of the *observation time* t_{acq}. A good estimate of this time is $t_{acq} \approx 5\,T_2^*$, the time it takes for the FID to decay to about 1% of its initial value. The signal from several scans is averaged until the *signal-to-noise ratio* of the signal sum is high enough, e.g. 100:1, to identify differences in signal amplitudes of a factor of 100. From scan to scan, the phases of the rf pulses and the receiver are changed in increments of 90° following the CYCLOPS *phase cycle* (Section 2.7).

Although sensitivity and resolution of ^{1}H NMR spectroscopy at 40 MHz are low compared to spectroscopy at high field (Fig. 3.4.1a), spectra useful for chemical analysis can be measured at such low frequency. For example, the spectrum of neat *acetylacetic acid* in a standard 5 mm diameter sample tube acquired with only 8 scans shows a signal-to-noise ratio of better than 32,000 : 1 and a resolution of 0.04 ppm (Fig. 3.4.4a). In fact, the sensitivity is so good that even the ^{13}C *satellites* of the ^{1}H resonances can be seen in the baseline of the spectrum (Fig. 3.4.4b). These small peaks arise from the indirect coupling of ^{1}H nuclei to the 1% ^{13}C nuclei in natural abundance, so that 1% of the ^{1}H resonances split into doublets by the hetero-nuclear *J-coupling* between ^{1}H and ^{13}C, i.e., the ^{13}C satellites have 0.5% of the intensities of the lines from the protons which are bound to ^{12}C.

While the single-pulse experiment is simple, some parameters may be set wrong after all (Tab. 3.4.2). The relative integrals of the multiplets provide the ratios of the protons in the respective chemical groups, but only if the *recycle delay* is chosen large

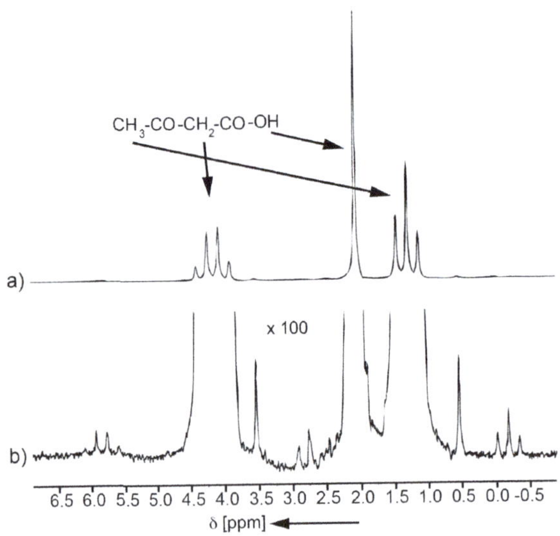

Figure 3.4.4. 40 MHz NMR spectrum of acetylacetic acid measured in 8 scans. (a) The multiplet structure is readily assigned to the chemical groups of the compound. (b) Baseline of the spectrum magnified 100 times. The ^{13}C satellites can be identified due to a signal-to-noise ratio of better than 32.000 : 1.

Table 3.4.2. Common issues encountered in proton NMR spectroscopy

- The recycle delay is too short so that signal integrals are no longer representative of the number of protons in a chemical group
- The transmitter frequency is not centered in the acquisition frequency window, and lines in the spectrum may be folded into the spectral range and appear at wrong frequencies
- The acquisition time is too short, so that the lines in the spectrum are widened
- The acquisition time is too long and noise is acquired and processed, which degrades the signal-to-noise ratio in the spectrum
- The phase cycle is faulty and signals subtract partially in the signal sum upon averaging
- The number of scans is inappropriately chosen for the signal-to-noise ratio needed

enough to allow recovery of each magnetization component to thermodynamic equilibrium. If the *transmitter frequency* is not set to the middle of the NMR spectrum, peaks that would fall out of the spectral width set with the *sampling interval* Δt may be attenuated by the internal filters of the spectrometer and folded back into the frequency window of the spectrum so that they appear at wrong positions in the spectrum. If the *acquisition time* is improperly chosen, the spectral lines may appear too wide or the noise level is too high in the spectrum. A faulty phase cycle typically leads to signal cancellation when accumulating signal instead of signal addition.

3.4.8 Advanced level measurements

There is an abundance of different ways to measure NMR spectra, which provide selected additional information not contained in simple *single-pulse spectra* or which help to disentangle the information provided by spectra crowded with many lines. Such methods are typically employed in studies of larger molecules, for example, proteins. For reasons of sensitivity, investigations of large molecules are unlikely to be conducted with compact magnets at low fields of one or two Tesla. Nevertheless, advanced NMR methods may be useful even at low field, for example, when the chemical shifts cannot unambiguously be extracted from the 1D spectrum due to complex multiplet structures.

The concept underlying most advanced NMR methods is that the excitation makes use of the existence of interactions between spins. These interactions are the *J-coupling* or *indirect coupling* and the *direct coupling* or *dipole-dipole coupling*. Either coupling may act between like spins, such as between two protons, or between unlike spins, such as between ^{1}H and ^{13}C. Indirect coupling is well observed in liquids by small line splittings, while in solids the splittings from the J-coupling are often buried under the lines broadened by the orders of magnitude stronger *dipole-dipole interaction*. The dipole-dipole interaction is typically reduced to zero in liquids by the rapid isotropic rotation of the molecules unless the molecules are partially aligned by embedding them in a liquid crystal or in the anisotropic pores of a strained gel. Then, a *residual dipole-dipole interaction* remains from the anisotropic rotational motion, and lines split from both, the indirect and the direct spin couplings.

The most prominent type of advanced NMR methods is two- or multidimensional NMR. Two pulse sequence diagrams are given in Fig. 3.4.3b and c. Instead of exciting the nuclear spins with one pulse, they are excited with two or more pulses. These pulses can be applied to the same or to different types of nuclei. The basic two-dimensional, homonuclear ^{1}H NMR experiments are the *COSY* experiment with two 90° excitation pulses (Fig. 3.4.3b), and the *exchange experiment* (*EXSY*) and *multi-quantum* (MQ) NMR experiment with three 90° excitation pulses (Fig. 3.4.3c). The hetero-nuclear version of the COSY spectrum is called *HetCor* for *hetero-nuclear correlation*. A HetCor spectrum helps to assign the lines in the ^{1}H NMR spectrum through the ^{13}C NMR spectrum, where the chemical shift range of 250 ppm for ^{13}C provides far more chemical detail than the ^{1}H chemical shift range of only 12 ppm. The ^{1}H NMR signal is detected indirectly via the modulation of the ^{13}C NMR signal which is acquired during the *observation time* t_{acq} for different evolution periods of the ^{1}H magnetization. The 2D spectrum is produced by two-dimensional *Fourier transformation* of the magnetization acquired as a function of the evolution period t_1 and the *detection period* $t_2 = t_{acq}$.

The ^{1}H NMR spectrum of *ethanol* (CH_3-CH_2-OH) exhibits signals at three *chemical shifts*, which correspond to the methyl (low δ), methylene (intermediate δ), and the OH (large δ) groups. *Indirect couplings* are active between all groups in neat ethanol, while in ethanol diluted by water the *J-coupling* to the OH group is suppressed by fast chemical exchange. In a *COSY* spectrum (Fig. 3.4.5a) both frequency axes denote the same chemical shifts, so that the 1D spectrum appears along the diagonal. The COSY spectrum reveals the couplings between groups at different chemical shifts in terms of *cross peaks* or *multiplets*, which appear off the main diagonal. At low field, the J-coupling between the OH group and the CH_3 group of neat ethanol is not resolved, so that no peaks are observed at the cross-coordinates of the associated two chemical shifts.

A hetero-nuclear variant of COSY is HetCor (Fig. 3.4.5b). In a 2D *HetCor spectrum* only the cross peaks between two different nuclei in a chemical group are observed. Often these nuclei are ^{1}H and ^{13}C. Both nuclei are irradiated with rf pulses. Initially the ^{1}H magnetization precesses in the *evolution period* t_1, while at the same time the indirect coupling of ^{1}H to ^{13}C is suppressed by the 180° echo pulse on the ^{13}C magnetization midway during the evolution period. Subsequently, ^{1}H magnetization is transferred to ^{13}C via the hetero-nuclear J-coupling during the mixing period. To properly align the peaks in the resultant spectrum, dephasing and rephasing periods 1(2J) and 1(3J) are part of the *mixing period* in addition to the 90° pulses on ^{1}H and ^{13}C. Because the hetero-nuclear coupling requires the presence of ^{1}H and ^{13}C in the same chemical group, only 1% of the protons is detected in HetCor as compared to 100% in COSY. The signal is then acquired during the detection period t_2 for all values of t_1 required to cover the chemical shift range of ^{1}H with the appropriate spectral resolution.

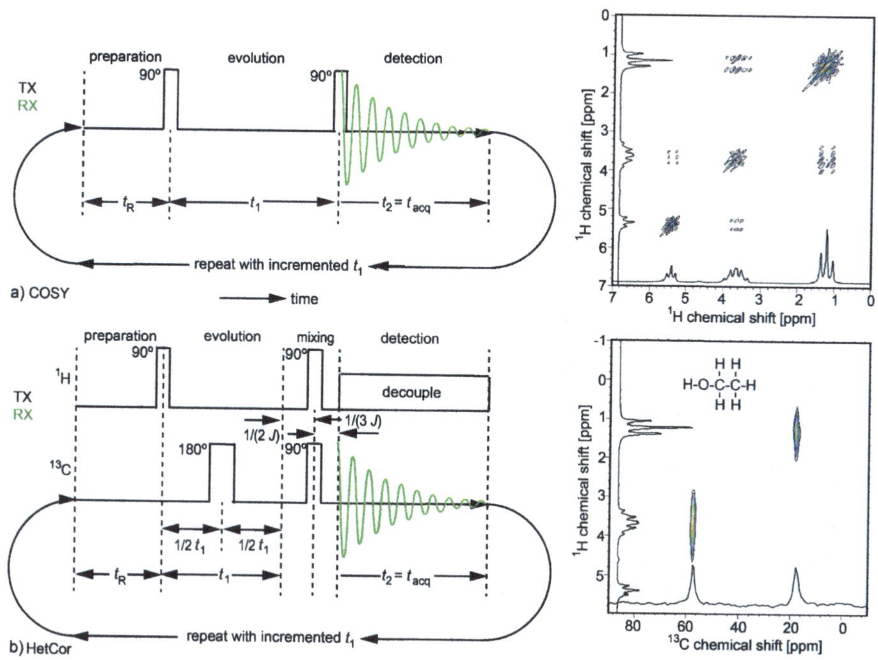

Figure 3.4.5. Basic pulse sequences (left) and spectra (right) of neat ethanol for 2D correlation NMR spectroscopy. (a) ^{1}H COSY NMR. The rf pulse sequence consists of two 90° pulses (Fig. 3.4.3b). The 2D spectrum provides the ^{1}H chemical shift on both axes. The 1D spectrum appears along the diagonal. Off-diagonal peaks connect lines that belong to multiplets with the same homonuclear J-coupling constant. (b) The HetCor experiment is the hetero-nuclear version of the COSY experiment with ^{13}C detection. Radio frequency pulses are applied to ^{1}H and ^{13}C nuclei. Only the signal of ^{1}H bound to ^{13}C is detected. The axes of the 2D spectrum are labeled by the ^{1}H and the ^{13}C chemical shifts. Peaks in the 2D HetCor spectrum identify the ^{1}H and ^{13}C resonances from the same chemical groups.

Two-dimensional NMR methods that generate spectra with cross peaks that cannot be observed in simple 1D spectra fall in the class of 2D *correlation spectroscopy*. There are other 2D methods that do not provide additional signals but simply spread the lines of a 1D spectrum into a second dimension. They belong to the class of 2D *separation spectroscopy*. Such techniques are helpful when analyzing crowded spectra, which are more often encountered at low field where the frequency range of the chemical shift is lower than at high field. The classic representative of this type of experiment is 2D *J-resolved spectroscopy*. The pulse sequence consists of a simple *Hahn echo* (Fig. 3.4.6a). The property of the Hahn echo is that the effects of magnetic field inhomogeneity and chemical shift are eliminated in the echo maximum, while the coupling information is maintained. During the free evolution both chemical shift and coupling modulate the signal. Thus when using the *echo time* as the evolution period t_1 in a 2D experiment and sampling the echo signal during the acquisition time t_2, the

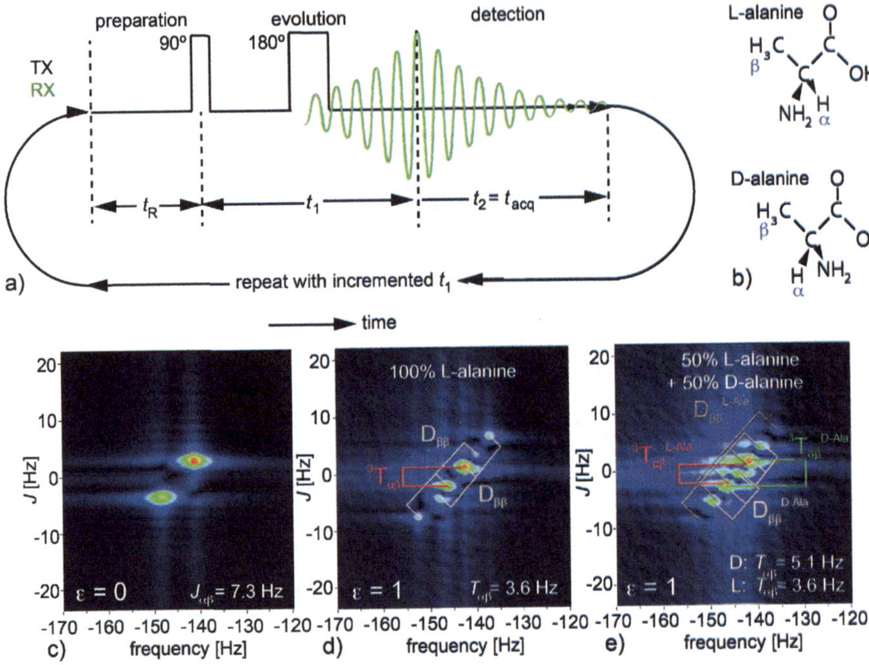

Figure 3.4.6. 2D J-resolved spectroscopy. (a) Pulse sequence. This is the Hahn echo pulse sequence. The second half of the echo is recorded, and the echo time t_1 is varied systematically through a range of values so that the frequency range of the homonuclear coupling between the observed nuclei is covered. (b) Chemical structures of the chiral molecules D-alanine and L-alanine. The indirect coupling between the α and β protons is the same for both structures in isotropic environment ($\epsilon = 0$). (c) 2D J-resolved spectrum of the methyl protons (β) for L-alanine and for D-alanine at 43 MHz. The coupling is observed in both dimensions, the chemical shift only in the directly detected (horizontal) dimension. The methyl resonance is split by the coupling to the α-proton. (d) 2D J-resolved spectrum for L-alanine in anisotropic environment ($\epsilon = 1$). The direct coupling is observed in addition to the indirect coupling. The resultant coupling between α and β protons is denoted by $T_{\alpha\beta}$. The methyl resonance is further split by the homo-nuclear direct coupling $D_{\beta\beta}$. (e) 2D J-resolved spectrum for a mixture of 50% L-alanine and 50% D-alanine in anisotropic environment ($\epsilon = 1$). The direct couplings differ in both compounds, so that chiral molecules can be discriminated with the help of the residual dipole-dipole interaction in anisotropic environment.

signal in the indirectly detected dimension is modulated only by the spin coupling, while the directly detected signal is modulated by spin coupling and chemical shift. Then a multiplet, which appears at a particular chemical shift in the 1D spectrum, appears rotated by 45° in the 2D J-resolved spectrum. By shearing, such a 2D spectrum can be transformed into a 2D spectrum where the multiplets appear in one dimension and the chemical shifts only in the other.

The name 2D *J-resolved spectroscopy* goes back to the early days of 2D NMR spectroscopy when direct coupling was not observed for molecules dissolved in liquids. However *residual dipole-dipole couplings* have since been made observable by confining the solute molecules in an anisotropic environment. An ingenious way of doing so is to incorporate the solution into a gel which fills a rubber tube. When the rubber tube is stretched by changing its original length L_0 by $\epsilon = (L - L_0)/L_0$, the formerly isotropic pores of the gel become elongated and the molecules in the solution no longer tumble isotropically. As a result they experience a *residual dipole-dipole interaction*, which adds to the indirect coupling and is also detected in 2D J-resolved spectroscopy.

Such residual dipolar couplings further split the lines in high-resolution NMR spectra. These splittings are helpful parameters when deriving conformations of complex molecules like proteins in solution. They can also be used to discriminate chiral molecules, such as D- and L-*alanine* (Fig. 3.4.6b). These give the same spectra in isotropic solution (Fig. 3.4.6c). Yet with the gel-stretching technique, residual dipole-dipole interactions can be introduced and these differ for the two mirror-image molecules. The 2D J-resolved spectra of pure L-alanine (Fig. 3.4.6d) and a mixture of D- and L-alanine (Fig. 3.4.6e) clearly reveal these differences even in low NMR field, whereby the spectral resolution is improved in the indirectly detected dimension of the pure J coupling.

There are many more spectroscopic experiments that can be done at low field. Because the technology for low-field liquid-state NMR spectroscopy is still young at the time of writing, examples are few, and many spectroscopic experiments useful for low-field instruments will become available. One particularly interesting 2D experiment is *diffusion-ordered spectroscopy* (*DOSY*, Fig. 4.1.5), whereby the 1D spectra of solutions with mixtures of molecules can be disentangled by resolving the signals at each chemical shift according to the distribution of *diffusion coefficients*. For DOSY measurements, *pulsed field gradients* are needed, which can comfortably be generated inside permanent magnets such as *Halbach magnets* without pole shoes.

3.4.9 Data processing

Phase correction

The *impulse response* at frequency ω_0 induced in the receiver coil of the spectrometer (Fig. 3.1.1a) is mixed and filtered inside the spectrometer with a sine and a cosine wave at frequency ω_{rf} of the transmitter to produce a signal at the difference frequency $\Omega = \omega_0 - \omega_{rf}$, which is digitized and stored in the computer for further processing. For single-pulse excitation this signal is the product of the transverse relaxation function and a wave rotating with frequency Ω,

$$s(t) = s(0)\exp\{-t/T_2\}\exp\{i(\Omega t + \phi)\}. \qquad (3.4.2)$$

For a fully relaxed spin system, $s(0)$ is the thermodynamic equilibrium magnetization M_0. The phase ϕ arises from the particular settings of the spectrometer hardware and from the receiver dead time where the signal cannot be acquired (Fig. 3.1.1), so that $t = 0$ in eqn. (3.4.2) marks the beginning of the signal acquisition. While the phase contribution of the spectrometer hardware can be trimmed to zero by adjusting the *receiver phase* (Section 2.7), the dead time contribution depends linearly on the frequency Ω of the spectrum.

The spectrum $S(\Omega)$ is obtained by *Fourier transformation* (FT) of the signal $s(t)$,

$$S(\Omega) = \mathrm{FT}\{s(t)\} = U(\Omega) + \mathrm{i}V(\Omega) = [A(\Omega) + \mathrm{i}D(\Omega)]\exp\{\mathrm{i}\phi\}. \tag{3.4.3}$$

It is a complex quantity with a real part $U(\Omega)$ and an imaginary part $V(\Omega)$ (Fig. 3.4.7a). Both can be displayed together on the spectrometer screen. They are related to the *absorption spectrum* $A(\Omega)$ and the *dispersion spectrum* $D(\Omega)$ by the phase factor $\exp\{\mathrm{i}\phi\}$. The spectrum to be analyzed is the absorption spectrum $A(\Omega)$. It can be retrieved from the Fourier transform $S(\Omega)$ of the sampled data if $S(\Omega)$ is multiplied by $\exp\{-\mathrm{i}\phi\}$, so

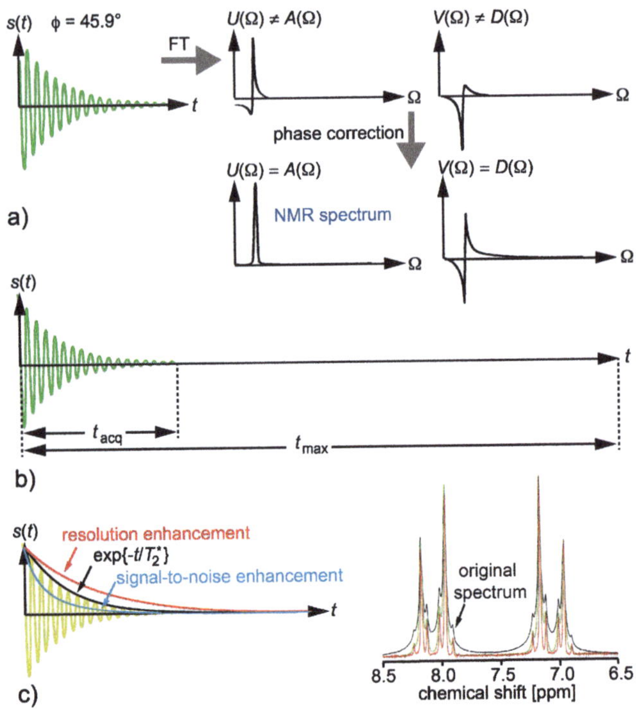

Figure 3.4.7. Data processing in high-resolution NMR spectroscopy. (a) Phase correction. The real part $U(\Omega)$ and the imaginary part $V(\Omega)$ of the Fourier transform of the FID are mixtures of the absorption signal $A(\Omega)$ and the dispersion signal $D(\Omega)$ unless the spectrum is phase corrected. (b) Zero filling. The acquired signal is appended with zeros prior to Fourier transformation to increase the digital resolution in the spectrum. (c) Apodization. Left: The acquired signal is multiplied with an exponential function to improve the signal-to-noise ratio at the expense of resolution or to improve the resolution at the expense of signal-to-noise ratio. Right: Aromatic region of the 42 MHz ^{1}H NMR spectrum of 500 mM 4'-hydroxypropriophenone dissolved in acetone-d_6 and acquired with 8 scans. The resolution gradually improves as the signal-to-noise ratio deteriorates when enhancing the resolution by multiplication of the FID prior to Fourier transformation with $\exp\{t/T_0\}$ as T_0 increases.

that

$$S(\Omega)\exp\{-i\phi\} = A(\Omega) + iD(\Omega), \quad (3.4.4)$$

where

$$\phi = \phi_0 + \phi_1\Omega. \quad (3.4.5)$$

This process is called *phase correction*. The quantities ϕ_0 and ϕ_1 are the constant and the linear phase correction parameters. Modern software will do the phase correction automatically, but a manual fine adjustment may be necessary so that the feet of the resonance lines in $A(\Omega)$ look symmetric also when magnified. Phase errors in the spectrum are also eliminated by computing the magnitude spectrum $|S(\Omega)| = [U^2(\Omega) + V^2(\Omega)]^{1/2}$, but the lines in the magnitude spectrum are broader than in the absorption spectrum so that magnitude spectra are avoided wherever possible.

Zero filling

The frequency interval $\Delta\Omega = 2\pi\Delta\nu$ with which the discrete spectrum is stored on the computer is determined by the length t_{max} of the data record which is Fourier transformed from the time domain into the frequency domain according to $\Delta\nu = 1/t_{max}$. To avoid that a line in a spectrum is represented by just one data point in the spectrum, it is helpful to set t_{max} to large values. On the other hand, the acquisition time t_{acq} should only be as long as the signal in $s(t)$ is larger than the noise to avoid recording noise and carrying it over into the spectrum by Fourier transformation. To accommodate both demands, the signal recorded for a duration t_{acq} is often appended with zeros to four or eight times the acquisition time prior to Fourier transformation (Fig. 3.4.7b). This process is called *zero filling*. It increases the *digital resolution* in the spectrum.

Apodization

If the signal is recorded not fully to the end but extended by zero filling, the lines in the high-resolution spectrum show rapid oscillations or wiggles right and left. To avoid these, the FID can be multiplied with an exponential function $\exp\{-t/T_0\}$, where T_0 is appropriately chosen to attenuate the signal, e.g. $5\,T_0 = t_{acq}$. This multiplication is called *apodization* (Fig. 3.4.7c), and functions other than exponential functions are also used, although the exponential function is the most common one in high-resolution NMR spectroscopy.

Exponential multiplication is also used when the signal is recorded for a time much longer than $5\,T_2^*$, so that the end of the signal record contains mostly noise and hardly any signal. To suppress the noise the best apodization parameter in this case is $T_0 = T_2^*$, and the spectrum is said to have been smoothed with a *matched filter*. This procedure, however, doubles the line width in the spectrum. Thus *signal-to-noise enhancement* is inherently associated with *line broadening*.

On the other hand, if the *signal-to-noise ratio* is good, the *line width* can be reduced and the spectral resolution enhanced by multiplying the FID with $\exp\{t/T_0\}$. This magnifies the tail of the FID and increases the noise, but lines closely separated can now better be resolved. In this way the instrumental resolution limited by the field inhomogeneity across the sample volume can be improved by data processing. This procedure is called *resolution enhancement*.

Two-dimensional NMR

The data processing steps of phase correction, zero filling and apodization apply also to *two-dimensional NMR* and multidimensional NMR data sets. However, phase twists particular to multidimensional Fourier transformation arise when the Fourier transform is not properly computed. If professional NMR data processing software is used, these issues are typically taken care of automatically.

4 Solutions, emulsions, and suspensions

Soft matter made from small molecules is divided in this chapter into solutions, emulsions, and suspensions. *Solutions* constitute a homogeneous phase throughout a mixture, and may be as simple as a solute mixed in a solvent. Such solutions are routinely encountered in chemical analysis by *high-resolution NMR spectroscopy* (Section 3.4). With the compact instrumentation of mobile NMR, the chemical analysis of solutions is readily extended to follow the progress of chemical reactions in *real time* under the *fume hood* [1]. Solutions can also be complex such as crude oil, and biological fluids. They exhibit not only more components but also different phases, and the primary task is to identify their constituents. To this end separation techniques are applied such as *gel permeation chromatography* (*GPC*) and *diffusion-ordered NMR spectroscopy* (*DOSY*). While high-field NMR spectrometers can well be used as GPC detectors [2], they require long feed lines so that the eluate peak broadens from laminar flow over long distances. This is avoided with compact low-field spectrometers at the expense of lower sensitivity and lower spectral dispersion (Fig. 3.4.1a). The perspective of low-cost NMR detection and portability is the motivation for exploring low-field NMR spectroscopy to identify GPC fractions in real time after all. But other than by separation time, fractions can also be separated by differences in their self-diffusion coefficients. This idea underlies DOSY NMR and its variants, where the peaks in the high-resolution NMR spectrum are resolved in a second dimension by the associated *distributions of diffusion coefficients*.

Emulsions are mixtures of immiscible fluids, which are phase separated on the micrometer scale. Examples are milk and *skin cream*. The droplets of the minor phase are suspended in the continuum of the major phase. The *droplet size distribution* is of critical importance for the properties of the emulsion. It is determined by the shear forces applied during production. NMR *diffusion* measurements are a standard technique to determine the droplet size distributions of emulsions.

Suspensions are mixtures of solid phase particles dispersed in a liquid phase. This heterogeneous mixture of two phases may exhibit sedimentation without mechanical agitation. Both phases have very different relaxation times, so that the *solid content* can conveniently be determined by NMR relaxation measurements. Fat is a suspension that is of practical interest in low-field NMR. Depending on temperature, fats consist of different fractions of solid fats and liquid fats or oils. The *solid fat content (SFC)* is the relative amount of solid fat in the total amount of fat. Among others it serves to assess the dietary value of the fat. *Pastes*, on the other hand, are also solid-liquid mixtures, but with a low concentration of the liquid phase, which serves to hold the solid particles together by capillary forces.

All three types of soft matter are usually analyzed by taking samples and inserting them into closed magnets, for example, *Halbach magnets* (Fig. 1.2.1, left). The demands on magnetic field homogeneity, however, are different. While the measurement of NMR spectra requires homogeneous magnetic fields, the relaxation and diffusion measurements needed for studying suspensions and emulsions can be performed in the moderately inhomogeneous fields of partially shimmed permanent magnets and even the strongly inhomogeneous stray fields of the *NMR-MOUSE*. The advantages of using the NMR-MOUSE instead of closed magnets are that closed packages can be inspected nondestructively, and that the sensitive volume is well defined so that concentrations can easily be determined from a component analysis of the NMR signal. The disadvantage is that the sensitive volume is small so the sensitivity is low.

4.1 Solutions

4.1.1 Introduction

A chemical *solution* is a homogeneous mixture of two or more components. The solutions encountered most frequently in NMR are liquid. During a bulk *chemical reaction*, the composition of the solution changes as *educt* concentrations decrease and *product* concentrations increase. In the course of the reaction, intermediate components may arise which are not present in either the initial educts or the final products. Such species are observable by *real-time NMR spectroscopy* in line with the chemical reaction vessel. The composition of a solution can be analyzed by different methods of *2D NMR spectroscopy*, in particular variants of *COSY*, where cross peaks are absent linking resonance lines that belong to different types of molecules, by physicochemical separation such as *GPC* with real-time detection of the eluate spectrum for chemical identification, and by *DOSY*, a 2D NMR method which separates the resonance lines observed in a conventional 1D spectrum according to differences in solute-solvent interactions such as molecular size and shape which give rise to different translational diffusion coefficients.

4.1.2 Objective

The objective in studying solutions is to identify their constituents. Real-time experiments are required in *size-exclusion chromatography* (*SEC*) and *reaction monitoring* where the composition of the solution changes with time. Here the measurement time must be short on the timescale of the concentration changes. If the solution concentrations are at equilibrium, more time is available, and 2D NMR methods can be employed to separate the chemical components not physically but only their NMR signals in a two-dimensional NMR map.

4.1.3 Further reading

Guthausen G. Produkt- und Prozesscharakterisierung mittels NMR-Methoden. Berlin: Logos-Verlag; 2012.

Dalitz F, Cudaj M, Maiwald M, Guthausen G. Process and reaction monitoring by low-field NMR spectroscopy. Prog Nucl Magn Reson Spectr. 2012; 60:52–70.

Callaghan PT. Translational Dynamics & Magnetic Resonance: Oxford: Oxford University Press; 2011.

Cudaj M; Guthausen G, Hofe T, Wilhelm M, SEC-MR-NMR: Online coupling of size exclusion chromatography and medium resolution NMR spectroscopy. Macromol Rapid Comm. 2011; 32:665–670.

Berger S, Brown S. 200 and More NMR Experiments. Weinheim: Wiley-VCH; 2004.

Nordon A, McGill CA, Littlejohn D. Evaluation of low-field nuclear magnetic resonance spectrometry for at-line process analysis. Appl Spectr. 2002; 56:75–82.

Johnson Jr CS. Diffusion ordered nuclear magnetic resonance spectroscopy: principles and applications. Prog Nucl Magn Reson Spectrosc. 1999; 34:203–256.

Maciel G. NMR in industrial process control and quality control. In: Maciel GE, Nuclear Magnetic Resonance in Modern Technology. Dordrecht: Kluwer Academic; 1994, pp. 295–275.

4.1.4 Theory

The constituents of a solution can be disentangled either by physical separation like *size-exclusion chromatography* and chemical *fractionation* or by measurement techniques which give different signals for different molecules. Depending on the complexity of the mixture, either approach is unlikely to achieve a complete separation and identification of the constituent molecules, so that a combination of different approaches is expected to improve the resolution with which similar molecules can be separated. Spectroscopic methods are sensitive to the chemical structure, and NMR spectroscopy in particular provides outstanding sensitivity towards structural details albeit at low mass sensitivity. Although NMR at the low field strengths of permanent magnets is far less sensitive than high-field NMR, it can nevertheless provide relevant information for identifying different molecules in mixtures through the *chemical shifts*, *multiplet splittings*, and *multiplet integrals* extracted from NMR spectra of liquids.

While this information is of purely chemical origin, it can be combined with NMR parameters of physical origin. These parameters are first of all the longitudinal and transverse relaxation times and the translational self-diffusion coefficient. For time-invariant samples, distributions of relaxation times and diffusion coefficients can be chemically resolved in a second dimension, so that the resonance lines of the same molecules can be identified by similar relaxation rates or common diffusion coefficients which can be read off from the second dimension of the NMR spectrum. This is a popular approach for *diffusion-ordered spectroscopy* (*DOSY*) but not for *relaxation-ordered spectroscopy* (*ROSY*). Because 2D NMR methods at low field require longer measurement times, this strategy is not feasible for *real-time monitoring* of solutions with changing compositions, and classical 1D single-pulse NMR spectroscopy is the default method of choice (Section 3.4.7).

4.1.5 Hardware

Most fluids can be filled into sample tubes or piped through tubes passing through closed magnets such as *Halbach magnets*. This is why closed magnets are preferred for fluid analysis, due to their superior sensitivity from a sensitive volume larger than that of the *NMR-MOUSE*. Nevertheless, one may want to use the NMR-MOUSE to inspect substances packaged in sealed containers without breaking the seal. Then the signal attenuation from diffusion in the large gradient of the order of 20 T/m of the stray field needs to be accounted for. The closed magnets used in solution studies are magnets with highly homogeneous magnetic fields suitable for NMR spectroscopy and relaxation measurements (Fig. 4.1.1a) and magnets with weakly inhomogeneous fields suitable for relaxation studies only (Fig. 4.1.1b). DOSY NMR experiments require *gradient pulses*. Such pulses tend to give rise to *eddy currents* in metallic parts close to the gradient coils inside the magnet. This is why magnets without iron pole shoes perform better in studies with pulsed gradient fields.

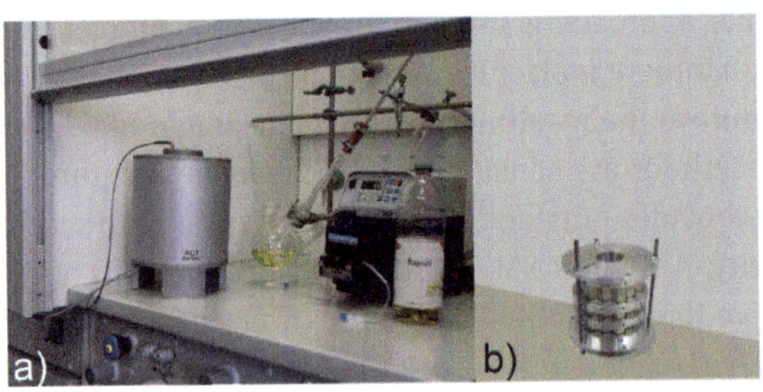

Figure 4.1.1. Desktop NMR instrumentation for solution studies. (a) 40 MHz Halbach magnet with shims from permanent magnets for in-line spectroscopy of solutions flowing through its center and for DOSY NMR. (b) Simple Halbach magnet for diffusion and relaxation studies.

4.1.6 Pulse sequence and parameters

Regular ^{1}H NMR spectra are measured with *single-pulse excitation* following the discussion in Section 3.4.6 with the pulse sequence of Fig. 3.4.3a. Depending on the rate of concentration changes in real-time NMR measurements, signal averaging may need to be kept to a minimum. But even one scan may provide a sufficiently high signal-to-noise ratio at high concentrations. Default values of the acquisition parameters applicable to 1D spectroscopy are listed in Tabs. 3.4.1 and 2.7.1 with and without digital filtering, respectively. For real-time applications, the *number* n_s *of scans* should be set to one if the *signal-to-noise ratio* is good enough.

4.1.7 Beginner's level measurements

A straightforward application of compact NMR spectroscopy is the monitoring of chemical reactions in real time by passing the reactor fluid in a tube through the magnet. For this purpose, the use of compact NMR spectrometers is attractive for two reasons [1–6]. One is that the equipment can be set up next to the reactor under the *fume hood*, so that hazardous reactions can be monitored more easily by NMR (Fig. 4.1.1a). The other is that shorter feeding tubes are needed than for high-field magnets, because compact magnets are smaller. Unless plug-flow conditions are assured, short tubing is essential for high time resolution, because for laminar flow the molecules in the center of the tube flow faster than those near the tube wall. Then, the longer the tube, the more a volume fraction pumped out of the reaction vessel into the tube is spread along the length of the tube until it arrives in the center of the NMR probe where it is detected.

An example for *real-time reaction monitoring* by ^{1}H NMR *spectroscopy* under the fume hood is the observation of the transesterification of vegetable oils to fatty acid alkyl esters, which are known as *biodiesel* (Fig. 4.1.2a) [7]. The reaction can be followed with single-scan spectra, taking one spectrum every 10 s (Fig. 4.1.2b). The reaction progress is traced by observing chemical shift and line intensities as a function of time. The chain length, degree of saturation, and concentration of fatty acid alkyl esters as well as the structure of the ester moiety derived from different feedstock affect the biodiesel properties such as its oxidation and storage properties. By means of low-field ^{1}H NMR spectroscopy a variety of different biodiesel structures can be discriminated (Fig. 4.1.2c). With the help of data mining strategies for spectral analysis the major constituents can be quantified [7] and with the help of chemometric methods the *conversion* can be determined from the low-field NMR spectrum.

Challenges encountered in reaction monitoring in addition to those common to 1D spectroscopy (Tab. 3.4.2) are summarized in Tab. 4.1.1. Reaction monitoring requires flow-through or stopped-flow operations with the reaction fluid passing through the magnet. When measuring flowing samples, the lines in the spectrum are

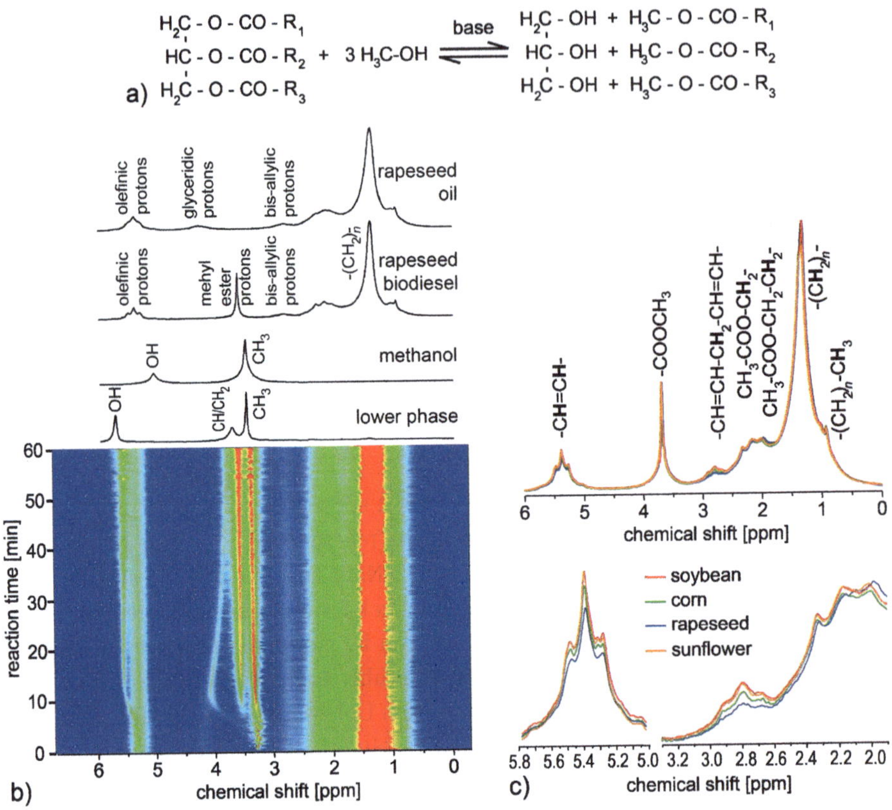

Figure 4.1.2. ^{1}H NMR spectroscopy of biodiesel (adapted from [7] with permission). (a) Transesterification reaction of a fat with fatty acid chains R_1, R_2 and R_3. The base-catalyzed reaction with methanol produces triglycerides (lower phase) and biodiesel (upper phase). (b) Evolution of ^{1}H NMR spectra during the transesterification reaction of rapeseed oil observed in real-time by single-scan spectra acquired every 10 s (bottom) together with educt and product spectra (top). (c) Spectra of biodiesel produced from soybean oil, corn oil, rapeseed oil, and sunflower oil. The different biodiesel sources can be discriminated in the spectra.

Table 4.1.1. Common issues encountered in real-time NMR spectroscopy

– The reaction proceeds too fast to be observed by flow-through NMR spectroscopy
– The concentrations of the educts, reaction intermediates, or products are too low to be detectable with a few scans
– The reaction mixture is passed through the magnet too rapidly leading to line broadening and magnetization lower than in thermodynamic equilibrium
– The recycle delay is longer than the time it takes to replace the sample volume from scan to scan so that it can be shortened to $t_R < T_1$

broadened in proportion with the flow velocity. The concentrations in the reaction mixture should change on timescales that are long compared to the measurement time for one spectrum including signal averaging. The recycle delay between scans can be shortened to the time it takes to replace the fluid in the measurement volume. But the number of scans per spectrum must be high enough to detect the educts, reaction intermediates, and products in question.

4.1.8 Advanced level measurements

Coupling of size-exclusion chromatography and ^{1}H NMR spectroscopy

The combination of *size-exclusion chromatography* and *high-resolution NMR spectroscopy* is attractive because different fractions in a *chromatogram* can be identified [2]. However, high cost and high maintenance discourage widespread use of this powerful combination. Because desktop ^{1}H NMR spectrometers with permanent magnets cost only a fraction of high-field machines and essentially do not require maintenance, they are being explored as detectors in *gel permeation chromatography* [8, 9].

In such a setup, the outflow of a *separation column* is led through the magnet of the NMR spectrometer, and ^{1}H NMR *spectra* are acquired while the *eluate* flows through the NMR coil. The challenge to face is the suppression of the *solvent* signal to a level sufficiently high to detect the ^{1}H NMR spectrum of the fraction in the time it takes to pass the coil, unless the flow is stopped for extended signal averaging. The solvent signal may be 1000 times stronger than the solute signal. Nevertheless, the solute signal can be detected if the flow rate, the pulse sequence, and the subsequent data processing procedures are optimized. For model polymer solutions, the principle of separating *molecular weight* by chromatography and identifying the chemical compound by low-field ^{1}H NMR spectroscopy has been demonstrated at 20 MHz [8, 9].

The eluate from a 20 mm × 300 mm preparative size-exclusion column was passed through PTFE tubing to a 13 mm long and 3 mm wide sensitive region of an NMR probe at a rate of 1 ml/min. Over a distance of 62 mm, the fluid was exposed to the magnetic field so that it could gain longitudinal magnetization. Because the longitudinal relaxation time T_1 of the solute was shorter than that of the solvent, the solute magnetization was close to maximum while the solvent magnetization was partially suppressed. The T_1 relaxation time of the solvent was enhanced by removing dissolved oxygen with argon gas. The total measuring time for poly(methylmethacrylate) (PMMA) concentrations of 10 g/l in $CHCl_3$ was about 60 min.

The pulse sequence for *SEC-NMR spectroscopy* (Fig. 4.1.3a) employs a *saturation recovery* sequence, which is commonly used for T_1 relaxation measurements (Fig. 3.2.2a). The longitudinal magnetization is first destroyed by a series of 90° pulses followed by a fixed *recovery delay* t_0. For t_0, a small value like 300 ms is chosen to allow the solute but not the solvent signal to recover. The resultant longitudinal magnetization is then converted to transverse magnetization with a 90° detection pulse for

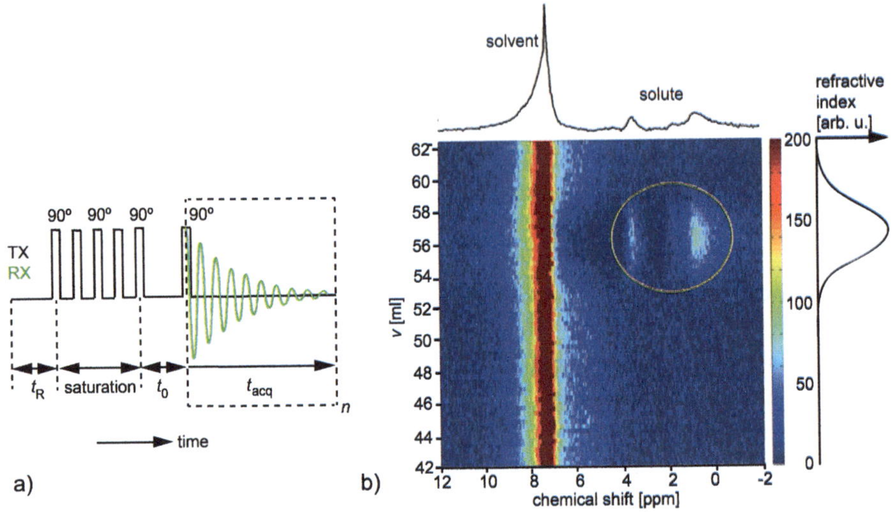

Figure 4.1.3. Coupling of size-exclusion chromatography and 20 MHz ^{1}H NMR spectroscopy (adapted from [9] with permision). (a) The excitation is a saturation recovery pulse sequence with recovery delay t_0, recycle delay t_R, and the acquisition of $n_s = 32$ FIDs in one saturation-recovery cycle for signal averaging. (b) Spectra of 10 g/l PMMA solute in $CHCl_3$ solvent. The number average molecular weight of PMMA was $M_n = 2.19$ kg/mol with a polydispersity index of 1.12. The PMMA signal arises at eluate volumes between 54 and 58 ml, where the change in the refractive index of the eluate assumes its maximum value.

acquisition of the FID. Thirty-two FIDs were acquired and averaged at each saturation recovery step. If the relaxation time T_1 of the solute is known, the flip angle of the detection pulse can be set to the *Ernst angle* (eqn. (2.7.1)). The solvent signal is further suppressed by a short recycle delay ($t_R = 100$ ms).

When properly optimized the *eluate* can indeed be identified by low-field NMR-spectra (Fig. 4.1.3b) in the volume range between 54 ml and 58 ml, the aliphatic and methoxy protons of PMMA are clearly visible on an amplitude scale which covers the bottom 25% of the residual solvent peak amplitude. With digital filtering of the acquired data, the solvent signal can be further suppressed. This demonstrates that in favorable cases, low-field NMR spectroscopy is sufficiently sensitive to be employed for detection of species at low concentrations in real time.

Diffusion-ordered spectroscopy: DOSY

Diffusion is routinely measured in homogeneous fields with *pulsed field gradients.* At least two gradient pulses are needed, one which marks the initial positions of the molecules and another one a time Δ later which marks the final positions of the molecules. With gradient pulses providing effective field gradients G and $-G$, each for duration δ in the two halves of the echo time of a Hahn echo or a stimulated echo, the echo amplitude is attenuated not only by *relaxation* but also by the *diffusion* of molecules along the gradient direction in the time Δ between the pulses (Fig. 3.2.5f). The signal attenuation by relaxation within the echo time can often be neglected, so that only the signal attenuation by diffusion remains,

$$s(G) = s(0)\exp\{-(\gamma G\delta)^2 D(\Delta - \delta/3)\}. \tag{4.1.1}$$

By recording many scans with different gradient amplitudes G, the diffusion coefficient can be extracted from the attenuation of the echo amplitude with the help of eqn. (4.1.1). Alternatively the distribution of diffusion coefficients can be obtained by inverse *Laplace transformation* of the decay. With DOSY, such decay is recorded for every line in the spectrum. The 1D Laplace transformation of the 2D data set is not trivial and different solutions have been published in the literature.

The DOSY experiment is obtained from the single-pulse spectroscopy experiment (Fig. 3.4.3a) by preparing the recorded magnetization with a *diffusion filter* and repeating the pulse sequence with *signal averaging* and *phase cycling* for different settings of the diffusion encoding gradient pulses for a range of gradient values. There are many different variants of DOSY pulse sequences. One commonly employed is based on the stimulated echo (Fig. 3.2.5f) with echo time t_E and pulsed field gradients of duration δ (Fig. 4.1.4a). The initial and final positions of the molecules diffusing for a time Δ are encoded each by a field gradient pulse, and the signal is recorded during the *observation time* t_{acq} following a delay t_z during which the echo is temporarily stored as longitudinal magnetization to accommodate time for the *gradient ringing* to die down.

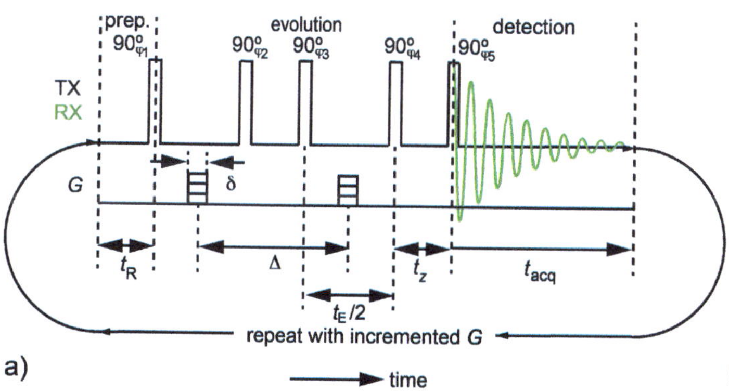

b)

φ_1	φ_2	φ_3	φ_4	φ_5	φ_{acq}
x	x	x	x	x	x
x	x	x	-x	x	-x
x	-x	x	x	x	-x
x	-x	x	-x	x	x
x	x	-x	-x	-x	-x
x	x	-x	x	-x	x
x	-x	-x	-x	-x	x
x	-x	-x	x	-x	-x
x	x	y	y	y	-y
x	x	y	-y	y	y
x	-x	y	y	y	y
x	-x	y	-y	y	-y
x	x	-y	-y	-y	y
x	x	-y	y	-y	-y
x	-x	-y	-y	-y	-y
x	-x	-y	y	-y	y

Figure 4.1.4. DOSY NMR. (a) Pulse sequence. Diffusion is encoded with a stimulated echo and two pulsed field gradients of equal area marking initial and final positions of the diffusing molecules. The delay t_z is introduced to accommodate time for the gradient ringing to die down. The experiment is repeated with different amplitudes of the pulsed gradient G to probe the evolution of the magnetization under the impact of diffusion. Note that in a homogeneous field, constant amplitude rf pulse can be employed, so that the 180° pulse is twice as long as the 90° pulse. (b) Minimum phase cycle.

Table 4.1.2. Parameters for measuring ^{1}H NMR DOSY spectra of liquids

Parameter	Value
transmitter frequency ν_{rf}	42 MHz
transmitter attenuation for 90° pulse	−6 dB at 10 W
duration t_p of 90° pulse	10 μs
dwell time Δt	2 ms
recycle delay t_R	6 s
number n_s of scans	16
echo time t_E	5 ms
observation time t_{acq}	4 s
number n_{acq} of data points	2000
diffusion time Δ	20 ms
gradient pulse width δ	2 ms
number n_G of gradient steps	32
maximum gradient G_{max}	1 T/m

Parameters applicable to DOSY are summarized in Tab. 4.1.2. The minimum phase cycle of 16 steps (Fig. 4.1.4b) requires at least $n_s = 16$ scans for each gradient setting. Of the order of 16 to 32 gradient settings are typically probed. The maximum gradient value G_{max} depends on the diffusion time Δ. The larger Δ, the smaller is G_{max} to attenuate the acquired signal to a few percent of its value without diffusion encoding.

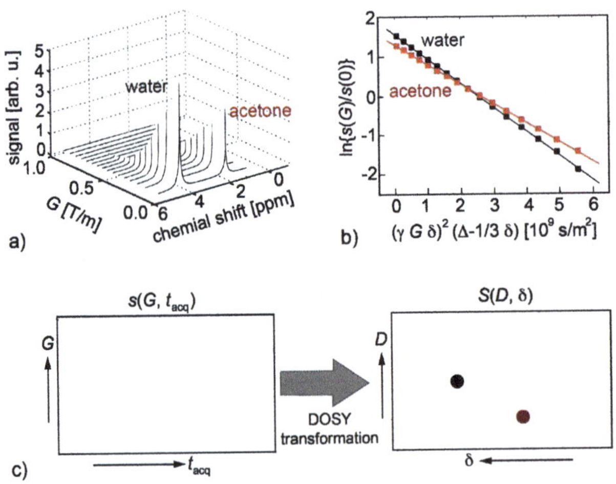

Figure 4.1.5. Principle of DOSY NMR. (a) Diffusion-filtered NMR spectra of water and acetone acquired with the stimulated echo pulse sequence and pulsed field gradients (Fig. 4.1.4a) with $t_z = 0$. (b) Logarithm of the peak integrals from the NMR spectra versus the square of the gradient strength G. The slope provides the diffusion coefficient D. (c) Illustration of the DOSY transformation by which the experimental data are converted into a DOSY spectrum.

The signal decay with increasing strength of the gradient pulses is illustrated in Fig. 4.1.5a by a series of spectra from a mixture of water and acetone. The integrated signal amplitude for each peak decays according to eqn. (4.1.1), so that the logarithm of the normalized signal amplitude is a linear function of the diffusion coefficient D (Fig. 4.1.5b), which can then be determined from the slope of the function. With *DOSY NMR*, this procedure is generalized, and raw data acquired as a function of observation time t_{acq} and *gradient strength* G are transformed into a two-dimensional map of chemical shift δ versus diffusion coefficient D (Fig. 4.1.5c), so that components with overlapping chemical shifts but different diffusion coefficients can be separated.

4.1.9 Data processing

The NMR data acquired in real-time applications and sometimes also those acquired in DOSY experiments are processed in the same way as regular free induction decays are processed into spectra (Section 3.4.9). The main steps are *Fourier transformation* and *phase correction*. The *chemical shifts* of the peaks in the spectrum, the *multiplet splittings*, and the *multiplet integrals* provide the information from which the *chemical structure* of the solute molecules is derived. This analysis is made difficult by the small chemical shift differences at low field and line widths broadened by the flow of spins during signal acquisition. Yet in monitoring applications of NMR spectroscopy, many similar data sets are available which have been acquired at different times, and the information is contained in their differences and similarities. To compare the data sets, statistical data analysis techniques are applied. In the context of chemical analysis these procedures are referred to as *chemometrics* [10].

Contrary to similar spectra from different samples, the raw DOSY data refer to one sample, and the different recordings are correlated via common diffusion processes. The raw DOSY data are converted to a DOSY spectrum by the *DOSY transformation*. This transformation is the conversion of the transverse magnetization s acquired as a function of time t_{acq} after the detection pulse and the amplitude G of the pulsed gradient into the DOSY spectrum S, which is a function of frequency or *chemical shift* δ and *diffusion coefficient* D (Fig. 4.1.5c),

$$s(G, t_{acq}) \rightarrow S(D, \delta). \tag{4.1.2}$$

The *DOSY spectrum* constitutes a set of one-dimensional NMR spectra, where each spectrum refers to a different diffusion coefficient. Although this transformation appears to be straightforward by using a *Fourier transformation* along the time axis t_{acq} and an inverse *Laplace transformation* along the gradient axis G, the instability of the inverse Laplace transformation in the presence of experimental noise and the two-dimensional correlation of the data as a result of the underlying molecular structure and diffusion have led to advanced DOSY transformation algorithms [11–13].

4.1.10 References

[1] Guthausen G, Garnier A, Reimert R. Investigation of hydrogenation of toluene to methylcyclohexane in a trickle bed reactor by low-field Nuclear Magnetic Resonance Spectroscopy. Appl Spectr. 2009; 63:1121–1127.
[2] Bayer E, Albert K, Nieder M, Grom E, Wolff G, Rindlisbacher M. On-line coupling of liquid chromatography and high-field nuclear magnetic resonance spectrometry. Anal Chem. 1982; 54:1747–1750.
[3] Küster SK, Casanova F, Danieli E, Blümich B. High-resolution NMR spectroscopy under the fume hood. Phys Chem Chem Phys. 2011; 13:13172–13176.
[4] McGill CA, Nordon A, Littlejohn D. Comparison of in-line NIR, Raman and UV-visible spectrometries, and at-line NMR spectrometry for the monitoring of an esterification reaction. Analyst. 2002; 127:287–292.
[5] Nordon A, Meunier C, Carr RH, Gemperline PJ, Littlejohn D. Determination of ethylene oxide content of polyether polyols by low-field ^{1}H nuclear magnetic resonance spectrometry. Anal Chim Act. 2002; 472:133–140.
[6] Vargas MA, Cudaj M, Hailu K, Sachsenheimer K, Guthausen G. Online low-field ^{1}H NMR spectroscopy: Monitoring of emulsion polymerization of butyl acrylate. Macromolecules. 2010; 43:5561–5568.
[7] Garro Linck Y, Killner M, Danieli E, Blümich B. Mobile low-field NMR spectroscopy for biodiesel analysis. Applied Magn Reson. 2013; 44:41–53.
[8] Cudaj M, Guthausen G, Hofe T, Wilhelm M. SEC-MR-NMR: Online coupling of size exclusion chromatography and medium resolution NMR spectroscopy. Macromol Rapid Commun. 2011; 32:665–670.
[9] Cudaj M, Guthausen G, Hofe T, Wilhelm M. Online coupling of size exclusion chromatography and low-field ^{1}H-NMR spectroscopy. Macromol Chem Phys. 2012; 18:1933–1942.
[10] Alam TM, Alam MK. Chemometric analysis of NMR spectroscopy data: A review. Ann Rep NMR Spectr. 2005; 54:41–80.
[11] Morris KF, Johnson CS. Diffusion-ordered two-dimensional nuclear magnetic resonance spectroscopy. J Am Chem Soc. 1992; 114:3139–3141.
[12] Nielson M, Morris GA. Speedy component resolution: An improved tool for processing diffusion-ordered spectroscopy data. Anal Chem. 2008; 80:3777–3782.
[13] Stilbs P. Diffusion studied using NMR spectroscopy. In: Encyclopedia of Spectroscopy and Spectrometry. Lindon JC, Tranter GE, Holmes JL, editors. London: Academic Press; 1999. pp. 369–375.

4.2 Emulsions

4.2.1 Introduction

Emulsions are phase-separated mixtures of immiscible liquids. They are encountered in foods, in cosmetics, and in industrial processing operations related to chemistry, pharmacy, and oil recovery. The *flow* properties, *texture*, and *shelf life* are important properties of emulsions, which are inherently connected to the *droplet size distribution* of the discontinuous phase dispersed in the continuous phase. *Microemulsions* have droplets with diameters below 100 nm and *macroemulsions* have droplet sizes between 100 nm and 100 μm. Microemulsions are thermodynamically more stable than macroemulsions, because the droplets in macroemulsions tend to coalesce. This process is called *Ostwald ripening*. It eventually leads to two phase-separated liquids. The *ripening* process, however, can be slowed down by emulsifiers or surfactants, which stabilize the interface between the two phases. Macroemulsions are more common in industry than microemulsions, and *oil-in-water* (o/w) as well as *water-in-oil* (w/o) *emulsions* are encountered most frequently. If clean drops of one phase are dispersed in the other, the emulsions are referred to as *single emulsions*. In *multiple emulsions*, the droplets themselves are complex systems of two or more phases.

4.2.2 Objective

The prime objective in studying emulsions by NMR is to estimate the *droplet size distribution*. While optically transparent emulsions can be studied by laser diffraction and photon correlation spectroscopy techniques, most emulsions are opaque and need to be studied by ultrasound techniques, microscopy, or NMR. *Pulsed field gradient (PFG) NMR* is a standard technique to study single emulsions and derive the droplet size distribution function.

4.2.3 Further reading

Guthausen G. Produkt- und Prozesscharakterisierung mittels NMR-Methoden. Berlin: Logos-Verlag; 2012.

Hardy EH. NMR Methods for the Investigation of Structure and Transport. Berlin: Springer; 2012.

Bernewitz R, Guthausen G, Schuchmann PH. NMR on emulsions: Characterisation of liquid dispersed systems. Magn Reson Chem. 2011; 49:93–104.

van Dynhoven J, Voda A, Vitek M, Van As H. Time-domain NMR applied to food products. In: Webb GA, editor. Ann Reports NMR Spectroscopy 69. Burlington: Academic Press; 2010. pp.145–97.

Voda A, van Duynhoven J. Characterization of food emulsions by PFG NMR. Trends in Food Science & Techn. 2009; 20:533–543.

Johns M, Hollingsworth KG. Characterization of emulsions systems using NMR and MRI. Progr Nucl Magn Reson Spectr. 2007; 50:51–70.

Pena AA, Hirasaki GJ. Enhanced characterrization of oilfield emulsions via NMR diffusion and transverse relaxation experiments. Adv Coll Interf Sci. 2003; 105:103–150.

4.2.4 Theory

The *droplet size distribution* $P(r)$ of most technical emulsions can well be modeled by a *logarithmic Gaussian distribution* of the droplet radius r with mean $\langle r\rangle$ and *standard deviation* σ,

$$P(r) = 1/(r\sigma(2\pi)^{1/2})\exp\{-(\ln[r/\langle r\rangle])^2/(2\sigma^2)\}. \tag{4.2.1}$$

Other distributions may also be considered. The droplet size distribution is determined by comparing the experimentally observed signal attenuation with the one calculated for the model distribution.

Given that the signal decay for restricted diffusion in a single pore with diameter $d = 2r$ is described by eqn. (4.1.1) with a *diffusion coefficient* D that depends on the pore size, the normalized decay for a distribution $P(r)$ of pores is given by

$$\begin{aligned} &s_d(\delta, G, \Delta, D, \langle r\rangle, \sigma)/s_d(\delta, 0, \Delta, D, \langle r\rangle, \sigma) \\ &\quad = \int_0^\infty P(r)s(\delta, G, \Delta, D, \langle r\rangle, \sigma)r^3\mathrm{d}r \Big/ \int_0^\infty P(r)s(\delta, 0, \Delta, D, \langle r\rangle, \sigma)r^3\mathrm{d}r, \end{aligned} \tag{4.2.2}$$

where the pulse-sequence parameters are the *gradient pulse* width δ, the gradient amplitude G, and the *diffusion time* Δ. By fitting the expression (4.2.2) to the observed, normalized signal decay in a diffusion experiment, the parameters $\langle r\rangle$ and σ of the model distribution (4.2.1) and thus the distribution itself are determined. Assumptions underlying this approach are that the diffusion proceeds in spherical pores, that the pore walls are impermeable and do not relax the magnetization, and that the magnetization phases exhibit a Gaussian distribution. For narrow distributions, diffraction effects can be observed in the signal decay.

4.2.5 Hardware

Emulsions are routinely studied by observing *relaxation decays* encoded by molecular *self-diffusion*, because these measurements do not require highly homogeneous magnetic fields. Commonly used instruments are equipped with magnets that provide a moderately homogeneous magnetic field and *pulsed field gradients* (Fig. 4.2.1a). Also, simple *stray-field sensors* like the *bar magnet* NMR-MOUSE can be employed (Fig. 4.2.1b), where diffusion is probed with constant field gradients [3], as well as spectrometers equipped with magnets that provide highly homogeneous fields to resolve the chemical shift in ^{1}H NMR spectra (Fig. 4.2.1c). The latter offer the most convenient way to determine droplet size distributions by means of measuring a set of diffusion-weighted NMR spectra.

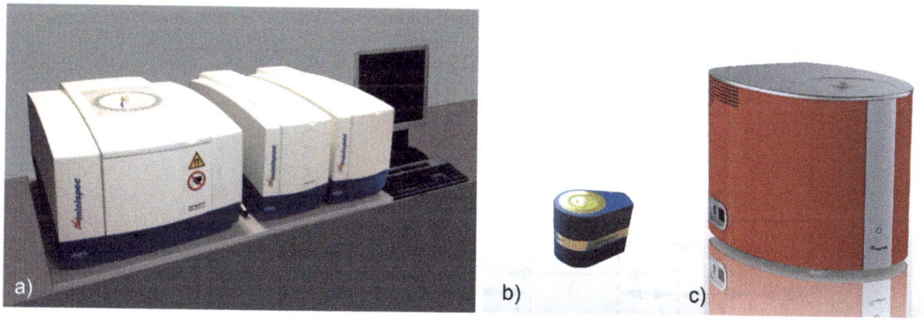

Figure 4.2.1. Equipment for measuring droplet size distributions. (a) Minispec (Bruker) with a temperature-controlled magnet with moderate field homogeneity, a gradient unit, and a radio frequency unit. (b) NMR-MOUSE for inspection of packaged goods. (c) Compact spectrometer (Magritek) with a temperature-controlled magnet with high field homogeneity for chemical shift resolved ^{1}H NMR spectroscopy.

4.2.6 Pulse sequences and parameters

The optimum pulse sequence for determining *droplet size distributions* is the *DOSY sequence* (Fig. 4.1.4), with which a diffusion curve is acquired for each resonance in the ^{1}H NMR spectrum, and the compounds in the different phases of the emulsion can be identified by their chemical shifts in the NMR spectrum. But many low-field NMR instruments do not have magnets suitable to resolve the *chemical shift*, so that the NMR signal from the continuous phase needs to be suppressed by the use of relaxation or diffusion filters before the start of the diffusion measurement.

A T_1 *relaxation filter* works well for *water-in-oil emulsions* (Fig. 4.2.2a). To use it, the longitudinal relaxation time T_{1o} of the continuous oil phase must be determined first (Section 3.2.6). The recovery time t_0 in an *inversion recovery* sequence is set to the value at which the longitudinal magnetization of the oil phase crosses the zero line, i.e. to $t_0 = T_{1o} \ln 2$. Because the longitudinal relaxation time T_{1w} of the water droplets differs from T_{1o}, longitudinal water magnetization remains, and the *diffusion* of the associated molecules is subsequently probed with *pulsed field gradients* and a *stimulated echo*. The basic parameters for the diffusion part of the pulse sequence are those for *DOSY* measurements (Tab. 4.2.1). Usually magnets with moderately inhomogeneous fields are employed, and a *CPMG echo train* with short echo time t_{E2} is used for detection. When measuring *oil-in-water emulsions*, the water signal needs to be suppressed. In this case the T_1 *relaxation filter* does not perform well and can be replaced by a *diffusion filter* (Fig. 4.2.2b) or a T_2 *relaxation filter* (Fig. 4.3.1b). Typical measurement times for food emulsions are of the order of 10 minutes per sample.

When the *chemical shift* can be resolved, the signal of the continuous phase does not necessarily need to be suppressed, because the signals from both immiscible phases arise at different chemical shifts. Also, a simple *FID* is acquired instead of a CPMG echo train, so that the *DOSY sequence* (Fig. 4.1.4) can be employed. The data are

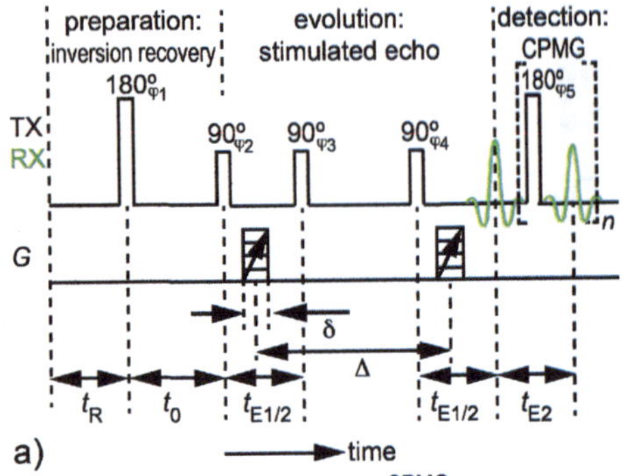

φ_1	φ_2	φ_3	φ_4	φ_5	φ_{acq}
x	x	x	x	y	x
x	x	-x	x	y	-x
x	-x	x	x	y	-x
x	-x	-x	x	y	x
x	x	x	x	-y	x
x	x	-x	x	-y	-x
x	-x	x	x	-y	-x
x	-x	-x	x	-y	x

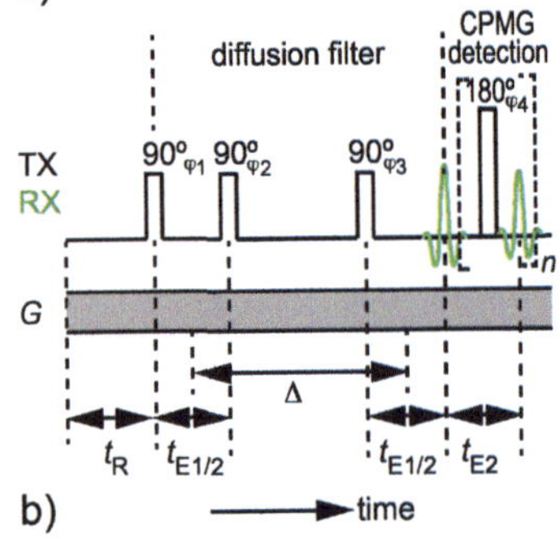

φ_1	φ_2	φ_3	φ_4	φ_{acq}
x	x	x	y	x
x	-x	x	y	-x
x	x	x	-y	x
x	-x	x	-y	-x

Figure 4.2.2. Pulse sequences and minimum phase cycles for determining droplet size distributions of emulsions with diffusion encoding and CPMG detection. (a) Pulsed field gradient scheme for water in oil emulsions. The signal of the continuous phase is suppressed in a preparation period by an inversion recovery sequence with the recovery time t_0 adjusted to the zero crossing of the signal to be suppressed. CPMG decays are detected for different amplitudes of the pulsed field gradients G applied in the evolution time. (b) Constant field gradient scheme for oil in water emulsions. The signal of the continuous phase decays rapidly due to unrestricted diffusion in the constant gradient. Diffusion is encoded in the CPMG detected signal by varying the diffusion time Δ.

then simply processed into sets of spectra versus gradient strength (Fig. 4.1.5a), so that the decays of the peak integrals with increasing gradient strength can be analyzed according to eqn. (4.2.2).

4.2.7 Beginner's level measurements

When droplet size distributions are to be determined from diffusion curves without spectroscopic resolution, the signal of the continuous phase must be suppressed by a *relaxation filter* or *diffusion filter*, which exploits the differences in relaxation or diffusion of the molecules in the two phases (Fig. 4.2.2). The parameters of these filters strongly depend on the sample and cannot be set without investigating the particular product prior to the *diffusion* measurement for determination of the *droplet size distribution*.

Table 4.2.1. Parameters for diffusion measurements with an inversion recovery filter

Parameter	Value
Parameter	Value
transmitter frequency ν_{rf}	42 MHz
transmitter attenuation for 90° pulse	−6 dB at 10 W
duration t_p of 90° pulse	10 μs
dwell time Δt	1 μs
acquisition time t_{acq}	16 μs
recycle delay t_R	10 s
number n_s of scans	4
recovery delay t_0	$T_{1o} \ln 2$
echo time t_{E1}	5 ms
echo time t_{E2}	0.1 ms
number n_E of echoes	5000
diffusion time Δ	20 ms
gradient pulse widths δ	2 ms
number n_G of gradient steps	32
maximum gradient G_{max}	1 T/m

In the NMR spectrum, however, the signals from both phases are usually easy to separate, because chemically very different compounds tend to have signals at very different chemical shifts. In *oil-in-water emulsions*, the oil and water signals are well separated in the ^{1}H NMR spectrum (Fig. 4.2.3a), so that either one can be integrated and the resultant relative amplitudes be plotted versus the strength of the field gradient pulses in a stimulated echo experiment with spectroscopic resolution (Fig. 4.2.3b). The signals from the oil droplets in various food emulsions decay slower with increas-

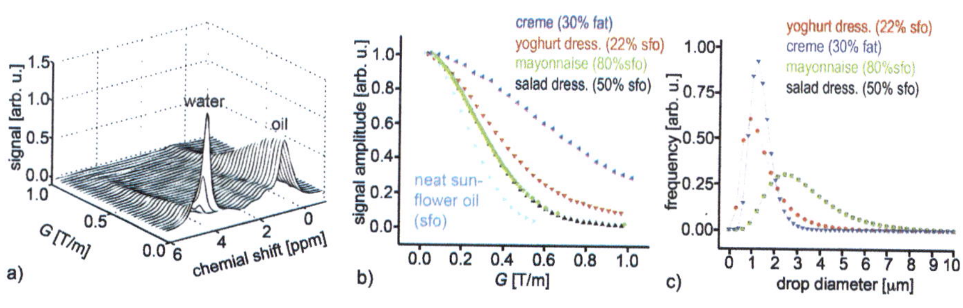

Figure 4.2.3. Determination of droplet size distributions for oil-in-water emulsions by diffusion-resolved ^{1}H NMR spectroscopy at 42 MHz. (a) Diffusion-weighted ^{1}H NMR spectra for a salad dressing versus the amplitude G of the field gradient pulses applied in a stimulated echo sequence with a diffusion time of $\Delta = 0.5$ s. Each spectrum was acquired with two scans. (b) Decays of the integrated oil resonances for different emulsions versus gradient amplitude G. (c) Droplet size distributions of the oil phases derived from the decay curves based on the validity of the logarithmic Gaussian model.

Table 4.2.2. Common issues encountered in measuring emulsions

– The droplet concentration is low and the number of scans needs to be high
– The sample temperature changes during the diffusion measurement
– The gradient pulses are not balanced
– The echo time t_E is too short to accommodate the gradient-pulse ring-down at high gradient amplitude or to probe the diffusive displacement across the pore diameter
– The recycle delay is set too short for the longitudinal magnetization of the disperse phase to recover between scans

ing gradient amplitude G than the signal from pure oil, where the diffusion length is not confined to the size of a pore.

When determining droplet size distributions by measuring diffusion-weighted NMR spectra, the *echo time* needs to be chosen large enough to accommodate the gradient pulses and their ring-down time (Tab. 4.2.2). Furthermore, the gradient pulses need to be well balanced, i.e. have the same time integrals and effectively have opposite signs. During the measurement, the *temperature* of the emulsion needs to be kept constant, because the diffusion coefficient is sensitive to temperature.

4.2.8 Advanced level measurements

Double emulsions

Emulsions may have more complex structures than droplets of one fluid embedded in the continuum of another fluid. *Double emulsions* are those where the droplets themselves are *microemulsions*. Then small droplets of the continuous phase from the macroemulsion or of yet another phase are embedded in droplets of the dispersed phase from the macroemulsion. This scheme can be carried on to *multiple emulsions* [1]. If only two fluids are involved, molecules from the continuous phase may *exchange* with molecules from the microdroplets embedded in the other phase. Without exchange, the attenuation of the signals from the continuous phase and the microdroplets is different due to unrestricted and confined diffusion, respectively. Because confined diffusion gives rise to less signal attenuation in magnetic field gradients, the signal from the continuous phase can be excluded from the analysis by discarding the initial part and analyzing the longer tail of the measured signal. With exchange of molecules between two environments, these environments may become indistinguishable, and the average signal decay will follow a single exponential function when the exchange is fast on the timescale of the measurement [2].

Studying emulsions with the NMR-MOUSE

The determination of droplet size distributions is based on the observation of the diffusion of the molecules in the dispersed phase of the emulsion. Because *stray-field NMR* facilitates the measurement of diffusion, single-sided NMR sensors like the *NMR-MOUSE* can also be employed for studies of emulsions. To this end, the time-invariant gradient of the sensor must be known from a calibration experiment with a liquid like water for which the diffusion coefficient is known. Because this gradient is very large, typically of the order of 20 T/m, diffusion dominates the signal decay and not relaxation, and the free induction signal or the echo amplitude decays faster for the component with the higher diffusion coefficient than for the component with the lower diffusion coefficient. Thus the signal of pure water decays faster than the signal of pure oil in a strong field gradient (Fig. 4.2.4b) [3]. This is different from the signal decay in homogeneous field, where relaxation and not diffusion dominates the signal decay, and the oil signal decays faster than the water signal (Fig. 4.2.4a) [3].

The pronounced difference in signal attenuation by diffusion can be exploited to separate the oil signal from the water signal in oil-in-water emulsions simply by cutting off the rapidly decaying initial part of the diffusion encoded signal and analyzing the remaining, slowly decaying part (Fig. 4.2.4c). A simple stimulated echo applied in the presence of the constant field gradient of the NMR-MOUSE suffices to encode dif-

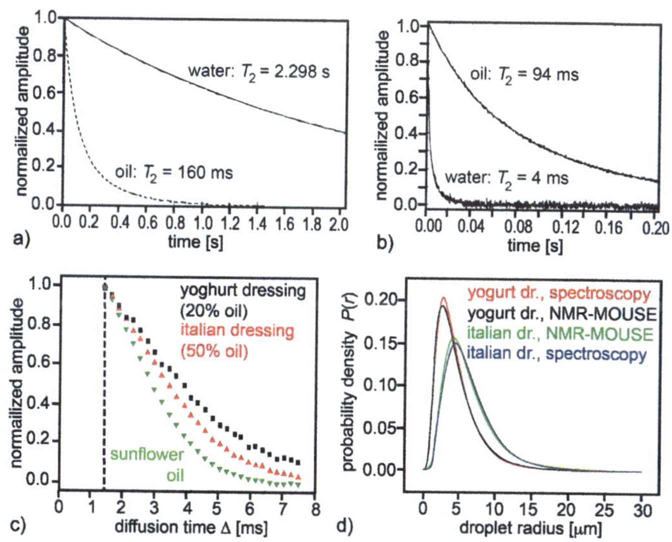

Figure 4.2.4. Droplet size distributions by stray-field NMR. (a) Free-induction decays of oil and water in (a) the homogeneous field and (b) the inhomogeneous field of the NMR-MOUSE (adapted from [3] with permission). (c) Normalized signals acquired with the NMR-MOUSE using a stimulated echo with varying diffusion time Δ and CPMG detection. The signal from the continuous phase decays in the initial part of the curve with $\Delta < 1.3$ ms. (d) Comparison of droplet size distributions measured by spectroscopic detection in homogeneous field and with the NMR-MOUSE.

fusion by varying the diffusion time Δ (Fig. 4.2.2b). The diffusion encoded stimulated echo is subsequently detected with a *CPMG sequence* using a short echo time. The droplet size distributions of Italian and yoghurt *salad dressings* determined in this way with the NMR-MOUSE agree well with those determined in homogeneous field (Fig. 4.2.4d) using pulsed field gradients and spectroscopic detection with a DOSY-type pulse sequence.

4.2.9 Data processing

The signal across the resonance of a disperse phase in the NMR spectrum is integrated and extracted from the NMR spectra as a function of the gradient amplitude G. These functions are normalized to their initial values at zero applied gradient, and the logarithm of this function is modeled with decays calculated for logarithmic Gaussian distributions according to eqn. (4.2.2). From the matching model, the mean $\langle r \rangle$ and the standard deviation σ of the *droplet size distribution* (4.2.1) are obtained so that the distribution can be reconstructed for each emulsion (Fig. 4.2.3c). A comparison of Figs. 4.2.3b and 4.2.3c shows that the signals from emulsions with large droplets decay faster than those from emulsions with small droplets. Chemometric methods are used to group similar signals in large data sets so that different products and processing methods can be identified.

4.2.10 References

[1] Wolf F, Hecht L, Schuchmann HP, Hardy EH, Guthausen G. Preparation of W1/O/W2 emulsions and droplet size distribution measurements by pulsed-field gradient nuclear magnetic resonance (PFG-NMR) technique. Eur J Lipid Sci Tech. 2009; 111:723–729.

[2] Guan XZ, Hailu K, Guthausen G, Wolf F, Bernewitz R, Schuchmann HP. PFG-NMR on W1/O/W2-emulsions: Evidence for molecular exchange between water phases. Eur J Lipid Sci Tech. 2010; 112:828–837.

[3] Pedersen HT, Ablett S, Martin DR, Mallett MJD, Engelsen SB. Application of the NMR-MOUSE to food emulsions. J Magn Reson. 2003; 165:49–58.

4.3 Suspensions

4.3.1 Introduction

Suspensions are liquids consisting of two or more phases, where at least one phase consists of small solid particles suspended in a continuous liquid phase. Because the relaxation times of solids and liquids are very different, both phases can readily be discriminated and their amounts determined by dissecting the NMR signal according to the contributions with different *relaxation times*. More complex structures are unraveled by exploring differences in *diffusion coefficients* and correlating *distributions of diffusion coefficients* and relaxation times.

4.3.2 Objective

Suspensions are frequently encountered in processed foods and fat. Fats are triglycerides, that is, triesters of glycerol with any of several fatty acids. Depending on the fatty acids, they are solid or liquid at room temperature. The food industry is interested in determining *solid fat content* (*SFC*) of foods, and NMR is a well-established method for that purpose. In fact, it is the official method Cd 16b-93 of the American Oil Chemists' Society AOCS, and the SFC method is defined in the norms ISO 8292-1 and ISO 8292-2. Also, *desktop NMR spectrometers* were first developed for determining the amount of protons from solid fat relative to the total amount of protons, which is referred to as SFC. Other mixtures of solids like pastes, viscous food products, and intact plant materials can be analyzed in similar ways.

4.3.3 Further reading

Guthausen G. Produkt- und Prozesscharakterisierung mittels NMR-Methoden. Berlin: Logos-Verlag; 2012.

Voda A, van Duynhoven J. Characterization of food emulsions by PFG NMR. Trends in Food Science & Techn. 2009; 20:533–543.

Song YQ. A 2D NMR method to characterize granular structure of dairy products. Prog Nucl Magn Reson Spectr. 2009; 55:324–334.

Trezza E, Haiduc AM, Goudappel GJW, van Duynhoven JPM. Rapid phase compositional assessment of lipid-based food products by time domain NMR. Magn Res Chem. 2006; 44:1023–1030.

Todt H, Burk W, Guthausen G, Guthausen A, Kamlowski A, Schmalbein D. Quality control with time-domain NMR. Eur J Lipid Technol. 2001; 103:853–840.

Maciel G. NMR in industrial process control and quality control. In: Maciel GE, editor. Nuclear Magnetic Resonance in Modern Technology. Dordrecht: Kluwer Academic; 1994. pp. 295–275.

Gribnau MCM. Determination of solid/liquid ratios of fats and oils by low-resolution pulsed NMR. Trends in Food Science & Technology. 1992; 31:186–190.

4.3.4 Theory

In liquids the *dipole-dipole interaction* among spins is averaged to zero by rapid isotropic motion of the molecules. In solids the motion is often anisotropic like that of the cross-link chains in *elastomers* or slow due to the confinement by close neighbor molecules. For spins that do not experience a dipole-dipole interaction, the *free induction decay* is slow in homogeneous magnetic field, and the apparent relaxation time T_2^* is long. For those that couple with others by the dipole-dipole interaction, the *FID* decays fast, and T_2^* is short. The *single-pulse response* or the FID of a sample that contains both solid and liquid fractions, is the sum of the solid component with a fast decay and the liquid component with a slow decay. The amplitude a_{fast} of the fast decay is proportional to the number of protons in the solid, and the amplitude a_{slow} of the slow decay is proportional to that of the liquid. The relative amplitude $a_{\text{fast}} / (a_{\text{fast}} + a_{\text{slow}})$ of a fat determines the solid fat content SFC.

Solids are easily discriminated from liquids by analyzing the NMR signal in response to excitation pulses in the time domain because the transverse relaxation times of both are vastly different. *Mixtures* of phase-separated liquids can be analyzed in the same way, and, for example, the oil content in a salad dressing can be determined from a relaxation analysis (Section 4.2). But also plasticizers in *polymers*, oil in seeds, the components in phase-separated polymer blends, and the *crystallinity* of *semi-crystalline polymers* can be quantified by a *relaxation analysis*.

4.3.5 Hardware

The hardware required for studying suspensions is the same as that required for studying emulsions. All three types of magnets can be employed, closed magnets with homogeneous fields for *spectroscopic analysis* (Fig. 4.2.1c), closed magnets with weakly inhomogeneous fields for *relaxation analysis* (Fig. 4.2.1a), and the *NMR-MOUSE* for inspection of packaged goods (Fig. 4.2.1b). A simple stray-field sensor can be used without particular specifications of the shape of the sensitive volume as long as the relevant volume section inside the package encloses the sensitive volume of the sensor. *Temperature control* is needed when measuring melting curves, i.e. the solid-fat content as a function of temperature.

4.3.6 Pulse sequences and parameters

Component amplitudes of mixtures, for example, the solid fat content but also the oil content in oil-in-water emulsions are mostly measured on samples inside closed magnets (Fig. 4.2.1a,c). Depending on the homogeneity of the magnetic field, two approaches are taken. In a homogeneous field, the *FID* is measured (Fig. 4.3.1a), and

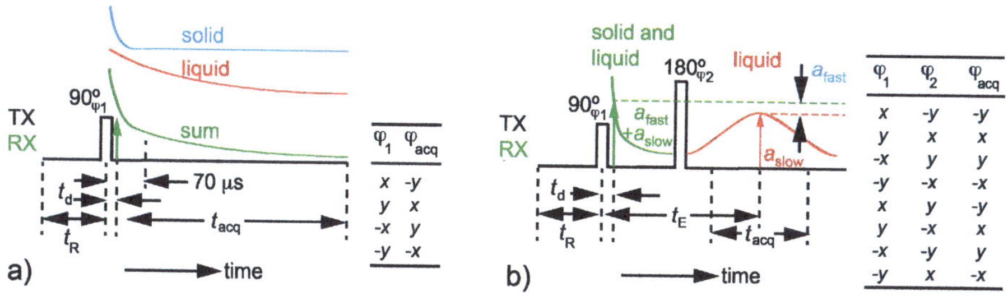

Figure 4.3.1. Pulse sequences for component analysis by the direct method based on differences in the transverse relaxation times. (a) Single-pulse excitation and FID as the sum of a rapidly and a slowly decaying component signal. (b) Hahn echo pulse sequence and full phase cycle for measurement in weakly inhomogeneous field. The echo time t_E is adjusted large enough so that only the slowly decaying signal component gives rise to an echo.

the signal is subsequently decomposed into two decays, a fast and a slow one corresponding to the signal contributions from solids and liquids, respectively, or oil and water. Even simpler, the signal amplitude is measured at just two times, once directly after the rf pulse when all protons in the sample contribute to the signal and a second time some time later when the signal from the solid components has relaxed and only the signal from the protons of the mobile liquid phase remains. In a weakly inhomogeneous field, the initial amplitude and an echo are measured instead with an echo time adjusted to suppress the signal from the solid or more viscous component by relaxation within the first half of the echo time t_E (Fig. 4.3.1b).

The problem with either scheme is that the amplitude at time zero, which is given by all protons, i.e. by the sum of the protons in the solid and in the liquid phases, cannot be measured accurately because the solid signal decays appreciably during the receiver *dead time*. There are different ways out. One is to estimate the signal amplitude at time zero by suitable extrapolation of the signal based on experimental data (Section 3.1.9). Another is to define a fixed time delay t_d after the pulse at which the sum signal is measured and establish a calibration curve with similar samples of known composition that corrects for the underestimation of the rapidly relaxing component extracted from a straightforward analysis of the experimental NMR data. Depending on the instrument, t_d = 70 µs is often chosen to quantify the oil content in *oil-in-water emulsions* (Fig. 4.3.1b). After t_E = 3.5 ms the oil signal has decayed and only the water signal remains. In *suspensions*, the signal from the solid component decays much faster than the signal from the liquid, often within less than 50 µs, so that t_d should be chosen as short as possible, e.g. t_d = 11 µs to catch at least some part of the signal from the solid component, and the amplitude of the liquid is sampled from the impulse response in a weakly inhomogeneous field at 70 µs (Fig. 4.3.1a).

The schemes described so far, which require the signal of the solid component to be determined at time zero, are known as the *direct method*. Two signal amplitudes are sampled, one shortly after the first pulse, which scales with the amplitude sum

Table 4.3.1. Parameters to determine the solid fat content

Parameter	Value
NMR frequency ν_{rf}	42 MHz
transmitter attenuation for 90° pulse	−6 dB at 10 W
duration t_p of 90° pulse	5 μs
dwell time Δt	10 μs
acquisition time t_{acq}	1 ms
recycle delay t_R	10 s
number n_s of scans	8
acquisition delay t_d for the total signal amplitude	11 μs
acquisition delay of the FID amplitude from the liquid	70 μs

$a_{fast}+a_{slow}$, and one at later time, which corresponds to the amplitude a_{slow}. From these two values the solid fat content is calculated following correction of the signal sum for its decay during the receiver dead time. The *indirect method* avoids extrapolation of the signal to zero time by melting the sample completely at an elevated temperature of typically 60° and determining the signal amplitude of the entire sample corresponding to $a_{fast}+a_{slow}$ at 70 μs after the rf pulse.

Table 4.3.1 lists the pulse sequence parameters applicable to both the FID and the echo techniques shown in Fig. 4.3.1. The echo technique with direct detection of the FID can only be applied in weakly inhomogeneous fields. The detection sensitivity can be enhanced when measuring not just one echo but a *CPMG echo train* following the concept of the pulse sequences in Fig. 4.2.2, which employ a diffusion filter. The *phase cycle* for the echo sequence is the same as for the Hahn echo or CPMG echoes (Fig. 2.7.3b), and that for the single-pulse excitation is the CYCLOPS phase cycle (Fig. 2.7.3a).

4.3.7 Beginner's level measurements

Single-pulse excitation is the simplest measurement procedure in NMR, so that the determination of the *solid content* from *suspensions* and *pastes* by NMR is robust. The free induction decay of a fat sample (Fig. 4.3.2a) shows a rapidly decaying component, which vanishes in less than 25 μs from the solid fraction, and a slowly decaying component from the liquid fraction. The NMR signal amplitudes a_{fast} and a_{slow} are determined right after the pulse and at 70 μs when the signal from the solid fraction has essentially decayed to zero, respectively. With increasing temperature, more and more of the solid fraction melts, and the amplitude of the liquid component increases. The solid fat content computed from a series of FIDs taken at different temperatures and calibrated with data determined for samples with known composition define the *melting curve* (Fig. 4.3.2b,c). Different fats have different melting curves, depending

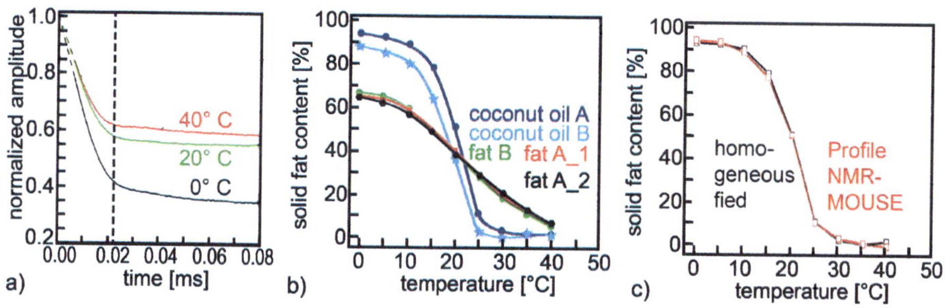

Figure 4.3.2. Determination of solid fat content. (a) Free induction decays of a fat sample measured at different temperatures in homogeneous magnetic field with single-pulse excitation. The signal from the solid components has decayed after about 25 μs. (b) Solid fat content determined with the indirect method as a function of temperature for different fats and reproducibility of the measurement for fat A. (c) Comparison of melting curves derived from single-pulse excitation in homogeneous field and Hahn echo excitation in the inhomogeneous field of the NMR-MOUSE.

on their composition. The reproducibility of these curves is excellent even when determined with different methods, for example, from free induction decays of samples in a homogeneous field (Fig. 4.3.2b) and from measurements through the packaging with the *NMR-MOUSE* in an inhomogeneous field using *Hahn echoes* (Fig. 4.3.2c).

Depending on the investigated *suspension* or *paste*, more than two components may contribute to the signal, e.g. water, oil, and solids in seeds. Then the default parameters specified in Tab. 4.3.1 need to be carefully checked, so that relevant signal contributions are measured and the others duly suppressed. A critical issue in determining solid fat content is the thermal history of the sample, which has to be tightly controlled according to ISO 8292-1 and ISO 8292-2 when absolute values are to be reported. These and further considerations to be addressed in executing the experiment are summarized in Tab. 4.3.2.

Table 4.3.2. Common issues encountered when determining solid fat content

- The sample temperature is not at equilibrium with the environment or the thermal history of the sample does not follow standard procedures
- The time t_{acq} when acquiring the amplitude of the liquid signal in the FID measurement or the echo time t_E in echo measurements needs to be adjusted to suppress unwanted signals in particular in multi-component mixtures
- The recycle delay is too short and the signal from the liquid phase with long T_1 is partially suppressed
- The number of scans is insufficient
- The number of echoes is set too high so that the rf circuit of the probe heats up and eventually the sample, and the rf power amplifier may fail
- The initial signal from the solid and the liquid components is not correctly extrapolated to zero, e.g. by calibration with a wrong set of samples

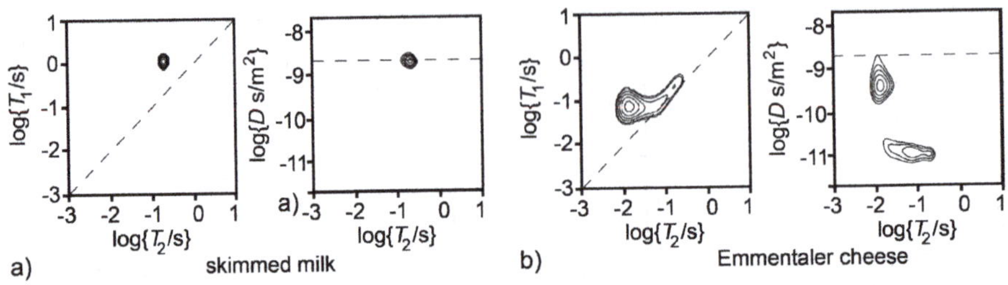

Figure 4.3.3. 2D correlation maps of T_1 and T_2 relaxation time distributions (left) and of diffusion and T_2 distributions (right) (adapted from [3] with permission). The dashed lines indicate the $T_1 = T_2$ limit (left) and the diffusion constant from bulk water (right). (a) Skimmed milk. The signal is from water only. (b) Emmental cheese with signals from water and from fat.

4.3.8 Advanced level measurements

The standard time domain NMR methods of Fig. 4.3.1 with which the solid content of emulsions and similar materials can accurately be quantified with uncertainties well below 1 % are typically employed with desktop instruments equipped with magnets that provide moderately homogeneous magnetic fields. This requires samples to be collected from batches. The same information can be obtained by nondestructive inspection through packed goods using stray-field sensors like the *NMR-MOUSE* with strongly inhomogeneous fields [1, 2]. Because the sensitive volume of the NMR-MOUSE is well defined, absolute concentrations of components can easily be measured.

A more detailed fingerprint of the composition of multi-component emulsions and pastes as well as their structure can be gained from 1D and 2D *Laplace distributions* of relaxation times and diffusion coefficients (Figs. 3.2.4 and 7.1.6) as has been demonstrated for *dairy products* [3, 4]. Dairy products are emulsions of milk fat in an aqueous phase, which also contain casein proteins and lactose. In milk, the casein proteins form micelles, while in cheese they form a continuous porous network. Their aggregation and consistency define the mechanical properties of the product and are determined by the *processing* conditions. The milk fat is contained in spherical globules. At body temperatures, it is liquid, while at lower temperature it starts to solidify.

T_1-T_2 *distributions* of milk and cheese have been measured with an inversion recovery sequence using about 30 values of the recovery delay t_0 logarithmically spaced between 1 ms and 10 s and detecting 8000 CPMG echoes at an echo time of 298 μs (Fig. 4.3.3) [3]. The D-T_2 maps were measured with a stimulated echo sequence (Fig. 3.2.7a) using a diffusion time of 40 ms with the same *CPMG* detection as for the inversion recovery preparation. Diffusion was encoded in a constant field gradient by changing the echo time of the stimulated echo in 64 steps. Because the signal from the aggregated casein proteins decays during the *echo time* of the CPMG detection sequence, it is not detected and only the signals from the liquid components in the dairy products are measured.

Generally, two distinct components are observed in the distribution functions of dairy products (Fig. 4.3.3) [3], a narrow one that corresponds to the watery phase and a broad one that corresponds to the liquid fat. In the T_1-T_2 correlation map, the water peak appears near a characteristic T_1/T_2 ratio of 4, where the exact value depends on the sample composition. The relaxation times of the water signal are significantly lower than those of free water, because the water molecules rapidly exchange between free water and water bound to proteins. On the other hand, the diffusion of the water is close to that of free water as indicated by the dashed line. While skimmed milk produces only one signal from water (Fig. 4.3.3a), another strong signal from fat is observed in cheese (Fig. 4.3.3b). It is characterized by wide distributions of T_1, T_2 and the diffusion coefficient D. The signal from liquid fat has a much lower diffusion coefficient than the water signal, and both signals are clearly separated in the *D-T_2 distributions*, while they overlap in the T_1-T_2 maps and in 1D distributions of relaxation times and diffusion coefficients. This separation of signals can be used to quantify water and liquid fat content in dairy products. Different dairy products lead to characteristically different 2D Laplace maps, which can be used for *product control* and for *quality control.*

4.3.9 Data processing

To determine the solid fat content from component amplitudes of a sum signal it is not necessary to sample a complete FID or CPMG decay and decompose it into different contributions with the help of suitable fit functions (Section 3.1). Rather it has been shown that it is sufficient to measure the signal amplitudes at two points in time. With the *direct method*, the amplitude $a_{\text{fast}} + a_{\text{slow}}$ of the solid and the liquid components together is measured right after the first excitation pulse in homogeneous or weakly inhomogeneous field and extrapolated to time zero either numerically or by means of a scaling factor determined in calibration experiments. Furthermore, the amplitude a_{slow} of the liquid component is determined at a time after the signal from the solid component has decayed to zero. With the *indirect method*, two measurements are needed, one at the temperature of interest determining the amplitude a_{slow} as above, and the other in the same way but at a temperature when all solid fat is molten. This avoids the extrapolation to time zero and gives the true solid fat content from the measured signal according to

$$SFC_{\text{indirect}} = a_{\text{fast}}/(a_{\text{fast}} + a_{\text{slow}}), \tag{4.3.1}$$

where a_{fast} is the difference of the signal amplitudes determined at the two sample temperatures. Accounting for the underestimation of a_{fast} with the direct method by a factor f, the apparent solid fat content determined with the *direct method* relates to the true solid fat content determined with the indirect method following

$$SFC_{\text{direct}} = fa'_{\text{fast}}/(fa'_{\text{fast}} + a_{\text{slow}}), \tag{4.3.2}$$

where a'_{fast} is the difference of signal amplitudes determined with the direct method at 11 μs immediately after the excitation pulses and at 70 μs. The scaling factor f can be determined by comparison of measurements with the direct and the indirect methods on representative samples.

4.3.10 References

[1] Petrov OV, Hay J, Mastikhin IV, Balcom BJ. Fat and moisture content determination with unilateral NMR. Food Res Int. 2008; 41:758–764.
[2] Veliyulin E, Mastikhin IV, Marble AE, Balcom BJ. Rapid determination of the fat content in packaged dairy products by unilateral NMR. J Sci Food Agri. 2008; 8:2563–2567.
[3] Hürlimann MD, Burcaw L, Song YQ. Quantitative characterization of food products by two-dimensional D-T_2 and T_1-T_2 distribution functions in a static gradient. J Coll Interf Sci. 2006; 297:303–311.
[4] Song YQ. A 2D NMR method to characterize granular structure of dairy products. Prog Nucl Magn Reson Spectr. 2009; 55:324–334

5 Polymers and elastomers

NMR is well suited to investigate *polymer* and *elastomer* materials because they are rich in protons and often soft like *biological tissue*, where *MRI* works well and provides superior relaxation and diffusion *contrast*. Most polymer and elastomer materials are made from synthetic macromolecules as the main constituent and additives like *fillers*, pigments, and processing agents. They are typically water repellant. Biological tissue is made from nature-made macromolecules. Their natural environment is typically water compatible.

Macromolecules are large molecules. They often look like chains where the chain segments are repeat units with particular chemical structures (Fig. 5.0.1a). In the melt and in solution the chains of most synthetic polymers assume the shape of random coils that penetrate each other, while many natural polymers like proteins assume well-defined shapes in their watery environment that relate to their function. *Elastomers* are melts in which the macromolecules are tied to each other via knots called *cross-links* (Fig. 5.0.1b). Solid polymer materials are either glasses, i.e. melts frozen

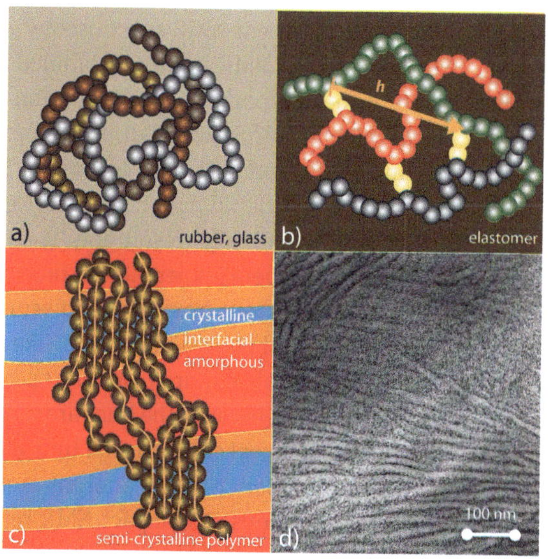

Figure 5.0.1. Most synthetic polymers and elastomers are made from thread-like macromolecular chains. Each bead represents the chemical structure of the repeat unit. (a) In the melt and in glasses the macromolecular chains are disordered and entangled. (b) In elastomers the macromolecules are in the molten state but interconnected by cross-links. Between cross-link points, the macromolecular chains wobble around their end-to-end vectors ***h***. One of these vectors is shown for one cross-link chain. (c) Semi-crystalline polymers consist of amorphous and crystalline domains. The repeat units in the interface are less ordered than in the crystalline domains and less mobile than in the amorphous domains. (d) Many polymers like polyethylene reveal a lamellar structure of crystalline and amorphous domains under the transmission electron microscope.

at temperatures below the *glass transition temperature*, whereby the macromolecules maintain their random coil *conformation*, or they are semi-crystalline materials, in which random domains and ordered crystallites coexist with different sizes, defining the polymer *morphology* (Fig. 5.0.1c,d).

The chemical and physical properties of these materials depend on three factors: (1) The *chemical structure* of the compounds, which can be analyzed in solution by *liquid-state NMR spectroscopy*, (2) the *cross-link density* in the case of elastomers, which is reported by *relaxation* measurements, and (3) the packing of the polymer chains in the solid state or the *morphology*, which is probed by relaxation and advanced *spin diffusion* measurements. For *semi-crystalline polymers* up to three transverse relaxation times are observed. They report on the details of the morphology that are assigned to the repeat units in the *amorphous domains*, the repeat units at the *interface* between amorphous and crystalline domains, and the repeat units in the *crystalline domains*. The *crystallinity* corresponds to the relative amount of crystalline domains. It can be determined from the amplitudes of signal components with different relaxation times. The thickness of these domains can be determined by spin-diffusion NMR experiments with a model-based numerical analysis of the experimental data.

The *chemical analysis* of synthetic polymers needs NMR spectroscopy with high resolution typically of ^{1}H and ^{13}C. High resolution in NMR spectra is obtained in liquid-state NMR spectroscopy by dissolution of the compound in a solvent and in solid-state NMR spectroscopy by *magic angle spinning* (MAS) and high-power decoupling. Analysis in the solid state is not yet available for compact NMR equipment and may remain in the near future in the domain of conventional high-field NMR for reasons of sensitivity and spinning speed. On the other hand, high-resolution ^{1}H liquid-state NMR spectroscopy is already available, and the general experimental procedures for solutions from macromolecules follow those of solutions from small molecules. But to dissolve macromolecules, elevated temperature and particular solvents like aromatic hydrocarbons and chlorinated solvents are commonly required, and the spectra often have lower resolution than those of small molecules because the lines are wider due to slower molecular motion and distributions of chemical shifts caused by a wide range of conformations. Liquid-state NMR spectroscopy for chemical analysis is treated in Sections 3.4 and 4.1, while the following applications of compact NMR refer to the characterization of material properties of solid polymers by unilateral *stray-field NMR* with the NMR-MOUSE, by NMR *relaxometry* of samples inside magnets with relatively homogeneous fields, and by NMR *imaging*.

Tires, conveyor belts, tubes, and fittings are common objects made from *elastomers*. In most cases they are easy to analyze with the NMR-MOUSE because the transverse relaxation time T_2 is long, i.e. larger than 1 ms and a good number of echoes can be acquired in a CPMG train. From a *relaxation analysis* local information can be obtained nondestructively on *cross-link density*, *strain*, state of *curing*, *aging*, and *solvent ingress*. Quantitative information relies on referencing the data from unknown objects to those from known objects. A particularity of most technical elastomer materials is that they appear heterogeneous on the space scale of the sensitive volume of the NMR-MOUSE. NMR relaxation measurements turn out to be so sensitive that the statistical nature of cross-link, chain length, and filler distributions on the course-grained scale of the detection volume of the NMR-MOUSE becomes visible in the NMR signal. Representative relaxation times and signal amplitudes must be derived as *mean values* from several measurements at equivalent positions of the object. The *standard deviation* and the *coefficient of variation* of the NMR data can be used to assess the homogeneity of the material, which depends on the processing conditions.

The same situation is encountered with semi-crystalline materials. Here the crystallite sizes are distributed and lead to local differences of average relaxation times and amplitudes from the material in the sensitive volume when this is shifted across the sample. *Semi-crystalline polymers* and *polymer glasses* are more demanding to measure due to transverse relaxation times much shorter than those encountered in *elastomers*. A short *dead time* of the NMR instrument is essential for investigating such materials, so that the accessible depth is restricted to a few millimeters. Nevertheless, the *phase composition* of the *morphology* can often be traced by a component analysis of transverse relaxation curves to study the effects of *mechanical deformation*, thermal treatment such as *annealing*, *crystallization*, and *aging*. The *solvent ingress* by *inter-diffusion* can be followed when measuring *depth profiles* as a function of exposure time. On the other hand, the Brownian motion of solvent molecules in elastomers and in the amorphous domains of polymers is quantified via the *self-diffusion coefficient* of the solvent, which can be determined with ease and great accuracy with the NMR-MOUSE and related to the solvent and material properties.

5.1 Elastomers

5.1.1 Introduction

Most NMR studies of *elastomer* materials concern *relaxation* measurements. These provide information on *cross-link density*, *strain*, strain anisotropy, *aging*, and *heterogeneity*. Quantitative information is extracted from measured NMR data with the help of network models or by comparison with data from samples with known properties. Because temperature stability is critical, the measurements have to be conducted under strict *temperature control* with temperature variations less than 0.1° C near room temperature or at elevated temperature where NMR parameters vary less with temperature. Advanced measurement methods explore *multi-quantum* effects, which arise when two interacting spins are excited coherently by the rf excitation. In solids and soft materials, the dominant interaction between spins is the *dipole-dipole interaction*, for example, the magnetic attraction and repulsion of the two protons in a methylene group of the polymer chain (Fig. 5.1.1).

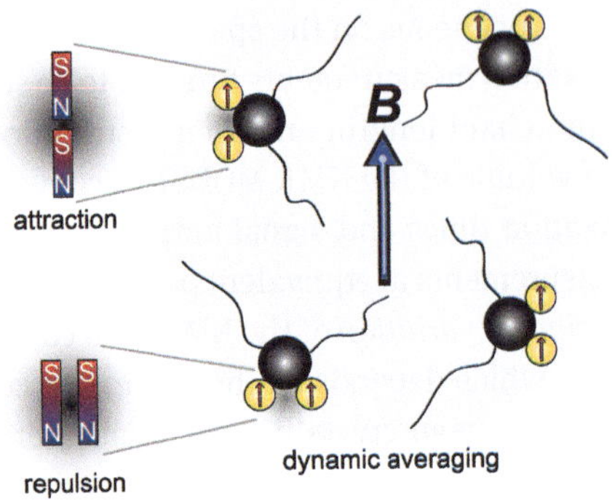

Figure 5.1.1. Attraction and repulsion of two magnetic dipoles close to each other. This is the situation of the protons in a methylene (-CH_2) group of a macromolecular chain. As the chain wobbles, the group rotates in the applied magnetic field ***B***. The field keeps the spins aligned, so that the magnetic interaction alternates between attraction and repulsion upon rotation. For rapid isotropic motion of a free chain, the dipole-dipole interaction vanishes. For rapid anisotropic motion of a chain constrained between two cross-links, a residual dipolar interaction remains.

The applied magnetic field is typically much stronger than the magnetic field one spin experiences from its neighbor spin, so that the applied field keeps the spins aligned along its direction. Then, as the methyl group rotates, the dipole-dipole interaction between the spins alternates between attraction and repulsion. Rapid isotropic motion of a free chain averages the dipole-dipole interaction to zero. Rapid anisotropic motion of a chain constrained between two cross-links (Fig. 5.0.1b), leaves a *residual dipole-dipole interaction*, which accelerates the *transverse relaxation rate* $1/T_2$. The more constrained the motion and the slower the motion, the stronger the residual dipole-dipole interaction and the faster the spins relax. Increasing *temperature* accelerates the motion and increases T_2 while increasing *cross-link density* and increasing *strain* restrict the motion further, leading to smaller T_2 due to larger residual dipole-dipole interactions from less motional averaging.

5.1.2 Objective

The objective in measuring elastomeric objects is to compare relaxation and amplitude data from samples with unknown properties with those from samples with known properties. If performed at the same temperature or extrapolated to the same temperature, such NMR data reveal differences in *cross-link density*, *formulation* defects, *strain*, *processing* history, sample *homogeneity*, and *aging*. Advanced measurements are conducted to test polymer network theories.

5.1.3 Further reading

Kimmich R. Principles of Soft-Matter Dynamics. Dordrecht: Springer; 2012.

Saalwächter K. Microstructure and dynamics of elastomers as studied by advanced low-resolution NMR methods. Rubber Chem Tech. 2012; 85:350-386.

Callaghan PT. Translational Dynamics and Magnetic Resonance. Oxford: Oxford University Press; 2011.

Casanova F, Perlo J, Blümich B, editors. Single-Sided NMR. Berlin: Springer; 2011.

Cheng HN, Asakura T, English AD. NMR Spectroscopy of Polymers: Innovative Strategies for Complex Macromolecules. Washington: American Chemical Society; 2011.

Blümich B, Casanova F, Perlo J. Mobile single-sided NMR. Prog Nucl Magn Reson Spectrosc. 2008; 52:197–269.

Saalwächter K. Proton multiple-quantum NMR for the study of chain dynamics and structural constraints in polymeric soft materials. Progr NMR Spectr. 2007; 51:1–35.

Blümich B. Essential NMR. Berlin: Springer; 2005.

Blümich B, Anferova S, Kremer K, Sharma S, Herrmann V, Segre A. Unilateral nuclear magnetic resonance for quality control, Spectroscopy. 2002; 18:18–32.

Litvinov VM, De PP, editors. Spectroscopy of Rubbery Materials. Shawbury: Rapra; 2002.

Blümich B. NMR Imaging of Materials. Oxford: Clarendon Press; 2000.

Becker ED. High Resolution NMR. New York: Academic Press; 2000.

Callaghan PT. Principles of Nuclear Magnetic Resonance Microscopy. Oxford: Clarendon Press; 1991.

Bovey FA. Chain Structure and Conformation of Macromolecules. New York: Academic Press; 1982.

5.1.4 Theory

Following the theory of rubber elasticity, the *shear modulus* G of an *elastomer* is proportional to the *cross-link density* n and the average *molecular weight* M_c of the chains between cross-links,

$$G = nRT = RT\rho/M_c, \tag{5.1.1}$$

where R is the gas constant, T the absolute temperature and ρ the material density. Note that there are two types of *cross-links*, chemical cross-links and physical cross-links from chain entanglements. Both are included in eqn. (5.1.1).

The *shear modulus* is measured at vanishingly small shear deformation where the relationship between shear stress and shear strain is linear. At higher deformation, the relationship becomes nonlinear. Mooney and Rivlin found a semi-empirical relationship between elongational stress σ and the *elongation ratio* $\Lambda = L/L_0$,

$$\sigma/(\Lambda - \Lambda^{-2}) = 2C_1 + C_2/\Lambda, \tag{5.1.2}$$

where C_1 and C_2 are elastic material constants which denote the chemical and the physical cross-link densities, respectively.

An elastomer consists of a proton fraction x_A of *cross-link chains*, a fraction x_B of dangling *chain ends*, and a fraction x_C of liquid and *sol molecules*. The normalized transverse proton magnetization of $M(t)/M_0$ which persists at time t after the excitation pulse is often approximated for fast chain motion at temperatures larger than about 120° C above the *glass transition temperature* by the *Anderson–Weiss model*

$$\begin{aligned} M(t)/M_0 = {} & x_A \exp\{-t/T_{2A} - qM_2\tau_c^2[\exp\{-t/\tau_c\} + t/\tau_c - 1]\} \\ & + x_B \exp\{-t/T_{2B}\} \\ & + x_C \exp\{-t/T_{2C}\}, \end{aligned} \tag{5.1.3}$$

where each fraction relaxes with its own relaxation time T_{2A}, T_{2B}, T_{2C}. The rapid but anisotropic motion of the cross-link chains leads to a *residual dipole-dipole interaction*, which is expressed in terms of a small fraction q of the *second moment* M_2 of the dipole-dipole interaction of the polymer chains constrained between cross-links and neighboring chains which move more slowly with correlation time τ_c.

The second moment is a quantity popular from the early days of NMR, because it can readily be extracted by integration of the wide lines observed in ^{1}H spectra of solids. If the relaxation rates $1/T_{2A}$, $1/T_{2B}$, and $1/T_{2C}$ are small compared to qM_2, the transverse magnetization decays according to $\exp\{-qM_2t^2/2\}$ so that an *effective transverse relaxation* rate can be defined as

$$1/T_{2\text{eff}} = (qM_2)^{1/2}. \tag{5.1.4}$$

Because qM_2 scales with the square of the residual dipole-dipole coupling, the effective transverse relaxation rate $1/T_{2\text{eff}}$ extracted from fitting a model function to the experimental transverse relaxation decay measures the *residual dipole-dipole coupling*. This coupling increases as the motion of the *cross-link chains* becomes more restricted either with increasing *cross-link density*, with increasing *strain*, or with decreasing *temperature*. Experimental evidence supports the notion that the transverse relaxation rate and the *residual dipole-dipole coupling* are both directly proportional to the cross-link density. Clearly, also chemical modification of the network chains due to material corrosion, i.e. chemical aging, temperature, and strain, affect the residual dipole-dipole interaction and thus $1/T_{2\text{eff}}$, so that analytical correlations can only be established if just one of these parameters varies [7].

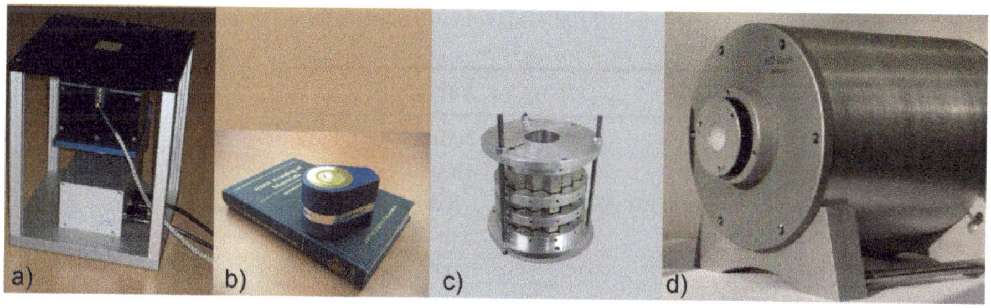

Figure 5.1.2. Magnets for measuring soft solids by ^{1}H NMR. (a) Profile NMR-MOUSE PM5 mounted on a lift for depth profiling of samples placed on the top plate of the device. (b) Bar magnet NMR-MOUSE. (c) 0.7 T Halbach magnet built from two coaxial rings of 12 hexagonal magnet bars each. (d) Compact 0.5 T MRI magnet with a 4 cm diameter bore.

5.1.5 Hardware

Usually *elastomers* are investigated by *relaxation techniques*, which do not require spectroscopic resolution. For elastomers, the transverse relaxation time T_2 is often longer than one millisecond, so that an instrument *dead time* of 50 μs can be tolerated. Most compact NMR instruments have dead times shorter than this, so that any NMR-MOUSE model (Fig. 5.1.2a,b) and Halbach magnets with relatively homogeneous fields (Fig. 5.1.2c) can be employed for studying elastomers and amorphous polymers above the glass transition temperature. *Imaging* experiments require a magnet with pulsed field gradients and good homogeneity for frequency encoding, for example, Halbach magnets (Fig. 5.1.2d). But the use of such closed magnets is limited to narrow samples like gaskets and fittings, which fit through the bore of the magnet.

Short *echo times* are needed to detect the non-exponential decay of the cross-linked chains (eqn. (5.1.3)) so that a few thousand echoes may need to be acquired before the *CPMG echo train* has decayed to zero. However, frequent pulsing of the transmitter heats up the rf coil. To avoid this heat being transferred to the sample, adequately long *recycle delays* are needed between scans. In particular for *elastomers*, variations of the sample temperature need to be avoided because the transverse relaxation near ambient temperature strongly depends on *temperature*, and it is mostly T_2 which is measured to gain information about sample properties like *cross-link density* and *strain*.

5.1.6 Pulse sequences and parameters

In homogeneous magnetic fields, the envelope of the response to a single excitation pulse can be recorded (Fig. 3.1.1a). The receiver *dead time* may be overcome with a simple Hahn echo. In inhomogeneous fields, the transverse relaxation curve is measured with a set of Hahn echoes acquired at variable echo time or with a *CPMG* sequence

Table 5.1.1. Acquisition parameters for ^{1}H NMR relaxometry of elastomers

Parameter	Value
magnet, probe	NMR-MOUSE PM5
transmitter frequency ν_{rf}	18.1 MHz
transmitter attenuation for 90° pulse	−8 dB at 80 W
duration t_p of 90° pulse	5 μs
dwell time Δt	1 μs
acquisition time t_{acq}	5 μs
echo time t_E	40 μs
number n_E of echoes	1000
recycle delay t_R	1.5 s
number n_s of scans	32

(Fig. 3.1.1b). Note that some magnetization components may be locked along the direction of the rf field when CPMG sequences are applied and that resonance off-set effects introduce contributions from stimulated echoes in the recorded echo train, so that T_{2eff} extracted from a CPMG measurement is often found to be longer than T_2 extracted from a set of Hahn echoes. The number of CPMG echoes should be kept as low as possible to reduce heating of the rf coil and thus of the sample from rf power dissipated in the coil. This heat can be removed in extended recycle delays at the expense of increased measurement time. The default excitation scheme uses the *CPMG* sequence and a long recycle delay. Typical acquisition parameters for investigations of elastomer materials with the NMR-MOUSE are summarized Tab. 5.1.1.

5.1.7 Beginner's level measurements

Transverse relaxation measurements provide a considerable amount of information on rubbery materials. Very often, the experimental decay curve is fitted with just a mono or a bi-exponential function when investigating technical *elastomers*. The most important application is the determination of *cross-link density*. It is proportional to *transverse relaxation* rate $1/T_{2eff}$, where the proportionality constant depends on the particular formulation and processing conditions. Within experimental error, often the *cross-link density* expressed in terms of the amount of cross-linker is found to scale linearly also with T_{2eff} (Fig. 5.1.3a) where slight differences are observed in the correlation with cross-link density values from shear (eqn. (5.1.1)) and stress-strain (eqn. (5.1.2)) experiments. The signal amplitude measures the total number of protons in the sensitive volume. Therefore, the NMR-MOUSE is a convenient tool to also determine *gravimetric density* (Fig. 5.1.3b) of elastomeric foams from the same data as *cross-link density*.

Strain experiments can be performed with simple devices (Fig. 5.1.3c). Because the NMR-MOUSE with a U-shaped magnet produces a magnetic stray field with the field

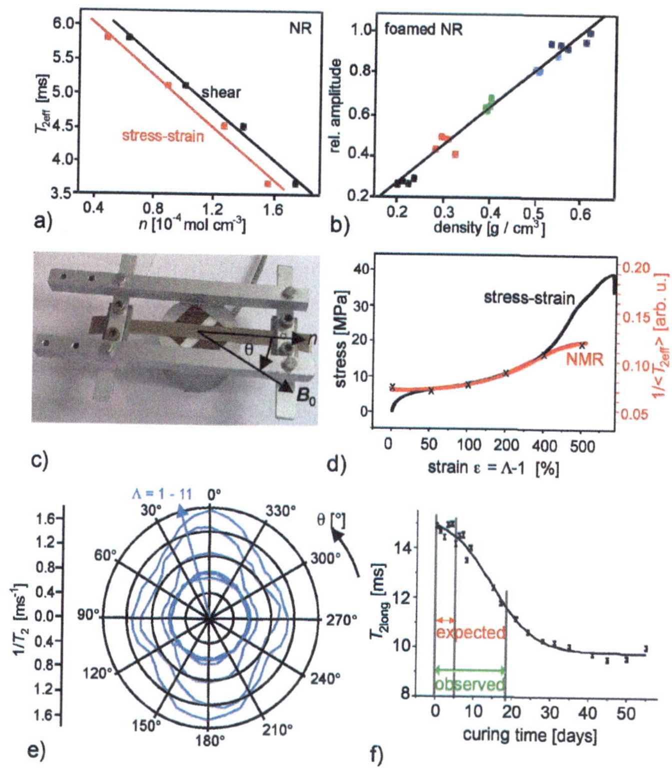

Figure 5.1.3. Applications of relaxation measurements to rubber materials (adapted from [1–4] with permission). (a) Dependence of the transverse relaxation time $T_{2\text{eff}}$ on the cross-link density of natural rubber as determined by shear and stress-strain measurements. (b) Dependence of the amplitude of the relaxation curve measured with the NMR-MOUSE on the gravimetric density of foams from natural rubber. (c) Simple frame for straining rubber bands suitable for relaxation measurements with the NMR-MOUSE. (d) Mechanical force-strain curve and dependence of average inverse relaxation time on the applied strain for natural rubber. (e) Anisotropy of the transverse relaxation rate from Hahn echoes for natural rubber strained to elongation ratios from 1 to 11. (f) Curing of a polyurethane adhesive used for repairing car windshields. The observed curing time dramatically exceeds the expected curing time.

vector pointing across the magnet gap and lying essentially parallel to the surface of the sensor, the angle dependence of the *transverse relaxation rate* $1/T_{2\text{eff}}$ can be determined by conducting CPMG measurements at different angles θ between the strain direction and the magnetic field direction. With increasing *elongation ratio* $\Lambda = L/L_0$, the relaxation rate increases and so does its *anisotropy* (Fig. 5.1.3e). Interestingly, the relaxation rate (at $\theta = 90^\circ$) follows the nonlinear dependence of the stress-strain curve up to strains of $\epsilon = (L - L_0)/L_0 = 400\%$ (Fig. 5.1.3d).

Relaxation measurements are fast, so that many *vulcanization* and *curing* reactions can be followed in real time. The moisture curing of a polyurethane adhesive used for sealing the windshield of a passenger car has been investigated with the NMR-MOUSE in a mock setup by recording CPMG decays of the reacting adhesive layer between magnetic sheet metal and glass windshield. The long relaxation time T_{2long} from a *bi-exponential fit* of the experimental CPMG curves decreased with increasing curing time, revealing a curing time of the order of 20 days instead of the expected 5 days (Fig. 5.1.3f).

While measurements of transverse relaxation curves are simple and straightforward, rubber and elastomers as such are tricky materials. In the vicinity of room temperature, T_{2eff} of most technical elastomer materials strongly depends on temperature, so that NMR data from different samples need to be compared at the same temperature either by measuring at the same temperature or by extrapolating the fit parameters to a common reference temperature. Also, materials that have been in use for a while have aged. Depending on the *aging* mechanism, the material may age homogeneously or develop surface-core layer structures (Fig. 5.1.4a).

Furthermore, elastomer materials exhibit random distributions of *cross-links*, *filler* particles, and may possibly contain filler aggregates and processing defects. The statistical nature of the material is revealed even on the scale of the sensitive volume of the NMR-MOUSE by variations of fit values extracted from experimental relaxation curves measured at different but equivalent positions. While the reproducibility of T_{2eff} values measured at one position may be better than 1%, the distribution of fit values is considerably broader when the measurements are taken at different positions (Fig. 5.1.4b). Mean values and their *standard deviations* need to be calculated to characterize a material property and the material homogeneity.

When the *standard deviation* is normalized to the mean value, the *coefficient of variation* is obtained. It is another measure of the material homogeneity when measurements are performed at different spots. This coefficient has been determined with the NMR-MOUSE in a factory at different stages in the *tire production process* (Fig. 5.1.4d). It essentially decreases with increasing number of processing steps from the master batch via the remill, the final mix, the extrudate, the unvulcanized tire to the vulcanized tire (Fig. 5.1.4c). The *rheometer* torque determined with the *rubber process analyzer* (RPA) on test samples reveals similar information, but these samples are larger than the sensitive volume of the NMR-MOUSE, so that the coefficients of variation from *RPA* are smaller than those from NMR. Moreover, RPA measurements are done on test samples, while stray-field NMR measurements can be done on the object of interest. This is why NMR can serve to control the quality of finished tires.

All of these measurements employ the *CPMG sequence* for data acquisition. Although this is a robust measurement scheme mistakes may be made by pulsing too fast leading to sample heating and choosing a larger than suitable value for the number of echoes. Common issues are summarized in Tab. 5.1.2.

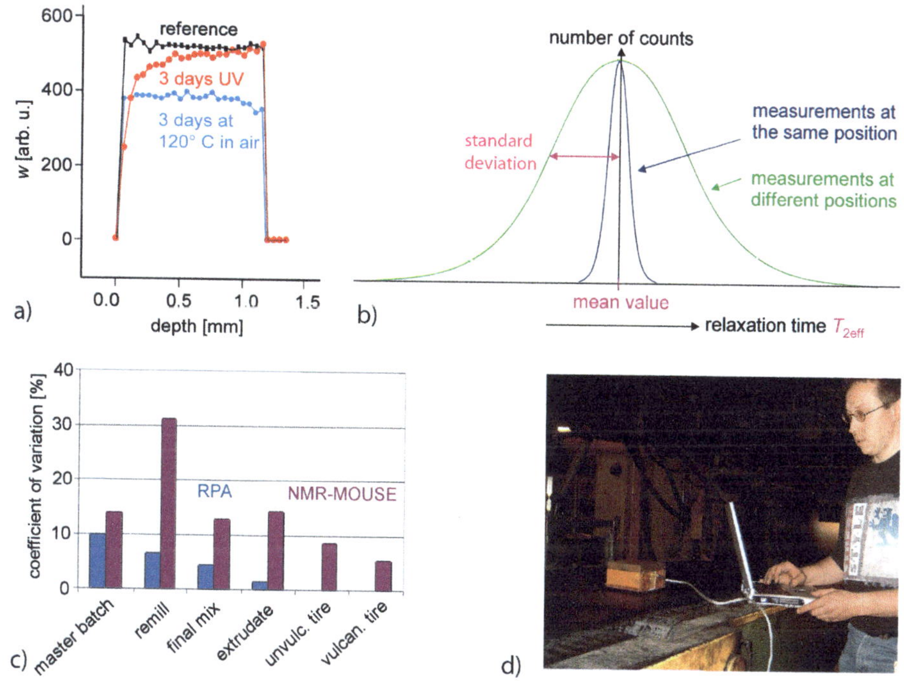

Figure 5.1.4. Inhomogeneity of rubber materials observed with the NMR-MOUSE (adapted from [1, 4–6] with permission). (a) Cross-sectional profiles through 1 mm thick sheets of natural rubber that have been aged in different ways. Thermal oxidative aging for three days lowers the relaxation-weighted spin density *w* homogeneously across the sheet while UV aging only affects the surface. (b) The scatter in determining T_{2eff} when measuring at the same position is much higher than when measuring at different positions. This indicates that on the scale of the sensitive slice of the NMR-MOUSE rubber is an inhomogeneous material. (c) Coefficient of variation for the rheometer torque measured with the rubber process analyzer (RPA) and T_{2eff} measured by the NMR-MOUSE for the different stages in the production of a car tire. The homogeneity improves with the number of processing steps. (d) Relaxation measurements with the NMR-MOUSE in a tire factory.

Table 5.1.2. Common issues ecnountered when measuring rubber with the NMR-MOUSE

- The recycle delay is too short so that signals with long relaxation T_1 times are suppressed
- The duration of the echo train exceeds the duration of the signal and the object may warm up from heat generated by the rf power dissipated in the coil
- The echo time is too short so that the experimental data are contaminated by receiver ringing
- The receiver phase is misadjusted and the signal phase needs to be adjusted after data acquisition
- Different samples are measured at different temperatures, so that the NMR data cannot be compared
- Measurements are only conducted at one spot or for one sample although the object is inhomogeneous

5.1.8 Advanced level measurements

Determining the initial signal decay

While the basic measurement aims at determining T_2 to correlate T_2 or $1/T_2$ with material properties via the *residual dipole-dipole coupling* between nuclear spins, advanced measurements are concerned with improving the accuracy of determining these couplings. Apart from improving the *signal-to-noise ratio* by signal averaging, the initial magnetization at zero detection time needs to be estimated to improve the accuracy of signal analysis by fitting model functions like the Anderson–Weiss expression (5.1.3) to the experimental data. Experimentally one cannot shorten the detection time to values below the *dead time* of the spectrometer after a single pulse (Figs. 2.7.1 and 3.1.1). Typically this is of the order of 10 μs. However, for some soft polymers like carbon black filled rubber, the signal from bound rubber at the surface of the *filler* particles may have decayed during this time as a result of the *dipole-dipole interaction* and enhanced relaxation by paramagnetic impurities in the carbon black filler.

Because the *dipole-dipole interaction* acts between two and more spins and not between one spin and the magnetic field, its effect on the signal decay cannot be undone by a simple *spin echo* or *Hahn echo* which consists of a 90° pulse followed by a 180° pulse (Figs. 3.2.3a and 5.1.5a). But a *solid echo*, which consists of two 90° pulses, which differ by a π/2 phase shift in their rf carrier signals (Fig. 5.1.5b), cancels the effect of the *dipole-dipole interaction* between two spins in the echo maximum. Of course, in organic matter there are more than two magnetic dipoles from the hydrogen nuclei that interact with each other. Therefore, the solid echo works well only at short echo

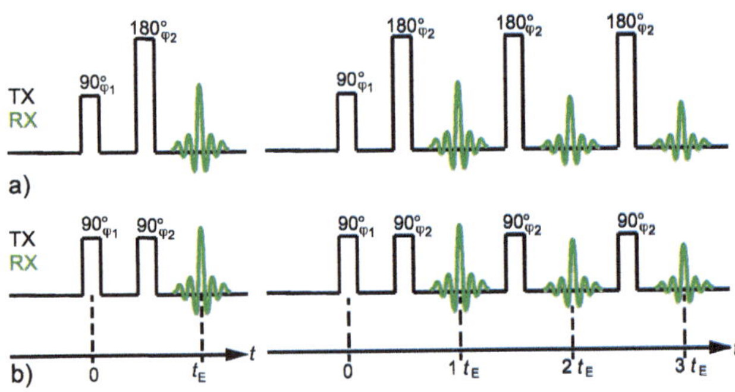

φ_1	φ_2	φ_{acq}
x	-y	-y
y	x	x
-x	y	y
-y	-x	-x
x	y	-y
y	-x	x
-x	-y	y
-y	x	-x

c)

Figure 5.1.5. Echoes (left) and echo trains (right). (a) Spin echo or Hahn echo. In the echo maximum the effects of field inhomogeneity and differences in resonance frequency or chemical shift of nuclei in different chemical groups are cancelled. The maximum echo amplitude is obtained when the pulse flip angles are 90° and 180°. The difference between the rf phases of the pulses is irrelevant. The multi-echo variant of the Hahn echo is the CPMG echo train. Significant pulse errors are compensated when the phase difference between the 90° pulse and the 180° pulses is π/2. (b) Solid echo. It eliminates the effects of the dipole-dipole interaction between two proton spins in the echo maximum. Both pulse flip angles are 90° and the phase shift between the first and the second pulse must be π/2. The multi-pulse variant is known as OW4 sequence. (c) Full phase cycle.

times. The so-called *magic sandwich echo* also takes care of multicenter dipole-dipole interactions, and the *mixed magic sandwich echo* is even better [7], because it eliminates the combined effects of field inhomogeneity and dipolar interactions on the magnetization decay. For *rubber* and many *polymer materials*, the value of the transverse magnetization at zero detection time can be extrapolated from several measurements of the *solid echo* with decreasing echo time (Fig. 3.1.4).

Double-quantum NMR

More direct access to *residual dipole-dipole couplings* than via transverse relaxation is provided by simultaneously exciting two magnetically coupled hydrogen spins. This requires two energy quanta, one for each spin. But from quantum mechanics it is known that such processes cannot be observed directly, unless a nonlinear process is invoked by high excitation power as in optics when generating green laser light. In modern NMR such *double-quantum* (2Q) phenomena are observed indirectly following the principles of *multi-dimensional* NMR. Here the measurement is repeated several times but with well-prepared initial conditions, which are varied systematically from scan to scan.

The *double-quantum* NMR pulse sequence (Fig. 5.1.6b) starts with the spin system in *thermodynamic equilibrium*. Then three pulses are applied in a time t_{MQ} that excite

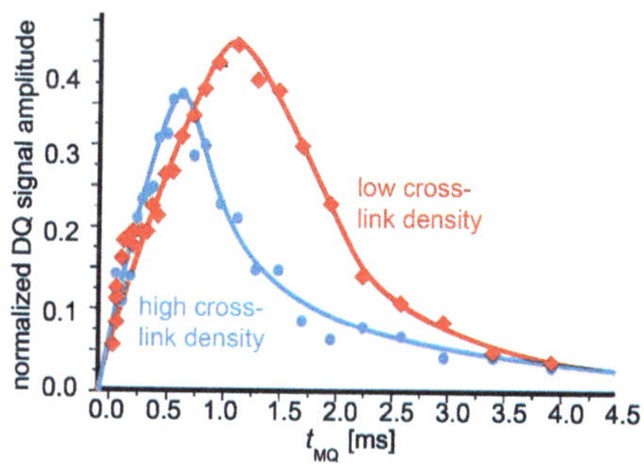

a)

φ_1	φ_2	φ_3	φ_4	φ_5	φ_{acq}
x	-x	-y	y	x	y
y	-y	-y	y	x	-y
-x	x	-y	y	x	y
-y	y	-y	y	x	-y

c)

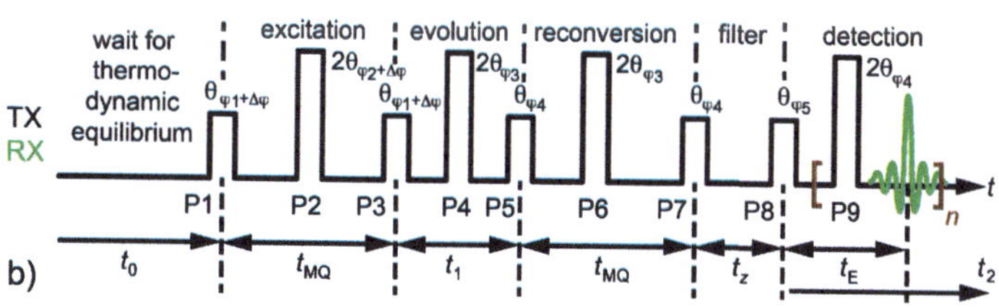

b)

Figure 5.1.6. Double-quantum NMR of elastomers with the NMR-MOUSE. (a) Build-up and decay curves of the double-quantum signal for strongly and weakly cross-linked natural rubber (adapted from [8] with permission). (b) Pulse sequence. Typical parameters are: pulse width $t_p = 2.5\,\mu s$, recycle delay $t_R = 0.5\,s$, multi-quantum conversion time t_{MQ} variable from 0 to 5 ms, multi-quantum evolution time $t_1 = 60\,\mu s$, filter time $t_z = 500\,\mu s$, echo time for detection $t_E = 200\,\mu s$. Note that the experiment is repeated three times for $\Delta\varphi = 90°, 180°, 270°$. (c) Phase cycle for selection of the double-quantum signal. This phase cycle is repeated three times to account for all values of $\Delta\varphi$ by shifting all phases with each repetition by 90°.

the coherent motion of a magnetically coupled spin pair, which evolves for a time t_1. A 180° pulse midway during this time compensates for the effects of field inhomogeneity on this motion. The double-quantum coherence is then converted into longitudinal magnetization by another series of three pulses, which, after a time t_z is observed with a Hahn echo or a *CPMG echo train*. This pulse sequence works in inhomogeneous fields with the NMR-MOUSE, where the pulse *flip angles* are not well defined. Ideally the flip angles are 90° and 180° instead of θ and 2θ. The 2θ pulses are only needed when measuring in inhomogeneous fields, e.g. with the NMR-MOUSE, and can be discarded when measuring in the more homogeneous field of a *Halbach magnet*. Nevertheless, the pulse phases need to be cycled through at least the set of values listed in Fig. 5.1.6c for each setting of the double-quantum build-up and decay time t_{MQ}.

The experiment is repeated for different values of t_{MQ} to obtain the double-quantum build-up and decay curves shown in Fig. 5.1.6a. The initial signal build-up is governed by the strength of the *dipole-dipole interaction* and the signal decay is governed by *relaxation*. From the normalized *double-quantum* signal amplitude

$$s_{2Q}(t_{MQ})/s_{2Q}(0) \propto \left(1 - \frac{3}{4}\varpi_D^2 t_{MQ}\right) \sin^4\theta \tag{5.1.5}$$

the value of the *residual dipole-dipole coupling* ϖ_D can be extracted by fitting the initial part of the experimental data (Fig. 5.1.6a) for short t_{MQ} to a quadratic function of the conversion time t_{MQ}. This procedure can be extended to probe spin modes other than the motion of two coupled spins, to extract distributions of dipole-dipole couplings, and to probe the theories of rubber elasticity and rubber networks.

NMR imaging

Elastomers are similar in consistency to *biological tissue*, where magnetic resonance imaging (*MRI*) provides excellent *sensitivity* and *contrast*. Objects from elastomers are tires, conveyor belts, fittings, tubes, and O-rings. Tires and conveyor belts are large and contain steel belts, so that they are best investigated with the NMR-MOUSE to scan pixel information by shifting the NMR-MOUSE from position to position across the surface or along the depth direction. Many fittings, seals, tubes and O-rings are small enough to fit into the hole of a tomograph for analysis by imaging. Linear fittings from the extruder can be passed through the MRI magnet for *process control* [9].

Most rubber products are inhomogeneous not only because *cross-links* and *filler* particles are randomly distributed, but also because the processing steps of mixing the rubber formulation from its different constituents, of mastication, calendaring and extrusion are only as perfect as needed. MRI is the method preferred over optical inspection because most *rubber* products are filled with *carbon black*, which gives little visual contrast [10]. Often aggregates of processing aides and fillers can be found in technical compounds. These are investigated by magnetic resonance imaging to improve the production process. Moreover, tubes, fittings and tires may consist of several layers from different rubber formulations. Their positions and curing efficiency can be inspected by *depth profiling* with the NMR-MOUSE. Optical and thermo-oxydative aging produce surface layers, which can be assessed by NMR depth profiling or imaging, and the location of inner surfaces of rubber gaskets and fittings can be inspected for *process control* in the production line (Figs. 3.3.5 and 3.3.7).

5.1.9 Data processing

Experimental transverse relaxation curves are fitted with model functions to extract the parameters of the function (Tab. 3.1.2). The *Anderson–Weiss formula* (5.1.3) has too many fit parameters to be of practical use for reliably analyzing noise-contaminated experimental data. This is why the formula is often approximated by other functions with fewer parameters. The lower the signal-to-noise ratio and the longer the receiver dead time, the simpler the approximation:

(a) Long *correlation time* τ_c, $T_{2A} = T_{2B} \equiv T_2$ and long relaxation time T_{2C},

$$M(t)/M_0 = x_A \exp\{-t/T_2 - qM_2t^2/2\} + x_B \exp\{-t/T_2\} + x_C. \tag{5.1.6}$$

(b) In the limit of long T_2, this equation reduces further to a simple *Gauss function*,

$$M(t)/M_0 = x_A \exp\{-qM_2t^2/2\} + x_{B,C}. \tag{5.1.7}$$

(c) If the initial Gaussian contribution cannot be resolved, the experimental data can be fitted with a *bi-exponential function* or some other function (Tab. 3.1.2),

$$M(t)/M_0 = x_A \exp\{-t/T_{2A}\} + x_B \exp\{-t/T_{2B}\}, \tag{5.1.8}$$

where the rapidly relaxing component is assigned to the chemically and physically cross-linked chains and the slowly relaxing component determines the mobile fraction of dangling chains and solvent molecules.

(d) When comparing different technical elastomers of the same type at low signal-to-noise ratio, the transverse relaxation curves can often be approximated by a *mono-exponential function*,

$$M(t)/M_0 = \exp\{-t/T_2\} \tag{5.1.9}$$

Because the relaxation curves of solids including polymer and rubber materials are inherently described by functions other than a simple exponential function, distributions of relaxation times and other parameters are extracted from the experimental data by transformations different from but similar in spirit to the inverse Laplace transformation [11].

The relaxation times or rates extracted from the transverse signal decay are subsequently employed in data banks together with acquisition and sample parameters

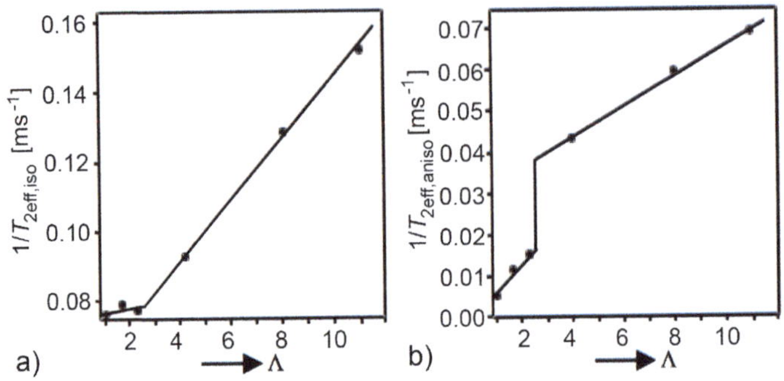

Figure 5.1.7. Relaxation rates versus elongation ratio extracted from modeling the relaxation anisotropy of strained natural rubber. Strain induced crystallization of cross-link chains is revealed at an elongation $\Lambda \approx 2.3$ by (a) a change of $1/T_{2\text{eff,iso}}$, which appears like a second-order phase transition and (b) a change of $1/T_{2\text{eff,aniso}}$, which appears like a first-order phase transition.

like the type of NMR device, the echo time, and the sample temperature for reference and comparison with the data from other samples or objects. By means of analytical models the NMR data can also be converted to material properties. In particular, the *transverse relaxation rate* and the *residual dipole-dipole coupling* are proportional to the *shear modulus* and the *cross-link density*.

For example, the way to utilize the angle-dependent relaxation rates $1/T_{2\text{eff}}$ of the strained rubber bands from Fig. 5.1.3e is to model the angle-dependence in terms of an angle-independent, isotropic relaxation rate $1/T_{2\text{eff,iso}}$ and an angle-dependent, anisotropic *relaxation* rate $1/T_{2\text{eff,aniso}}$,

$$1/T_{2\text{eff}}(\theta) = 1/T_{2\text{eff,iso}} + 1/T_{2\text{eff,aniso}} \int_0^{-\pi} P(\theta - \theta')[3(\cos^2\theta' - 1)/2]^2 d\theta', \qquad (5.1.10)$$

where $P(\theta)$ is a Gaussian distribution function, which accounts for a spread in the alignment angles of the vectors connecting both ends of the cross-link chains (Fig. 5.0.1b) with the direction of the applied strain. At an *elongation* Λ of about 2.3, natural rubber tends to form crystallites. This *crystallization* phenomenon is revealed in graphs of the *relaxation rates* $1/T_{2\text{eff,iso}}$ and $1/T_{2\text{eff,aniso}}$ extracted from eqn. (5.1.10) versus the *elongation* ratio Λ. While the change of $1/T_{2\text{eff,aniso}}$ at this elongation ratio appears to indicate a first-order *phase transition* (Fig. 5.1.7b) that of $1/T_{2\text{eff,iso}}$ appears like a second-order phase transition (Fig. 5.1.7a).

5.1.10 References

[1] Blümich B, Casanova F, Perlo J. Mobile single-sided NMR. Prog Nucl Magn Reson Spectrosc. 2008; 52:197–269.

[2] Blümich B, Anferova S, Kremer K, Sharma S, Herrmann V, Segre A. Unilateral nuclear magnetic resonance for quality control. Spectroscopy. 2002; 18:18–32.
[3] Hailu K, Fechete R, Demco DE, Blümich B. Segmental anisotropy in strained elastomers detected with a portable NMR scanner. Solid State Nucl Magn Res. 2002; 22:,327–343.
[4] Kolz J, Martins J, Kremer K, Mang T, Blümich B. Investigation of the elastomer-foam production with single-sided NMR. Kautschuk Gummi Kunststoffe. 2007; 60:179–183.
[5] Blümich B, Buda A, Kremer K. Non-destructive testing with mobile NMR. RFP. 2006; 1:34–37.
[6] Goga N, Demco DE, Kolz J, Ferencz R, Haber A, Casanova F, et al. Surface UV aging of elastomers investigated with microscopic resolution by single-sided NMR. J Magn Reson. 2008; 192:1–7.
[7] Fechete R, Demco DE, Blümich B. Chain orientation and slow dynamics in elastomers by mixed magic Hahn echo decays. J Chem Phys. 2003; 118:2411–2422.
[8] Wiesmath A, Filip C, Demco DE, Blümich B. NMR of multipolar spin states excitated in strongly inhomogeneous magnetic fields. J Magn Reson. 2002; 154:60–72.
[9] Danieli E, Berdel K, Perlo J, Michaeli W, Masberg U, Blümich B, Casanova F. Determining object boundaries from MR images with sub-pixel resolution: Towards in-line inspection with a mobile tomograph. J Magn Reson. 2010; 207:53–58.
[10] Blümler P, Blümich B. NMR imaging of elastomers: A review. Rubber Chem Tech. 1997; 70:468–518.
[11] Chalcea RI, Fechete R, Culea E, Demco DE, Blümich B. Distributions of transverse relaxation times for soft solids measured in strongly inhomogeneous magnetic fields. J Magn Reson. 2009; 196:179–190.

5.2 Amorphous polymers

5.2.1 Introduction

Amorphous polymers cover a wide range of properties. They are materials from disordered macromolecules that lack crystalline domains in which repeat units are arranged on crystal-lattice points (Fig. 5.0.1a). Above the *glass transition temperature* these materials are soft like *rubber* unless they are highly cross-linked such as polyester resins. Below the glass transition temperature, these materials are brittle like *glass*. Because light is scattered by crystalline domains, polymer glasses are transparent like Plexiglas or polycarbonate unless they are filled with talcum and other fillers like many *polyvinylchloride* (*PVC*) materials, or with glass and carbon fibers like polyester resins.

5.2.2 Objective

Like any other material, amorphous polymers may corrode and swell upon exposure to chemicals. Many polymer materials, *PVC* in particular, are plasticized, and the *plasticizer* may be washed out in adhesive joints or by excessive thermal exposure. Mechanical *stress* will impose local strain on the macromolecular chains and eventually lead to *stress whitening* through the formation of microcracks or *crazes*. All of these material changes affect the segmental mobility of the macromolecules and in turn the NMR *relaxation* times. Comparative relaxation measurements of amorphous polymers can therefore be used to identify localized regions of stress, aging, solvent ingress and plasticizer loss. Parts made from amorphous polymers exposed to critical loads can be inspected by stray-field NMR relaxation measurements with the NMR-MOUSE for material change in different processing steps and during their service life. Attention needs to be paid to *carbon fiber* filled materials because such fibers are electrically conducting. If the filler fibers are long and show multiple orientations inside the sensitive volume, the radio frequency field of the coil may be shielded from entering the sensitive volume, and NMR measurements may not be possible at all.

5.2.3 Further reading

Blümich B, Casanova F, Perlo J. Mobile single-sided NMR. Prog Nucl Magn Reson Spectrosc. 2008; 52:197–269.

5.2.4 Theory

The basic theory needed to analyze amorphous polymers by mobile NMR refers to the measurement of transverse and longitudinal relaxation phenomena (Section 3.2). Transverse relaxation curves need to be integrated to obtain relaxation weights or fitted with model functions to extract fit parameters like signal amplitudes and relaxation rates (Section 3.1.9). A common fit function for the CPMG echo train decay of amorphous polymers is the exponential function, but a Gaussian function is more appropriate for rigid amorphous polymers.

5.2.5 Hardware

With respect to compact NMR, products from amorphous polymers are preferably investigated with the NMR-MOUSE. Due to short relaxation times in most rigid, amorphous polymer materials, an NMR-MOUSE with a short dead time needs to be employed, e.g. a *Profile NMR-MOUSE* with a 3 mm or 5 mm depth range (Fig. 5.1.2a) or a *bar magnet NMR-MOUSE* (Fig. 5.1.2b). Imaging experiments are hard to conduct on polymers below the *glass transition temperature* because of the short transverse relaxation time, and studies with a *Halbach magnet* (Fig. 5.1.2c) are suitable only when sample sections can be prepared to fit inside the magnet.

Table 5.2.1. Acquisition parameters for ^{1}H NMR relaxometry of rigid polymers

Parameter	Value
Magnet, probe	Bar-magnet NMR-MOUSE
transmitter frequency ν_{rf}	19.1 MHz
transmitter attenuation for 90° pulse	−8 dB at 80 W
duration t_p of 90° pulse	3 μs
dwell time Δt	0.5 μs
acquisition time t_{acq}	3 μs
echo time t_E	25 μs
number n_E of echoes	500
recycle delay t_R	0.5 s
number n_s of scans	2048

5.2.6 Pulse sequence and parameters

The standard pulse sequences for the NMR-MOUSE are multi-echo sequences with short echo times. For soft amorphous polymers the CPMG sequence is employed (Fig. 5.1.5a), and the acquisition parameters are similar to those for *rubber* (Tab. 5.1.1), while for rigid *amorphous polymers* the *multi-solid echo sequence* is employed (Fig. 5.1.5b) and the acquisition parameters are similar to those for *semi-crystalline polymers* (Tab. 5.2.1).

5.2.7 Beginner's level measurements

The impact of plasticizer loss and *mechanical stress* can be characterized by simple spin echo and *CPMG* measurements with the NMR-MOUSE, and products can be tested in critical areas for fabrication *defects*. Even if the acquisition of the NMR signal is limited to some tens of seconds, NMR is still a slow measurement method compared to Raman, IR and other spectroscopic methods in the optical regime. This is why larger components are unlikely to be screened completely by single-sided NMR, and the NMR analysis is restricted to known, localized regions such as areas of high mechanical impact and sections where flaws from manufacturing or use are suspected to arise. The benefit of a localized analysis is that the point in question and surrounding reference material can be measured to compare NMR parameters from both regions.

For example, in the vicinity of a *crack* in a PVC pipe, the transverse relaxation time has been found to be lower than in the surrounding material (Fig. 5.2.1a). This is explained by reduced mobility and increased motional anisotropy of the polymer chains when they are strained. Both, slower and more confined motion, increase the *residual dipole-dipole interaction* among the protons of the polymer chains and enhance the transverse relaxation rate. The data shown refer to the transverse relaxation time $T_{2\text{eff,long}}$ of the slowly relaxing component extracted from a bi-exponential fit of exper-

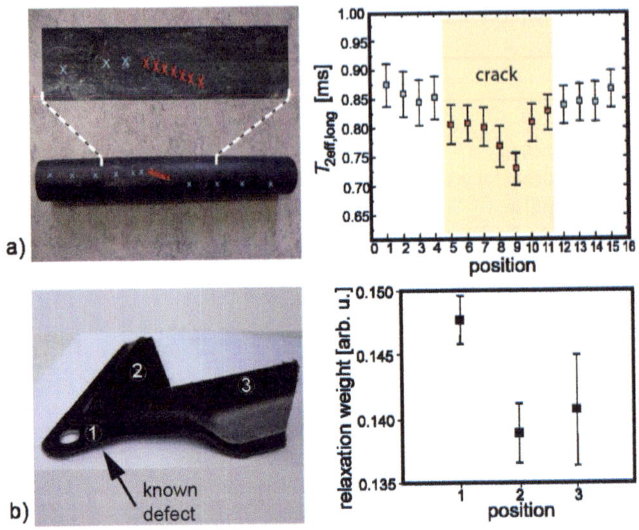

Figure 5.2.1. Relaxation measurements of defects in objects from amorphous polymers with the NMR-MOUSE. (a) Carbon black filled PVC pipe with a crack and relaxation times $T_{2\text{eff,long}}$ of the slowly relaxing component at different points along and in the vicinity of the crack. (b) Glass fiber reinforced polymer composite and relaxation weights at selected positions. Position 1 was known to have an invisible defect.

imental CPMG decays. It is often found that the slowly relaxing component is more sensitive to material change than the rapidly relaxing component.

A way of eliminating the spin density M_0 from experimentally determined echo trains other than by use of the w function (eqn. (3.1.5)) applies to *mono-exponential* relaxation. In this case the ratio of echo amplitudes acquired with two different echo times t_{E1} and t_{E2} defines a relaxation weight according to $M_0 \exp\{-t_{E2}/T_2\}/[M_0 \exp\{-t_{E1}/T_2\}] = \exp\{-(t_{E2} - t_{E1})/T_2\}$. This approach has been used for a quick estimate of relaxation properties, and a hidden defect could be discovered in this way in a section in question of a glass fiber reinforced polymer composite component from the automotive sector (Fig. 5.2.1b). This defect shows a higher relaxation weight in accordance with a shorter transverse relaxation time similar to the crack in the PVC pipe. About the same information is obtained with the *weight parameter* w (eqn. (3.1.5))

Table 5.2.2. Common issues encountered when measuring amorphouse polymers

– The echo time is too short, and receiver ringing covers the NMR signal
– The number of echoes is too high so that the echo train extends far beyond the signal
– The receiver phase is misadjusted and the signal phase needs to be adjusted after data acquisition
– The NMR signal may depend on the sample temperature
– The object may be inhomogeneous by nature and not from flaws

but at better *signal-to-noise ratio* because several echoes from a CPMG echo train are added. Some common issues that may be encountered when investigating amorphous polymers with the NMR-MOUSE are summarized in Tab. 5.2.2.

5.2.8 Advanced level measurements

Advanced measurements concern the acquisition of *double-quantum build-up curves* (Fig. 5.1.6) and the execution of *spin diffusion* experiments (Fig. 5.3.5) to interrogate the strength of *dipole-dipole couplings* and to gain information on morphological heterogeneity from differences in molecular mobility on the nanometer scale, respectively. Given the strong dipole-dipole interaction due to slower motion of chain segments in hard amorphous polymers than in rubber, a *multi-solid echo train* (Fig. 5.1.5b) may give higher echoes and slower echo envelope decays because the dipole-dipole interaction among the spins (Fig. 5.1.1) is partially averaged out by the solid echo. This increases the sensitivity in terms of the acquired signal energy but may decrease the sensitivity in terms of the *contrast* needed to differentiate between different materials.

5.2.9 Data processing

Experimental NMR data of *rigid polymers* are typically noisier than those from *soft polymers*, because a significant part of signal decays within the *dead time* of the instrument before the first echo of the echo train has been acquired. Also the number echoes may be as few as 10 or less instead of hundreds or thousands in the case of *elastomers*. This is why slight deviations from the correct receiver phase will be noticed much more strongly so that the phase of the complex transverse magnetization measured with a set of echoes or an echo train may need to be corrected by multiplying the acquired signal with a constant phase factor $\exp\{-i\phi\}$ in order for the signal to appear only in the real channel of the data record and for the imaginary channel to contain only noise. The signal from the real channel is then preferably integrated to compute the *relaxation weighted spin density* w (eqn. (3.1.5)) or fitted with model functions to extract relaxation rates and signal amplitudes (Tab. 3.1.2).

5.3 Semi-crystalline polymers

5.3.1 Introduction

Semi-crystalline polymers are unusual materials. They consist of *crystalline domains* with ordered polymer chains and *amorphous domains* with disordered chains (Fig. 5.0.1c,d). Their preferred temperature of use is above the *glass transition temper-*

ature, where the chains in the amorphous domains are mobile while the chains in the crystalline domains are rigid. In this case the *degree of crystallinity* can be determined by analyzing the NMR signal according to differences in chain mobility, i.e. according to differences in relaxation times in the time domain or correspondingly to differences in line shapes and widths in NMR spectra. The size of the crystalline domains is not uniform. It varies randomly within limits. This variation gives rise to a small but significant variation in amplitude of the NMR signal from small volume elements because the proton densities of crystalline and amorphous domains are different due to different packing densities of the polymer chain segments.

While the crystalline domains are conveniently studied by *X-ray diffraction*, signal from the amorphous domains is obtained by NMR, IR and Raman-scattering methods. An NMR *relaxation analysis* not only identifies signal from crystalline and amorphous domains but also often from the *interface* between both, which exhibits intermediate mobility (Fig. 5.0.1c). The relative amounts of protons in these regions can be determined from the signal amplitudes of the magnetization components with short, intermediate and long transverse relaxation times. The shortest diameter of each region, the so-called *domain size*, can be determined by experiments that monitor the diffusive spreading of longitudinal magnetization from the crystalline domains into the amorphous domains. These experiments are called spin-diffusion experiments.

Spin diffusion is transport of nuclear magnetization. It has to be discriminated from *self-diffusion* of small molecules, which is the *Brownian motion* of molecules in a liquid or of solvent molecules in polymer materials. Self-diffusion can well be measured by NMR (Section 3.2.8), and its measurement is particularly easy with the *NMR-MOUSE*. Solvent molecules diffuse in polymer materials mostly in the *amorphous domains*, where the chain packing is less dense than in the *crystalline domains*. The amorphous domains are also the ones that are affected first when semi-crystalline polymer material is deformed. When the polymer chains in the amorphous domains are deformed from the random coil conformation (Fig. 5.0.1a) into an elongated or compressed state, the chain mobility is slowed down and the motion is more confined. This increases the *residual dipole-dipole interaction* between the protons in the chain (Fig. 5.1.1) and enhances the transverse relaxation rate.

5.3.2 Objective

Semi-crystalline polymers are investigated by NMR, because the method is nondestructive, so that the sample can be used again for further analysis, and because information about the *amorphous domains* is obtained directly in addition to information about the crystalline and interfacial domains. Given the proton densities, the amounts of crystalline, interfacial and amorphous material can be calculated from proton signal amplitudes extracted from measured echo trains with the help of suitable *fit functions*. The fit parameters report about the state of the material as it is determined by the *thermal history* and the *mechanical history* of the sample or object. They are expected to be important for the estimation of the remaining *service life* of the material. The average domain sizes of these material components are determined in *spin-diffusion* experiments. *Solvent-diffusion* studies help to understand *aging* mechanisms and *barrier* properties of polymer containers. Most studies need to be conducted with spatial *depth resolution*, because *swelling* and *aging* processes start from the surface and produce layer structures even in originally homogeneous polymer material. High-quality *packaging* materials and polymer containers like gasoline tanks are made from laminated polymer sheets, and depth profiling is used to study the thickness and properties of the different layers and their changes in response to mechanical and chemical impact.

5.3.3 Further reading

Kimmich R. Principles of Soft-Matter Dynamics. Dordrecht: Springer; 2012.

Callaghan PT. Translational Dynamics & Magnetic Resonance. Oxford: Oxford University; 2011.

Blümich B, Casanova F, Perlo J. Mobile single-sided NMR. Prog Nucl Magn Reson Spectrosc. 2008; 52:197–269.

Hedesiu C, Demco DE, Kleppinger R, Vanden Poel G, Remerie K, Litvinov VM et al. Aging effects on the phase composition and chain mobility of isotactic poly(propylene). Macromol. Mat Eng. 2008; 93:847–857.

Hedesiu C, Demco DE, Kleppinger R, Adams-Buda A, Blümich B, Remerie K, Litvinov VM. The effect of temperature and annealing on the phase composition, molecular mobility and the thickness of domains in high-density polyethylene. Polymer. 2007; 48:763–777.

Strobl G. The Physics of Polymers, 3rd edition. Berlin: Springer; 2007.

Blümich B. NMR Imaging of Materials. Oxford: Clarendon Press; 2000.

Callaghan PT. Principles of Nuclear Magnetic Resonance Microscopy. Oxford: Clarendon Press, 1991.

5.3.4 Theory

Stray-field relaxometry

The *CPMG decay* from many materials including semi-crystalline materials is often approximated by a double or *triple exponential decay* when the NMR-MOUSE is employed,

$$s(t)/s(0) = x_r \exp\{-t/T_{2\text{eff,short}}\} + x_i \exp\{-t/T_{2\text{eff,inter}}\} + x_m \exp\{-t/T_{2\text{eff,long}}\}. \quad (5.3.1)$$

Here the indices r, i, m stand for rigid, interfacial, and mobile, respectively, and the relative signal amplitudes x_r, x_i, and x_m are the relative proton *spin densities* associated with the *relaxation rates* $1/T_{2\text{eff,short}}$, $1/T_{2\text{eff,inter}}$, and $1/T_{2\text{eff,long}}$, respectively. The first term scales with the X-ray *crystallinity*, while the magnitude of the interfacial component cannot be measured with methods other than NMR. All three densities listed in eqn. (5.3.1) sum up to 1. The fact that *exponential functions* suffice to model the CPMG decay has to do with the loss of signal in the dead time (Fig. 3.1.1b) and a partial averaging of the *dipole-dipole interaction* between protons from *spin-lock* effects associated with the CPMG sequence as well as a *flip angle distribution* because most spins are irradiated away from resonance in the inhomogeneous stray field of the NMR-MOUSE (Fig. 3.1.5).

Relaxometry in a largely homogeneous field

In a sufficiently homogeneous magnetic field like that of a *Halbach magnet*, the transverse magnetization decay can be determined more accurately because the entire sample can be excited with well-defined flip angles, and the signal sensitivity benefits from excitation of a sample volume larger than the sensitive volume of the NMR-MOUSE. To avoid the spin-lock effects typical for multi-echo trains, the transverse magnetization decay is measured step by step in many single-echo experiments with different echo times (Fig. 3.1.4a), which extends the measurement time. At long magnetization decay times, the *Hahn echo* (Fig. 5.1.5a, left) is employed to eliminate signal loss from the influence of an inhomogeneous magnetic field, and the magnetization decay largely follows an exponential function. At short times below about 100 μs, the signal from the rigid component dominates the magnetization decay. It is determined with *solid echoes* (Fig. 5.1.5b, left), and extrapolated to zero time (Fig. 3.1.4b). The decay of the rigid component follows the *Abragam function* (Tab. 3.1.2).

The complete transverse magnetization decay from semi-crystalline polymers determined in this way in a relatively homogeneous magnetic field is then modeled by the sum of an *Abragam function* and two *exponential functions* [7],

$$\begin{aligned} s(t)/s(0) = {} & x_r \exp\{-(1/2)(t/T_{2,\text{short}})^2\}(\sin\{at\})/(at) \\ & + x_i \exp\{-t/T_{2,\text{inter}}\} + x_m \exp\{-t/T_{2,\text{long}}\}. \end{aligned} \quad (5.3.2)$$

A fit of the experimental curve (Fig. 3.1.4a) with this model function produces the relaxation rate for each of the three domains and the associated amplitudes corresponding to the spin densities of the components.

5.3.5 Pulse sequences and parameters

The standard pulse sequence for investigating semi-crystalline polymers with the NMR-MOUSE is the *multi-solid echo sequence* (Fig. 5.1.5b). Because the *crystalline domains* are rigid, the *dipole-dipole interaction* is strong, and the initial signal decay is rapid. This is why a short *echo time* should be used. The acquisition parameters are the same as those for rigid amorphous polymers (Tab. 5.2.1). If small samples are investigated in a closed magnet with a comparatively homogeneous magnetic field, the transverse magnetization decay can be measured more accurately without the effects of resonance off-set from the presence of a strong field gradient. Then *Hahn echoes* (Fig. 5.1.5a, left) should be used for echo times larger than 100 μs and *solid echoes* (Fig. 5.1.5b, left) or just the *free induction decay* for shorter times (Fig. 3.1.4a). Alternatively, the CPMG sequence can be started with a magic echo, which reverses signal loss from multicenter dipole-dipole interactions, to get a more accurate estimate of the initial signal amplitude[1].

5.3.6 Hardware

NMR of semi-crystalline polymers needs hardware with short *dead times* so as to lose as little of the signal of the rigid crystalline domains as possible. As a rule of thumb, the larger the *depth range* of the NMR-MOUSE, the larger the diameter of the radio frequency coil and the lower the *NMR frequency*. Both facts engender larger *dead times*, so that measurements of glassy and semi-crystalline polymers by single-sided NMR devices are limited to low depth. Suitable instruments are the *Profile NMR-MOUSE* with depth ranges of 3 mm and 5 mm (Fig. 5.1.2a), the *bar magnet NMR-MOUSE* (Fig. 5.1.2b), and a closed magnet with a small diameter coil like the *Halbach magnet* of Fig. 5.1.2c. Imaging studies of objects from semi-crystalline polymers are rather uncommon due to the short transverse relaxation times of these materials.

5.3.7 Beginner's level measurements

Several phenomena of practical importance for the use of objects from semi-crystalline polymers can be investigated by *stray-field* relaxometry with the NMR-MOUSE when applying regular CPMG measurements with short echo times. The *crystallinity* x_r can be estimated from the relative signal amplitude of the rapidly decaying component of the transverse relaxation signal (eqns. (5.3.1) and (5.3.2), Fig. 5.3.1a). The crystallinities derived from a bi-exponential fit to the signals of a PE test sample measured with the NMR-MOUSE and by high-field NMR were in good agreement with 70.3% and 72.4%, respectively. Note that NMR crystallinity is defined through differences in molecular mobility, while X-ray crystallinity is defined through differences in molecular order. It

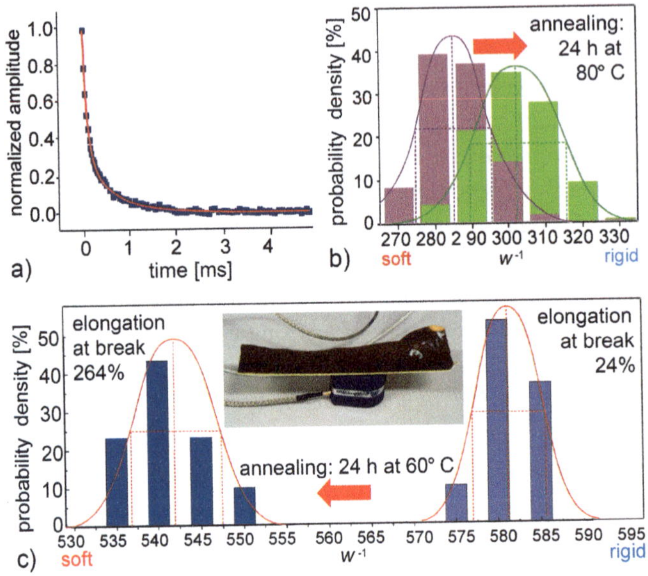

Figure 5.3.1. Relaxation measurements of polyethylene (PE) products (adapted from [2, 3] with permission). (a) CPMG decay fitted with a double exponential function. The relative amplitude of the rapidly decaying component defines the NMR crystallinity. (b) Distributions of the w-parameter measured inside a 10 cm diameter pipe from high-density polyethylene (HDPE) PE 100. Annealing shifts the distribution to the right as the material becomes more brittle. (c) Distributions of the w-parameter measured for an aged and brittle steel pipe shell from low-density polyethylene (LDPE). Annealing shifts the distribution to the left as the material becomes softer.

is well known that values for crystallinity obtained by different methods do not give the same results. But the values determined with the NMR-MOUSE and by high-field NMR agree well.

Similar to rubber materials (Fig. 5.1.4c), the CPMG decays from semi-crystalline materials measured with the NMR-MOUSE vary from spot to spot due to different average crystallinities in the sensitive volume. This *distribution* depends on the thermal and mechanical history of the semi-crystalline polymer material, because processing conditions and exposure to heat and mechanical stress affect the morphology. Distributions of the *w-parameter* were measured inside a 10 cm diameter pipe from *polyethylene* PE 100 before and after annealing at 80° C. *Annealing* of semi-crystalline polymers is known as *crystallization* in polymer physics. Its shifts the distribution to higher values of the contrast parameter as the material becomes more brittle (Fig. 5.3.1b). An aged steel pipe shell from low-density *PE* became elastic by annealing at 60° C. This can be seen by a shift of the contrast parameter from high values to lower values as the material becomes softer (Fig. 5.3.1c). Along with this shift, the *elongation at break* improved from 24% to 264%. This demonstrates how simple relaxation measurements with the NMR-MOUSE can be used to assess the properties of objects from semi-crystalline polymers. Effects of annealing can also be observed near room temperature right after production when storing products like pipes from semi-crystalline polymer material.

Aging and *annealing* affect the segmental mobility and thus the relaxation times just as *stress* and *strain* do. The relative tensile strength of differently aged *polyethylene* materials appears to affect T_{2eff} extracted from mono-exponential fits to experimental multi-echo train signals decays in a linear fashion: The more the PE material is aged, the lower the relative tensile strength and the shorter T_{2eff} (Fig. 5.3.2a). This observation links nondestructive NMR measurements at the site of the object to mechanical measurements on test samples in the laboratory. Bending a curved PE *pipe* section cut from a large role of pipe straight and even further over a pin in the center between both ends (Fig. 5.3.2b), increases T_{2eff} from the *amorphous domains* and the *interfacial domains* (Fig. 5.3.2c), but leaves T_{2eff} of the *crystalline domains* unaffected. In this case, the CPMG decay could be fitted with a *triple exponential decay* (eqn. (5.3.1)). On the other hand, from the build-up curve of the *longitudinal magnetization* only two T_1 relaxation times could be extracted with a bi-exponential fit function. They are assigned

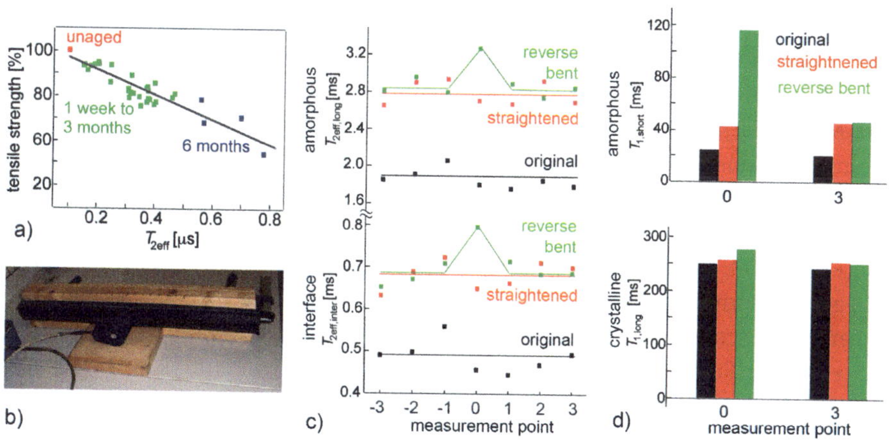

Figure 5.3.2. Relaxation of polyethylene objects measured with the NMR-MOUSE (adapted from [2] with permission). (a) Relative tensile strength of differently aged polyethylenes versus the T_{2eff} extracted from mono-exponential fits to the experimental CPMG decays. (b) Experimental set-up to measure a curved PE pipe section cut from a large role of pipe. The section was measured in the originally curved state, straightened by force against a wooden support and reverse bent with a pin between the pipe and the board in the middle between both ends of the pipe. (c) Transverse relaxation times extracted from a tri-exponential fit of CPMG data acquired at different spots displaced along the pipe axis on the outside of the pipe and the inside curvature. The relaxation times of the amorphous and the interfacial regions are consistently increased by the deformation, those of the crystalline regions do not change (not shown). (d) The longitudinal relaxation shows two components, one from the crystalline and one from the amorphous domains. The longitudinal relaxation time of the amorphous component is increased by the deformation while that of the crystalline domains remains unaffected.

Table 5.3.1. Common issues encountered when measuring semi-crystalline polymers

- The echo time is too short, and the receiver ringing covers the NMR signal
- The signal is too low, because a CPMG sequence is used instead of a multi solid-echo sequence
- The number of echoes is too high so that the echo train extends way beyond the signal
- The receiver phase is misadjusted and the signal phase needs to be adjusted after data acquisition
- The statistical nature of semi-crystalline has not been taken into account by averaging data from different positions of the sensitive volume
- The object may be inhomogeneous not only by nature but also from flaws

to the crystalline and the amorphous domains. Only the relaxation time T_1 from the amorphous domains changes upon deformation (Fig. 5.3.2d). These observations are consistent with the interpretation that aging and deformation affect primarily not the *crystalline domains* but the *amorphous domains* and the *interfacial domains*.

The *amorphous domains* are also the ones affected most when semi-crystalline polymers are exposed to *solvents*. Depending on the solubility, some solvents like water ingress into *polyethylene* only under extreme conditions like 6° C and 180 bar pressure from a nitrogen atmosphere (Fig. 5.3.3a). The *effective transverse relaxation* times $T_{2\text{eff}}$ extracted from fitting mono-exponential functions to the CMPG decays grows towards a saturation level following an exponential law with a time constant of 19 days for medium density polyethylene. Such measurements are easy to conduct with the NMR-MOUSE by just laying standard test samples on the sensor surface for measuring.

Instead of measuring at one depth only, *depth profiles* can be acquired as a function of time. In this way the ingress of water into a polymer sheet from the automotive sector dipped into a water bath has been traced at ambient pressure and temperature (Fig. 5.3.3b). The water ingresses from both sides, and about 65 days are required until an equilibrium concentration of 4% is reached across the 3 mm thick sheet. The concentration scale has been obtained by calibration of the NMR amplitude with a reference sample. From a time series of such depth profiles, the interdiffusion coefficient can be derived [4]. Similarly, the barrier properties of laminated walls from fluid containers are probed. A plastic *gasoline tank* consists of several layers, an outer wall for stability, an adhesive layer that binds a *barrier* layer to the outer wall, another adhesive layer between the barrier layer and the inner wall, and the inner wall. All these layers are identified in an NMR depth profile (Fig. 5.3.3c). Upon filling the tank with gasoline, the inner layer swells, but the progression of the *swelling* front is blocked by the barrier layer. The swelling process proceeds over a few days and can readily be resolved in time by successive measurements of depth profiles through the container wall. All of these measurements require only standard multi-solid echo sequences with short echo times. Some commonly encountered issues are summarized in Tab. 5.3.1.

5.3.8 Advanced level measurements

Molecular diffusion

The uptake of solvents by swelling is governed by *inter-diffusion*, that is, by the thermally activated translational motion of molecules along a concentration gradient in the direction from high to low concentration. Such *swelling* processes can readily be followed by repetitive measurement of depth profiles (Fig. 5.3.3b,c). The Brownian motion of molecules in liquids and of solvent molecules in solids at equilibrium is referred to as *self-diffusion* (Section 3.2.8).

In the time-invariant gradient field of a *stray-field* NMR device the information on *diffusion* is inscribed into the measured *CPMG* decay by a *stimulated echo* (Figs. 3.2.5b and 5.3.4a). The time-invariant, applied *gradient* is modulated in time by application of 180° pulses or two successive 90° pulses. This time modulation is expressed in terms of the *effective gradient* G_{eff}. The time integral of the effective gradient defines the *wave number* of the magnetization wave according to eqn. (3.3.4). Each time when the total time integral over G_{eff} is zero, a gradient echo is observed. This echo is attenuated not only by relaxation but also by the displacement of the molecules due to *diffusion*. For the *stimulated echo*, the signal attenuation is given by

$$s(t_E, \Delta)/s(0) = \exp\{-(\gamma G t_E/2)^2 D(\Delta + 1/3 t_E) - \Delta/T_1 - t_E/T_2\}. \qquad (5.3.3)$$

Given the strong gradient of the order of 20 T/m of the NMR-MOUSE, the relaxation terms can often be neglected so that the diffusion coefficient follows from the slope of a semi-logarithmic plot of $s(t_E, \Delta)/s(0)$ versus the cube of the echo time (eqn. (3.2.12)). Alternatively, the diffusion coefficient can be retrieved from the measured signal by fitting eqn. (5.3.3) to the experimental data (Fig. 5.3.4b). For diffusion encoding, the *echo time* t_E is varied at a fixed mixing time t_m, and CPMG echoes acquired in the detection period at short echo time and added to improve the signal-to-noise ratio (Fig. 5.3.4a).

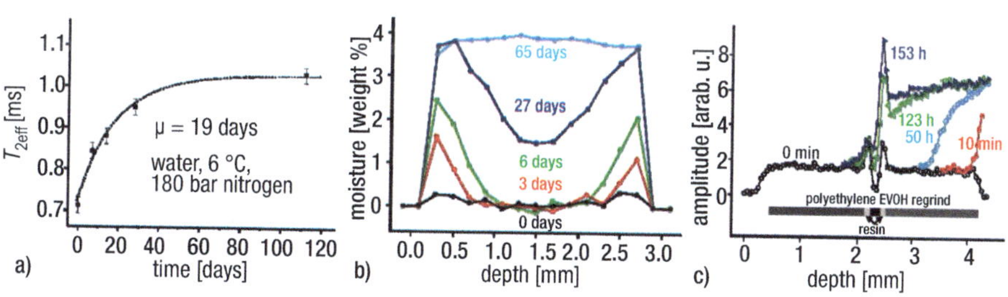

Figure 5.3.3. Solvent ingress into polymers (adapted from [5] with permission). (a) Ingress of water into a test sheet of medium density polyethylene at 6 ° C under 180 bar pressure in a nitrogen atmosphere followed in terms of the effective transverse relaxation time. (b) Ingress of water into a polymer sheet from the automotive sector at ambient pressure and temperature in terms of depth profiles and time. (c) Ingress of gasoline from the right-hand side into a laminated gasoline tank wall in terms of depth profiles and time. The wall consists of two polymer layers separated by a barrier layer, which is glued to either layer by a resin.

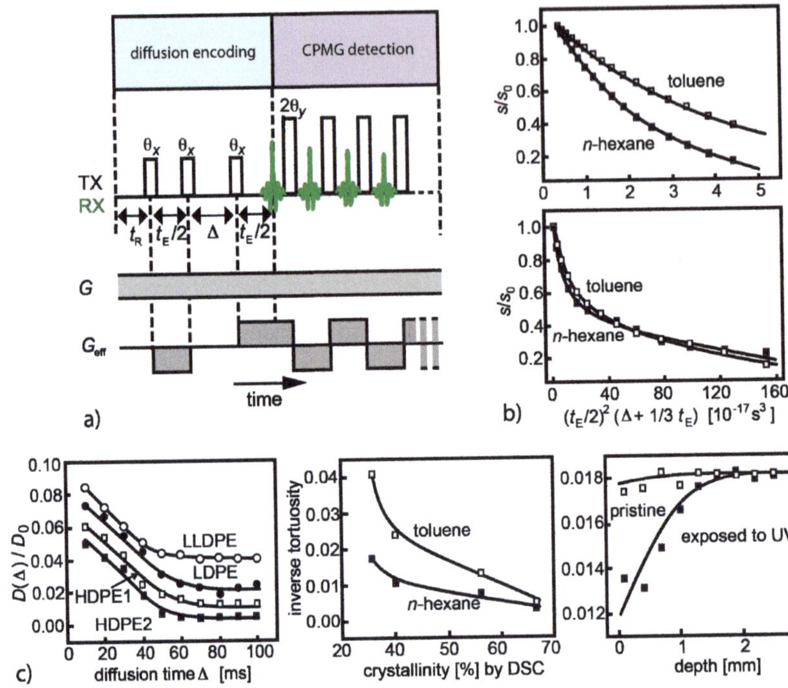

Figure 5.3.4. Measurement of translational self-diffusion with the NMR-MOUSE (adapted from [6] with permission). (a) The pulse sequence consists of a diffusion encoding period and a CPMG detection period. Ideally the pulse flip angle θ is 90°. While the gradient of the NMR-MOUSE is time invariant, the effective gradient is not because it changes sign with every 180° pulse and is suspended for longitudinal magnetization. For diffusion encoding, the echo time t_E is varied at a fixed diffusion time Δ, and the echoes acquired in the detection period are added to improve the signal-to-noise ratio. (b) Diffusion curves of pure solvents and the same solvents in polyethylene. When diffusion is restricted, the signal attenuation is less. (c) For long diffusion time Δ, the relative diffusion coefficient reaches the tortuosity limit. It scales with the crystallinity, which increases from LLDPE to LDPE, HDPE1 and HDPE2. For *n*-hexane the tortuosity changes across an aged PE pipe wall while across a pristine pipe section it does not. The lines are drawn as guides for the eye.

Typical parameters for PE are $T_1 > 0.25$ s, $T_2 > 1$ ms, $t_E = 32\,\mu s$ to $200\,\mu s$, $\Delta t_E = 12\,\mu s$, $\Delta = 50$ ms.

For pure fluids like *n*-hexane and toluene one diffusion coefficient is observed (Fig. 3.2.6 and 5.3.4b, top), while for the swollen samples, two coefficients are observed (Fig. 5.3.4b, bottom). One derives from the rapid signal decay and is assigned to free fluid wetting the sample, and the other one derives from the slow signal decay and is assigned to fluid molecules inside the polymer material, which are confined by the polymer chains.

The longer the *diffusion time* Δ, the more the solvent molecules experience the confinement of their diffusive motion by the polymer chains. As a result, the diffusion coefficient decreases with increasing diffusion time (Fig. 5.3.4c, left). At short diffusion time, the diffusion is limited to distances inside the amorphous domains while at long diffusion time the diffusion distance is constrained by the polymer morphology. The ratio of the diffusion coefficient D_0 of the pure solvent to that at long diffusion times defines the *tortuosity* τ,

$$\tau = \lim_{\Delta\to\infty} D_0/D(\Delta). \tag{5.3.4}$$

It turns out that the tortuosity scales with the *crystallinity* (Fig. 5.3.4c, middle); it increases from linear low density PE (*LLDPE*) to low density PE (*LDPE*), high density PE1 (HDPE1) and high density PE2 (HDPE2). This is strong evidence that the solvent molecules diffuse mainly in the mobile *amorphous domains*. Therefore, morphological changes induced by aging can also be observed by monitoring the solvent diffusion. This point is proven with plots of the tortuosity for n-hexane versus depth for a section from a pristine *PE pipe* and one that has been aged by *UV irradiation* from one side (Fig. 5.3.4c, right). In the aged region of the PE pipe wall the diffusion is more restricted than in the pristine material, indicating a more brittle material.

Spin diffusion

Spin diffusion is measured preferably in the more homogeneous magnetic fields of a *Halbach magnet* than in the highly inhomogeneous field of the *NMR-MOUSE*. The basic experiment involves three time periods (Figs. 5.3.5a and 5.3.6c), a *preparation period*, a *spin diffusion period* and a *detection period*. In the preparation period, longi-

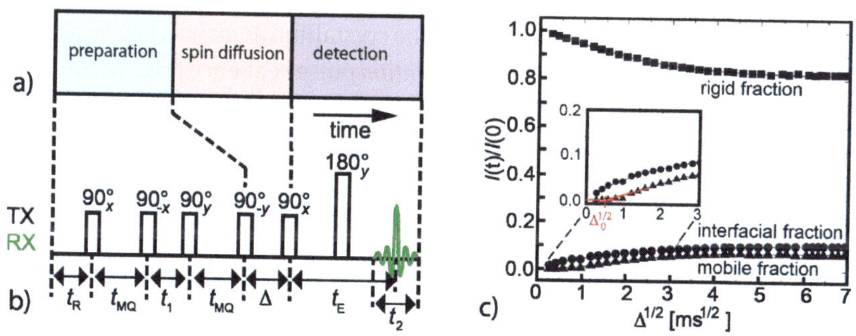

Figure 5.3.5. Timing diagram for measuring spin diffusion and experimental spin-diffusion curves for polyethylene. (a) The generic timing diagram consists of a preparation period, a spin diffusion period, and a detection period. (b) In the preparation period, maximum longitudinal magnetization is prepared in the rigid domains and zero magnetization in the other domains by establishing thermodynamic equilibrium magnetization first and then applying a double-quantum filter. During the spin diffusion period, the magnetization spreads from the crystalline domains through the interface into the amorphous domains. The transverse magnetization decay is acquired in the detection period with a Hahn echo or an echo train. (c) Spin diffusion curves for the mobile, interfacial and rigid fractions of HDPE at 100 ° C. The magnetization diffuses from the rigid domains through the interface into the mobile domains. The inset shows a magnification of the initial part. The time $\Delta_0^{1/2}$ is proportional to the size of the interface (adapted from [7] with permission).

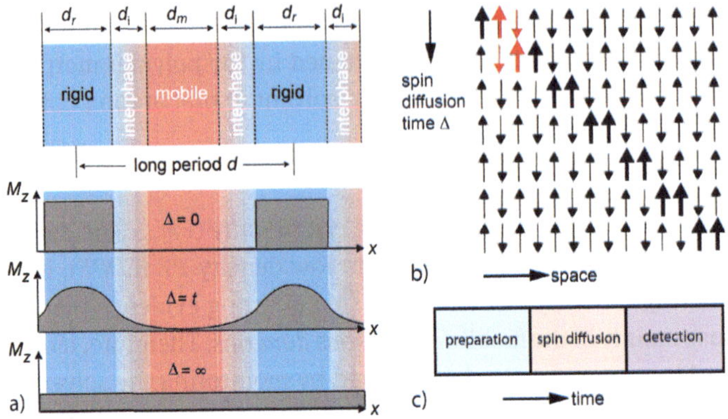

Figure 5.3.6. Spin diffusion in a semi-crystalline polymer with a lamellar morphology. (a) Drawing of stacked domains with rigid, interfacial, and mobile polymer chain segments. With the help of suitable pulse sequences, the longitudinal magnetization is localized initially only in rigid domains. The pulse sequence then allows for a spin diffusion time Δ during which the magnetization leaks from the rigid domains across the interface into the mobile domains. (b) The longitudinal magnetization (bold vectors) migrates through the object by mutual flip-flops of spin pairs (red vectors), which are coupled through the dipole-dipole interaction. (c) General timing scheme of spin diffusion experiments.

tudinal magnetization is prepared in a non-equilibrium state. Starting from thermodynamic equilibrium magnetization, which builds up in a waiting time t_R of duration 5 T_1, a pulse sequence which acts as a *magnetization filter* is applied and eliminates the magnetization from all domains other than the crystalline domains (Fig. 5.3.6a). A useful pulse sequence of this sort is a *double-quantum* pulse sequence (Fig. 5.3.5b). For short double-quantum conversion time t_{MQ} the magnetization from only the rigid domains with strong *dipole-dipole interaction* passes the filter, the other magnetization is blocked.

During the *spin diffusion time* Δ the magnetization spreads from the rigid *crystalline domains* through the interface into the mobile *amorphous domains* (Fig. 5.3.6a). This migration is caused by thermally induced flip-flop transitions of neighboring spins, which are coupled by *dipole-dipole interaction* (Fig. 5.3.6b): At room temperature, only very few spins form the longitudinal magnetization. All other spins have their magnetization cancelled by oppositely oriented neighbor spins. Wherever two neighboring spins point in the same direction, *longitudinal magnetization* is established in this volume element of the sample. If one of these spins exchanges its orientation with a neighbor spin pointing in the opposite direction in a flip-flop process, the magnetization changes its position. In this way the magnetization spreads from its source in the *rigid domains* to the sink given by the *interface* and the *mobile domains*. The spreading of magnetization follows the *diffusion equation*

$$\partial M_z / \partial \Delta = D \, \partial^2 M_z / \partial x^2 \tag{5.3.5}$$

in each direction of space, where D is the *spin diffusion constant*, Δ the spin diffusion time, and M_z is the *longitudinal magnetization*.

In the *detection period*, the transverse magnetization decay is acquired and analyzed with the help of eqn. (5.3.5), so that the magnetization fractions in the rigid, interfacial and mobile domains are determined as a function of the *spin diffusion time* Δ (Fig. 5.3.5c). With the help of a model of the polymer *morphology* (Fig. 5.3.6a), the *domain sizes* d_r, d_i, d_m are determined by fitting analytical solutions of the diffusion equation derived for the boundary conditions that describe the polymer morphology to the diffusion time dependent values of the magnetization components from the individual domains [8]. These solutions are rather complex expressions, which require as input the spin diffusion constants for each domain. These constants can be determined from wide-line NMR spectra, i.e. the Fourier transforms of the transverse magnetization of the different morphological domains.

Because the differences in segmental mobility are enhanced for polyethylene at elevated temperature, the measurements were conducted at 100° C. At this temperature the measurement time is restricted to short times to avoid annealing of the *morphology* which can be tested by monitoring the amplitudes of the mobile, interfacial and rigid components over time. By fitting the theoretical expressions for the longitudinal magnetization diffusing from the rigid domains through an interface into the amorphous domains for a lamellar morphology (Fig. 5.0.1d) to the experimental spin diffusion curves of Fig. 5.3.5c, the diameters of the rigid, interfacial, and mobile parts of the lamellae are determined to 17.0 nm, 1.3 nm, and 0.7 nm, respectively [8]. Note that the diameter d_i of the interface is proportional to the spin diffusion time $\Delta_0^{1/2}$ obtained by extrapolation of the spin diffusion curve of the mobile domains to zero amplitude (Fig. 5.3.5c).

5.3.9 Data processing

Depending on the *signal-to-noise ratio*, relaxation data acquired with the NMR-MOUSE are typically fitted by mono-, double, or triple exponential functions. In homogeneous fields, larger sample volumes can be acquired, so that the signal-to-noise ratio allows for a more precise analysis of relaxation decays following eqn. (5.3.2). In either case, the signal amplitude at zero time needs to be extrapolated from the measured signal to improve the estimation of NMR *crystallinity* from the rapidly relaxing component of the transverse magnetization.

Diffusion coefficients from *solvent diffusion* data are extracted from semi-logarithmic plots of the signal amplitudes in terms of the slope of a straight line (Fig. 3.2.6) or by fitting the theoretical expression to the experimental data (Fig. 5.3.4b). *Spin diffusion* curves are analyzed with dedicated programs in terms of a model of the polymer morphology to obtain the diameters of the crystalline, interfacial and amorphous domains. Without such programs relative differences in the thickness of the *interface* between crystalline and amorphous domains in a lamellar morphology can be determined from the extrapolated intercept of the spin diffusion curve of the amorphous domains with the axis of the square root of the spin diffusion time (Fig. 5.3.5c).

5.3.10 References

[1] Maus A, Hertlein C, Saalwächter K. A robust proton NMR method to investigate hard/soft ratios, crystallinity, and component mobility in polymers. Macromol Chem Phys. 2006; 207:1150–1158.
[2] Adams A, Adams M, Blümich B, Kocks HJ, Hilgert O, Zimmermann S. Optimierung der Umhüllung von Stahlrohren. 3R International. 2010; 49:216–225.
[3] Blümich B, Adams-Buda A, Baias M. Alterung von Polyethylen: Zerstörungsfreies Prüfen mit mobiler magnetischer Resonanz. GWF Gas Erdgas. 2007; 148:95–98.
[4] Reuvers NJW, Huinink HP, Fischer HR, Adan OCG. Quantitative water uptake study in thin nylon-6 films with NMR imaging. Macromolecules. 2012; 45:1937–1945.
[5] Blümich B, Casanova F, Perlo J. Mobile single-sided NMR. Prog Nucl Magn Reson Spectrosc. 2008; 52:197–269.
[6] Kwamen R, Blümich B, Buda A. Estimation of self-diffusion coefficients of small penentrants in semicrystaline polymers using single-sided NMR. Macromol Rapid Commun. 2012; 33:943–947.
[7] Hedesiu C, Demco DE, Kleppinger R, Adams-Buda A, Blümich B, Remerie K et al. The effect of temperature and annealing on the phase composition, molecular mobility and the thickness of domains in high-density polyethylene. Polymer 2007; 48:763–777.
[8] Buda A, Demco DE, Bertmer M, Blümich B, Litvinov LV, Penning JP. General analytical description of spin-diffusion for a three domain morphology. Application to melt-spun nylon-6 fibers. J Phys Chem B. 2003; 107:5357–5370.

6 Biological tissues

The most appreciated use of NMR is *magnetic resonance imaging* (*MRI*) for medical diagnostics. MRI provides images of soft biological tissues with contrast superior to that in X-ray images obtained with *computer tomography* (CT). Early medical MRI machines worked at a magnetic field strength of 0.5 T, about the same field strength the NMR-MOUSE provides. In an overwhelming number of cases, the contrast in MRI is generated by differences in *spin density*, *relaxation*, or molecular *self-diffusion*, none of which requires homogeneous magnetic fields. Therefore, it is to be expected that stray-field NMR will provide signal that is good for discrimination of medical tissues, and the vision of an NMR-based tricorder that diagnoses disease from a distance by pointing it at a person or limb haunts the fiction literature (Fig. 6.0.1a) [1]. The fact is that the depth range of today's stray-field NMR technology is limited to a few centimeters, so that leaves, bark, *skin* as well as near-surface *tendon*, *bone* and brain can be studied with the *NMR-MOUSE* in close contact with the object.

A stray-field sensor like the NMR-MOUSE collects signal from a slice parallel to the sensor surface at a given distance away from the surface (Fig. 6.0.1b). This slice corresponds to a *pixel* in an NMR image, and by shifting the slice, high-resolution depth profiles and low-resolution lateral images can be measured point by point. The in-

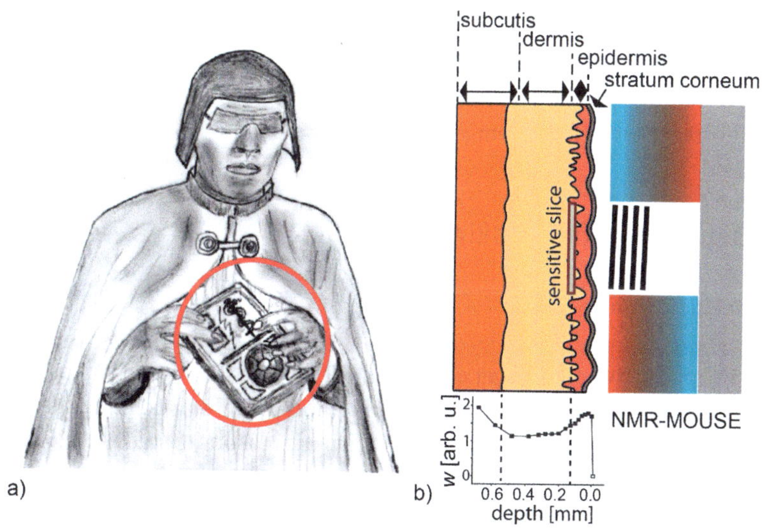

Figure 6.0.1. Fiction and reality associated with single-sided NMR in medicine. (a) A device like the tricorder® (trademark of CBS Studios Inc.) from Starship Enterprise is supposed to be able to diagnose illness remotely, whereas the NMR-MOUSE cannot collect NMR signal from far away [1]. (b) The Profile NMR-MOUSE has a sensitive volume in the shape of a thin flat slice capable of detailing the stratigraphy of skin, e.g. from the lower arm by shifting the sensitive slice through the depth range of interest.

plane pixel resolution of the NMR-MOUSE is rather crude and of the order of 1 cm^2, while the depth resolution is high, on the order of 10 μm, so that the resolution of the depth profile is determined by the flatness of the layers and the quality of alignment of the sensitive slice with the layers.

Biological tissues are found in animals and in plants. They are assemblies of cells that serve particular functions. The study of tissue is called histology. Conventional histological techniques are invasive, because they require the tissue to be embedded in a paraffin block to be stained, frozen, sliced, and inspected under a microscope. There are four basic types of *animal tissue*. These are (1) connective tissue such as cartilage, tendon, bone, adipose tissue, and blood, (2) muscle tissue, (3) nerves, and (4) epithelial tissue which lines the cavities and surfaces of structures in the body. *Plant tissues* are the (1) epidermis formed by the outer cells of leaves, (2) the vascular tissue defining xylem and phloem which transport the fluid and nutrients, and (3) the ground tissue which produces nutrients by photosynthesis and stores reserve nutrients. Plant tissue is also divided into (1) meristematic tissues with actively dividing cells and (2) permanent tissues without actively dividing cells.

NMR is of great interest for studies of biological tissues, because the investigation can be conducted in a nondestructive way. Moreover, mobile NMR allows the tissues to be studied at the site, i.e. in the medical doctor's office, senior citizens' home, greenhouse, the field or in the forest. The NMR-MOUSE is particularly well suited for characterizing extended and possibly layered tissues within a centimeter or two close to the surface. Examples are studies of *connective tissue* like *skin*, *tendon* and *bone*. On the other hand, portable MRI can provide images of small animals such as mice near the animal shelter and of *plants* such as cross sections through the stem for monitoring fluid and transport in the xylem and phloem.

6.1 Depth profiling of skin

6.1.1 Introduction

Skin is the protective outer layer of the body. It protects the body from pathogens and water loss and for many humans plays an important role in defining their perceived identity. Therefore, there are medical and cosmetic reasons to study skin. Skin consists of different layers. The most important ones are the *stratum corneum* formed from dead cells, the *epidermis*, which constitutes the barrier layer, the *dermis*, where the cells grow, and the *subcutis*, which is the first layer underneath the skin (Fig. 6.0.1b).

The thickness of these layers and their properties vary with gender, location of the skin on the body, age, skin type, sun exposure, and skin care. *Cosmetics* are classified to affect only the outer layers of the skin and not to transgress the skin barrier into the body. This is different for *drugs* like nicotine patches that deliver active ingredients across the skin from outside to inside, while *sweat* passes from inside to outside.

The morphology of skin and its function are difficult to image with conventional MRI unless dedicated hardware is employed to resolve the thin layers, which define the stratigraphy of skin. Stray-field NMR sensors like the NMR-MOUSE appear to be predestined for imaging skin because they perform best at low depth and provide superior resolution due to high gradients. However, because the sensitive slice extends laterally across several millimeters, the wavy interfaces between different skin layers limit the ultimate *depth resolution* that can be reached experimentally (Fig. 6.0.1b).

6.1.2 Objective

NMR investigations of skin aim at resolving its layer structure and at diagnosing its state. Such studies are motivated by medical conditions like skin grafting, wound healing, allergic reactions, transdermal drug administration, and skin *hydration* reflecting body function as determined by fluid uptake, exercise, and medical treatment like dialysis. Another motivation is the assessment of skin care treatments with lasers, cosmetic agents, and by other means. By choosing the right contrast in depth profiling, changes in the skin layers can be correlated with skin treatment and resolved in time, for example, to quantitatively analyze the ingress of *skin cream*.

6.1.3 Further reading

Danieli E, Blümich B, Single-sided magnetic resonance depth profiling in biological and materials science. J Magn Reson. 2013; 229:142–154.

Van Landeghem M, Perlo J, Blümich B, Casanova F, Low-gradient single-sided NMR sensor for one-shot profiling of human skin. J Magn Reson. 2012; 215:74–84.

Casanova F, Perlo J, Blümich B, editors. Single-Sided NMR. Berlin: Springer; 2011.

Ciampi E, van Ginkel M, McDonald PJ, Pitts S, Bonnist EY, Singleton S, Williamson AM. Dynamic in vivo mapping of model moisturiser ingress into human skin by GARField MRI. NMR Biomed. 2010; 24: 135–144.

Rähse W, Dicoi O. Produktdesign disperser Stoffe: Emulsionen für die kosmetische Industrie. Chemie Ingenieur Technik. 2009; 81:1369–1383.

Blümich B, Casanova F, Perlo J. Mobile single-sided NMR. Prog Nucl Magn Reson Spectrosc. 2008; 52:197–269.

Casanova F, Perlo J, Blümich B. Depth profiling by Single-Sided NMR. In: Stapf S, Han SI, editors. NMR Imaging in Chemical Engineering. Weinheim: Wiley-VCH; 2006, pp. 107–123.

McDonald PJ, Akhmerov A, Backhouse LJ, Pitts S. Magnetic resonance profiling of human skin in vivo using GARField magnets. J Pharm Sci. 2005; 94: 1850–1860.

6.1.4 Theory

Without spectroscopic resolution, NMR is capable of discriminating magnetization components with different *relaxation rates* and *diffusion coefficients*. Moreover, *component amplitudes* and associated relaxation rates and diffusion coefficients can be quantified as a function of depth and time. From such data models of skin can be developed that account for the stratigraphy of the skin, which can be resolved by NMR, and the layer function in terms of the uptake, release, and transport of components such as sweat as well as oil and water in creams. In this way stray-field NMR is unique in providing the essential information for establishing functional models of living skin and model skin in terms of material and transport properties.

6.1.5 Hardware

Typically clinical MRI is used to analyze biological tissue. Yet MRI machines are available not only for measuring humans, but also in compact form for desktop operation or as portable devices (Fig. 1.3.3c). The best approach for studying the stratigraphy of skin appears to be stray-field NMR, because such measurements are straightforward, and provide high-resolution *depth profiles* of the skin layers with information about spin density, relaxation and diffusion. One such device is the *GARField magnet* (Gradients At Right Angles to Field) [2], designed to produce high-resolution NMR depth profiles of objects placed in the large gap of a desktop magnet, which is open on three sides and exhibits a magnetic field with a constant gradient in an extended volume (Fig. 6.1.1c). This magnet, the NMR-MOUSE (Fig. 6.1.1a) and more recently also the *Fourier NMR-MOUSE* (Fig. 6.1.1b) are employed to study skin *in vivo* in terms of depth profiles.

Because skin is thin, it is measured best with an NMR-MOUSE with a low depth range such as 5 mm. When measuring humans or animals with stray-field NMR devices, the electrically conducting body may act as an antenna that feeds noise to the NMR receiver. To reduce the noise collected by the subject under study it proves helpful at times to wrap the NMR-MOUSE and a considerable part of the limb or the subject in electrically conducting *parachute silk*, which is brought into electrical contact with the grounded lift of the NMR-MOUSE or the case of the NMR spectrometer (Fig. 6.1.1a).

Depth is scanned with the *Profile NMR-MOUSE* by adjusting the distance between the sensor and the object (Fig. 6.0.1b). While the Profile NMR-MOUSE is well suited to resolve the layer structure of skin *in vivo* in terms of the *w-parameter* (eqn. (3.1.5)) and to follow changes in the skin with time at a given depth, the sensitivity is not good enough to acquire depth profiles as a function of time to track, for example, moisture uptake, swelling of layers and drying as a function of time after application of *skin cream*. Such experiments are preferably conducted with the *Fourier NMR-MOUSE* (Fig. 6.1.1b) and the GARField magnet (Fig. 6.1.1c) because with these devices a depth

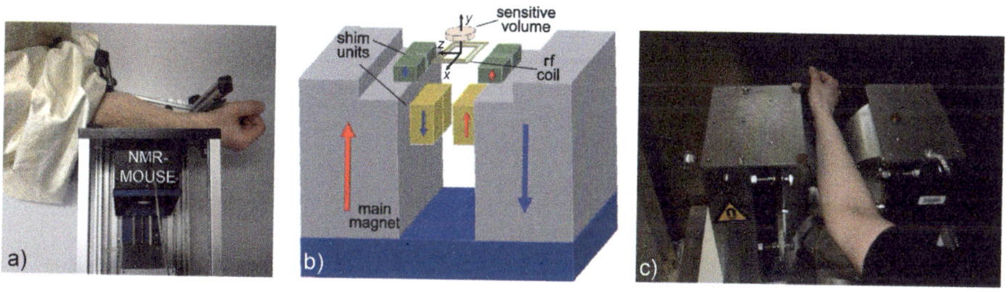

Figure 6.1.1. Stray-field sensors to measure depth profiles of skin. (a) Profile NMR-MOUSE mounted on computer-controlled lift. The arm is positioned above the NMR-MOUSE in an armrest. The subject is covered with electrically conducting parachute silk, which is grounded with the spectrometer to reduce far-field noise. (b) Fourier NMR-MOUSE. This is a Profile NMR-MOUSE with shim magnets, which are adjusted to produce a thick sensitive slice and a uniform gradient in the sensitive volume. (c) GARField magnet with a uniform gradient over an extended volume for 1D imaging of thin-layered samples (photo courtesy of Peter McDonald). With the GARField magnet and the Fourier NMR-MOUSE a depth profile can be acquired in a single shot in terms of the Fourier transform of an echo.

profile can be measured in a single scan without advancing the sensor step by step to shift the sensitive slice through the different layers of the skin. Instead, either device makes use of the *frequency encoding* technique from MRI, and the depth profile is obtained in terms of the Fourier transform of the echo. Depending on the gradient strength, signal across a range of depths can be excited with a narrow rf pulse. The GARField magnet provides a gradient of the order of 20 T/m, the same as the Profile NMR-MOUSE, and achieves high homogeneity of the gradient with accurately shaped pole shoes (Fig. 6.1.1c). But to cover a depth range of a millimeter or two, the magnetic field gradient has to be made about ten times smaller and homogeneous with the sensitive slice. To meet these requirements, shim magnets are accurately positioned in the magnet gap of the Fourier NMR-MOUSE (Fig. 6.1.1b).

6.1.6 Pulse sequences and parameters

Default measurements of skin with the Profile NMR-MOUSE use the *CPMG sequence* (Fig. 3.1.1b). More advanced measurements aim at determining the diffusion coefficient or *distributions of diffusion coefficients*. These are preferably conducted with the stimulated echo sequence (Figs. 3.2.5b and 5.3.4a). The parameters for a CPMG sequence to acquire the transverse magnetization decays of *elastomers* and of soft tissue like skin are similar (Tabs. 6.1.1 and 5.1.1). But opposed to elastomers, skin contains water and fat molecules, which can diffuse. The choice of the *echo time* therefore determines the *diffusion weight* of the signal acquired in the field gradient of the NMR-MOUSE. When the diffusion coefficient or the distribution of diffusion coefficients are to be determined, the CPMG sequence is preceded by a stimulated echo where the echo time t_E

Table 6.1.1. Default acquisition parameters for ^{1}H stray-field NMR of skin

Parameter	Value
Magnet, probe	NMR-MOUSE PM5
transmitter frequency ν_{rf}	17.1 MHz
transmitter attenuation for 90° pulse	–8 dB at 300 W
duration t_p of 90° pulse	5 µs
dwell time Δt	1 µs
acquisition time t_{acq}	20 µs
echo time t_E	50 µs
number n_E of echoes	1024
recycle delay t_R	1 s
number n_s of scans	32

is varied from scan to scan from 0.1 ms to 2 ms with exponential or linear time increments of Δt_E while the diffusion time is kept fixed at $\Delta = 5$ ms (Fig. 5.3.4a). Because the Fourier NMR-MOUSE is not a standard tool yet, acquisition parameters are discussed together with applications below in the section on advanced level measurements.

6.1.7 Beginner's level measurements

Depth profiles of the w-parameter (eqn. (3.1.5)) through human skin *in vivo* vary from person to person (Fig. 6.1.2a). These variations are well reproducible, even when a manual lift is employed. However, the profile shapes also differ between skin from different parts of the body (Fig. 6.1.2b). CPMG measurements with the NMR-MOUSE provide ample opportunity to study these differences and to correlate characteristic features of the skin NMR profile with personal data. The *contrast* in the *depth profile* can be adjusted via the particular choice of the *w-parameter* and via the *echo time*.

The w-parameter is defined as the amplitude sum of the last echoes of the train divided by that of the first echoes of the train (eqn. (3.1.5)). A good choice for skin is to split a train of 300 echoes into the first 30 echoes and the remaining 240 echoes for calculation of the w-parameter. The echo time t_E changes the *diffusion weight* (Fig. 6.1.2c). A long echo time enhances the *diffusion contrast*. The signal of the *epidermis* does not change much with t_E in contrast to that of the *dermis* and that of the *subcutis*.

While Profile NMR-MOUSE provides high depth resolution and can resolve features below 10 µm in perfectly planar structures, the acquisition of one profile demands several minutes. A 1 mm depth profile with a step width of 50 µm, 64 scans per depth position, and a recycle delay of 0.3 s takes about 8 minutes to acquire with the automatic lift. This time is too long to follow kinetic phenomena of *moisture ingress* and drying of skin in terms of depth profiles, but initial and final states of the skin can well be recorded. Depth profiles can be acquired more rapidly using the *Fourier NMR-MOUSE* described below.

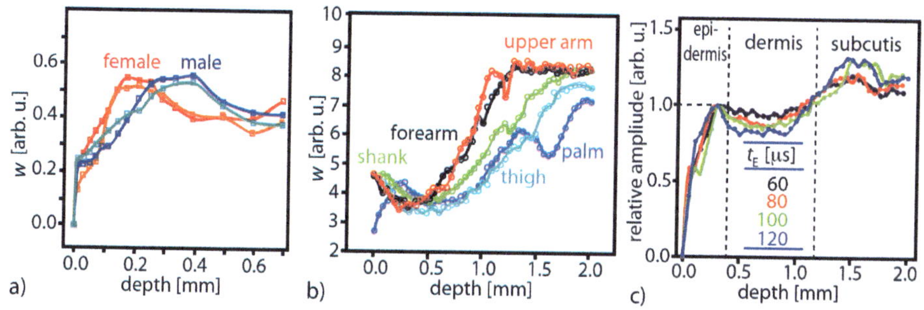

Figure 6.1.2. Depth profiles through skin. (a) Palm of the hand of a male and a female volunteer. Two scans are shown for each position demonstrating the quality of reproducibility when using a manual lift. (b) Skin at different parts of the body of a female volunteer (adapted from [3] with permission). (c) Normalized profiles through the palm of the hand acquired with different echo times to vary the diffusion weight. Diffusion is weak in the epidermis and strong in the dermis and the subcutis.

Measuring skin *in vivo* requires some practice to make sure that comparable data are obtained from different individuals (Tab. 6.1.2). The skin on the lower arm and the palm of the hand is measured most easily. The different skin layers are relatively thick so that they can be identified in the depth profiles. But the time for measuring a depth profile is of the order of 5 to 10 minutes. During this time the limb exposed to the sensor has to be absolutely still. Any change in *pressure* on the sensor may not only affect the layer dimensions of the skin but also shift the zero point of the profile by slightly bending the plate of the lift on which the limb rests. For studies involving many persons, a dedicated limb rest should be employed. The initial position of the depth profile can be marked with the signal from an adhesive strip attached to the lift plate above the rf coil. Furthermore, the skin of each individual needs to be conditioned to reproducible initial conditions, and the development of *sweat* on the skin exposed to the sensor must be avoided.

Table 6.1.2. Common issues encountered when measuring skin

– The recycle delay is too short so that the signal from water is partially suppressed
– The duration of the echo train is too short and the echo train is truncated
– The echo time is badly chosen so that the diffusion weight provides bad contrast
– The measurement time is too long and the skin under the sensor starts to sweat
– The pressure on the sensor changes during the measurement so that the zero position shifts
– The receiver phase is misadjusted and the signal phase needs to be adjusted after data acquisition
– The noise level is too high. The subject acts as an antenna and needs to be electrically shielded, e.g. by wrapping the subject with electrically grounded parachute silk

6.1.8 Advanced level measurements

The Fourier NMR-MOUSE

With the *Profile NMR-MOUSE* the position of the sensitive slice is advanced step by step to collect the integral signal from the sensitive slice at each step (Fig. 6.0.1b). This way of scanning a depth profile in real space is slow compared to the *frequency encoding* of space employed in *Fourier imaging* (Fig. 3.3.4), which requires the acquisition of an echo in the presence of a gradient. With the NMR-MOUSE echoes are measured in the presence of gradients in the depth direction y, but in contrast to spin-echo imaging, these gradients are very strong and cannot be turned off when rf pulses are applied. This is why even a short rf pulse selects only a magnetization slice in the object but never excites the magnetization in the entire object. In such a case, the duration of the echo is well approximated by the duration t_p of the rf pulse. In practice, the echo decay is sampled for the *acquisition time* t_{acq} until the echo signal is covered by noise. Following eqn. (3.3.5a), which applies when the resolution is not limited by relaxation, the *spatial resolution* with the Fourier NMR-MOUSE is then given by $1/\Delta y = k_{y,max}/(2\pi) = \gamma G_y t_{acq}/(2\pi)$.

In addition to relaxation, the NMR signal is attenuated by *diffusion*, which is considerable for low viscosity liquids freely diffusing in strong gradients. This signal attenuation is ameliorated when the gradient strength is reduced. Low gradient strength is also required for a large *depth range* $y_{max} = 2\pi/(\gamma G_y \Delta t)$ (eqn. (3.3.5d)) to be scanned by *frequency encoding*, whereby the interval Δt between data points sampled under the echo is typically of the order of 0.5 to 1 μs. In practice, the depth range is limited by the excitation bandwidth of the rf pulse. A 5 μs long rf pulse applied in a gradient G_y of 2 T/m gives access to a 2 mm depth range with a resolution of about $1/\Delta y \approx 1/(30\ \mu m)$. The Fourier NMR-MOUSE (Fig. 6.1.1b) has been designed to meet these specifications. The small magnets in the gap between the main magnets are adjusted to scale down the field gradient provided by the main magnets by a factor of 10 compared to that of the Profile NMR-MOUSE, and to provide a uniform gradient in the entire volume of the slice, which is 2 mm thick and covers the coil diameter.

With the Fourier NMR-MOUSE depth profiles can be measured fast, because a depth profile across the sensitive slice is the Fourier transform of an echo. The echoes of a CPMG train can be added to improve the *signal-to-noise ratio* or processed separately to calculate the *w-parameter*, to extract *relaxation times* and *component amplitudes*, or to calculate the *distribution of relaxation times*. Also, the CPMG train can be edited with a *stimulated echo* (Fig. 5.3.4a) to introduce a *diffusion weight*. From a set of CPMG data edited by stimulated echoes with different echo times, the distribution of diffusion coefficients can be extracted by *Laplace inversion*. Basically all experiments that can be done with the Profile NMR-MOUSE can also be done with the Fourier NMR-MOUSE albeit faster and at the expense of somewhat lower spatial resolution.

Time-resolved depth profiles

Instead of several minutes, useful depth profiles through skin can be acquired with the Fourier NMR-MOUSE in 30 s. This is fast enough to follow the uptake and release of *skin care products*. For example, the moisturizing effect of cream has been followed for the skin of the palm of the hand by acquiring depth profiles of the w-parameter for 30 seconds every three minutes (Fig. 6.1.3a). The dry *epidermis* of regular skin without cream gives low profile values while the moist *dermis* gives high values and the *subcutis* slightly lower values than the dermis. Both the outer layer of the epidermis, the *stratum corneum*, and the epidermis take up the skin cream at about the same rate, while the profiles remain unchanged in the regions of the dermis and the subcutis. With time increasing after application of the cream, the peak in the profile that marks the dermis shifts to slightly larger depths indicating that the epidermis swells as it absorbs more of the skin cream.

Relaxation-resolved depth profiles

Depth profiles that show the *distribution of transverse relaxation times* $T_{2\mathrm{eff}}$ at each depth can be measured in about 10 minutes by acquiring a CPMG echo train at high signal-to-noise ratio with 2048 scans at a repetition time of $t_R = 300$ ms. The data depicted in Fig. 6.1.3b were acquired with CPMG trains consisting of 340 echoes, where each echo was sampled with 128 complex points at a dwell time of $\Delta t = 2\,\mu$s. This sets the *depth resolution* to 1/(50 μm), a value slightly lower than the maximum resolution of the sensor, which is limited by the uniformity of the gradient in the sensitive slice.

The maxima of the logarithmic $T_{2\mathrm{eff}}$ distribution versus depth trace the w*-parameter* with high fidelity (Fig. 6.1.3b). This observation is in accordance with the fact that the w-parameter approximates the time integral of the normalized CPMG decay curve, which is an *average relaxation time* (eqn. (3.1.4)). Thus most of the information provided by depth-resolved *relaxation time distributions* is contained in the w-parameter

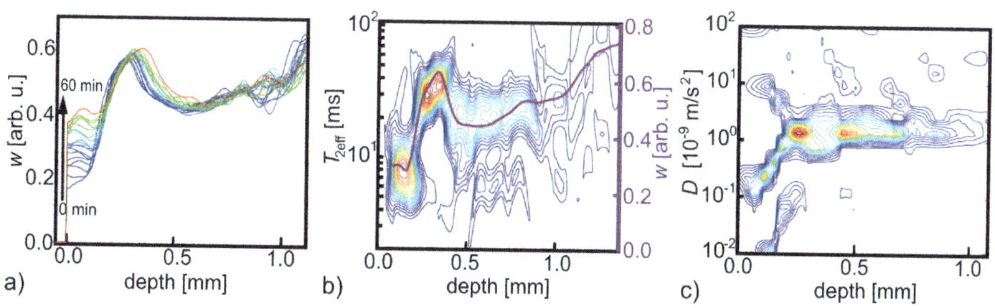

Figure 6.1.3. Depth profiles through the skin of the palm of the hand acquired with the Fourier NMR-MOUSE (adapted from [4] with permission). (a) Time resolved profiles acquired every three minutes following the application of skin cream. (b) Depth resolved distribution of transverse relaxation times. (c) Depth resolved distribution of diffusion coefficients.

depth profile, which can be obtained about 20 times faster. The depth profile of the w-parameter depicted in Fig. 6.1.3b was measured with 128 scans in 38 s.

Diffusion-resolved depth profiles

While the relaxation time distributions show mostly one peak in the different skin layers, which is traced by the w-parameter profile, up to three diffusion coefficients can be observed in skin layers, depending on the skin care history. The preferred pulse sequence consists of a *stimulated echo* for *diffusion encoding* followed by a *CPMG sequence* for detection (Fig. 5.3.4a). To record 32 points in the diffusion direction of a depth profile from the palm of the hand (Fig. 6.1.3c), 20 minutes measurement time were needed given 128 scans for each diffusion-encoding step. In this case the acquisition time of the CPMG echoes was reduced to half the value used for the depth-resolved *relaxation time distributions* (Fig. 6.1.3b) in order to shorten the echo time of the CPMG detection train to 160 μs for trading spatial resolution against higher signal-to-noise ratio which results from lower signal loss due to *diffusion*. The *diffusion time* was set to $\Delta = 5$ ms to probe a wide range of diffusion coefficients. Although such depth-resolved *distributions of diffusion coefficients* take too much time to resolve the changes in skin during uptake or release of agents, they provide detailed information on the state of skin, which cannot be obtained by any other method, because in one skin layer different diffusion coefficients can be identified and subsequently assigned to different skin structures or cosmetic ingredients.

6.1.9 Data processing

Depth profiles of the *w-parameter* are calculated from the real parts of the phase corrected, complex CPMG data by summing the amplitudes of the latter 90% of the echoes and dividing this sum by the sum of the first 10% of the echoes. On the other hand, if the signal-to-noise ratio is good enough, the entire CPMG train can be inverted by inverse Laplace transformation to obtain the *distribution of relaxation times*. This distribution and the w-parameter are calculated at each depth at which the Profile NMR-MOUSE has been positioned. *Distributions of diffusion coefficients* are measured in a two-dimensional fashion with indirect detection of the diffusion information by preparing the signal detected with a CPMG echo train with a stimulated echo sequence for which the echo time is varied from scan to scan. Such distributions are computed by *inverse Laplace transformation* along the indirectly detected diffusion dimension. The directly detected CPMG dimension can be treated in different ways: (1) Echoes are added to improve the *signal-to-noise ratio* and to obtain an *average relaxation time*. (2) A *relaxation-weighted spin density* corresponding to the *w-parameter* is calculated. (3) The *distribution of relaxation times* is calculated by a second Laplace inversion to correlate the distribution of relaxation times with the distribution of diffusion coefficients. Each type of processing requires different minimum quality of the experimental data. Data with a high noise level should be processed with echo summation, and high-quality data can be Laplace inverted. When diffusion data are acquired with the Fourier NMR-MOUSE, a 3D data set can in principle be extracted from the diffusion-edited experimental CPMG data, which correlates the distribution of diffusion coefficients, the distribution of relaxation times and depth.

6.1.10 References

[1] Kaku M. Physics of the Future. New York: Doubleday; 2011. pp. 60–62.
[2] Glover PM, Aptaker PS, Bowler JR, Ciampi E, McDonald PJ. A novel high-gradient permanent magnet for the profiling of planar films and coatings. J Magn Reson. 1999; 139: 90–97.
[3] Casanova F, Perlo J, Blümich B, editors. Single-Sided NMR. Berlin: Springer; 2011.
[4] Van Landeghem M, Perlo J, Blümich B, Casanova F. Low-gradient single-sided NMR sensor for one-shot profiling of human skin. J Magn Reson. 2012; 215: 74–84.

6.2 Anisotropy of tendon by relaxometry

6.2.1 Introduction

Tendon is the connective tissue between bone and muscle. The tissue connecting bone and bone is referred to as ligament while the tissue connecting muscle and muscle is called fascia. All three types of connective tissue are substantially made of collagen. *Collagen* is the most abundant protein material in mammals. At the molecular level it is composed of three amino acid chains arranged in a triple helix. These helices form the

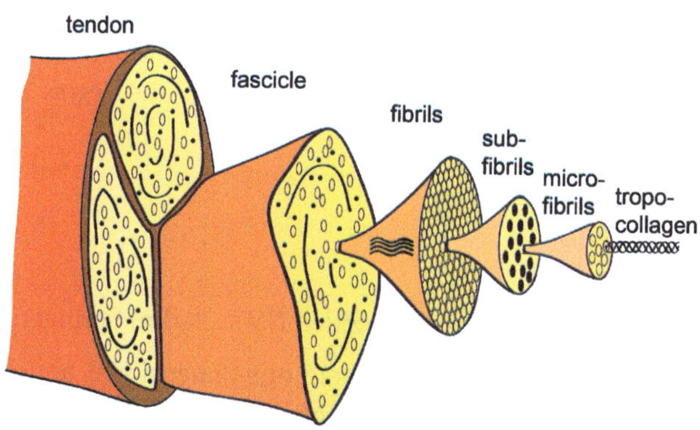

Figure 6.2.1. Structure of tendon (adapted from [1]). Tendon is mostly a highly ordered hierarchy of aligned bundles and fibrils. These are made from triple helices of polypeptides, which form the tropocollagen at the molecular level.

tropocollagen. The tropocollagen is arranged in an ordered hierarchy of elongated and highly ordered *fibrils*, similar to the wires in a steel rope of a suspension bridge that form the tendon ligament and provide mechanical strength. The tropocollagen forms microfibrils, which are bundled into subfibrils and then further into fibrils. These fibrils are crimped into wavy structures, which play a determining role for the *elasticity* of the tissue. The fibrils are arranged into *fascicles* together with the fibroplasts, which are the biological cells that produce the tendon. Bundles of fascicles then make up the tendon (Fig. 6.2.1).

6.2.2 Objective

The thickest and strongest tendon in the human body is the *Achilles tendon* of the heel. It has extensively been studied by MRI but can also be accessed by single-sided NMR. The anisotropy of the tendon structure revels itself by an *anisotropy of translational diffusion*, which can be measured with *pulsed field gradient* NMR by varying the gradient direction in MRI diffusion studies. While the gradient direction is easy to change in MRI machines, the direction of the magnetic field with respect the axis of the body, limb, or tissue is not. This complicates studies or *relaxation anisotropy* in ordered tissues by conventional MRI. In some cases a so-called *magic angle effect* is observed in NMR images of tendon and cartilage when the average *dipole-dipole coupling tensor* of the uniaxially ordered tissue is oriented at the magic angle of 54.7° with respect to the direction of the applied magnetic field $\boldsymbol{B}_0$. At this orientation the average *residual dipole-dipole interaction* is reduced so that the longitudinal and transverse relaxation times are longer and give rise to higher image intensities.

The orientation dependence of the relaxation rates can be studied with the U-shaped NMR-MOUSE, which has a magnetic field direction perpendicular to the magnet gap, so that the orientation of $\boldsymbol{B}_0$ can be changed by rotating the NMR-MOUSE while maintaining the position of the sensitive slice. The aim of studying tissue anisotropy with the NMR-MOUSE is to quantify the *isotropic relaxation rate* and the *anisotropic relaxation rate*, and to gain information about the *degree of order* of the tissue.

6.2.3 Further reading

Casanova F, Perlo J, Blümich B, editors. Single-Sided NMR. Berlin: Springer; 2011.

Rössler E, Mattea C, Mollava A, Stapf S. Low-field one-dimensional and direction-dependent relaxation imaging of bovine articular cartilage. J Magn Reson. 2011; 2013: 112–118.

Blümich B, Casanova F, Perlo J. Mobile single-sided NMR. Prog Nucl Magn Reson Spectrosc. 2008; 52:197–269.

Navon G, Eliav U, Demco DE, Blümich B. Study of order and dynamic processes in tendon by NMR and MRI. J Magn Reson Imag. 2007; 25: 362–380.

Blümich B. Essential NMR. Berlin: Springer; 2005.

Xia Y. Magic-angle effect in magnetic resonance imaging of articular cartilage: A review. Invest Radio. 2000; 35:602–621.

Haken R, Blümich B. Anisotropy in tendon investigation in vivo by a portable NMR scanner, the NMR-MOUSE. J Magn Reson. 2000; 144:195–199.

Blümich B. NMR Imaging of Materials. Oxford: Clarendon Press; 2000.

6.2.4 Theory

Tendon is a material in which the molecules are highly ordered although not as perfectly as in a crystal. The triple helices from amino acids in the tropocollagen are well aligned parallel to each other. Consequently their molecular motion is highly constrained, but so is the motion of the water molecules embedded in the tendon. Because the translational and rotation diffusion of the molecules is no longer isotropic, the magnetic dipole-dipole interaction between neighboring protons on a molecule is no longer averaged to zero, and a *residual dipole-dipole interaction* remains (Fig. 5.1.1). This interaction shortens the decay of the transverse magnetization. The more constrained the molecular motion is, the stronger the dipole-dipole interaction and the larger the relaxation rate of the transverse magnetization. This situation is analogous to the relaxation in strained *rubber*, where the molecules are aligned on average along the strain direction (Figs. 5.1.3c,e).

Depending on whether or not the transverse magnetization decay is measured with a set of Hahn echoes and variable echo time or with a CPMG sequence, different relaxation rates are observed in a mono-exponential fit of the experimental set of echoes measured with the NMR-MOUSE. The relaxation time $T_{2\mathrm{eff}}$ of the magnetization decay from a CPMG echo train with short echo time shows little attenuation from *diffusion* in the field gradient of the NMR-MOUSE, and the dipole-dipole interaction is partially averaged by locking magnetization components along the direction of the magnetic radio frequency field $\boldsymbol{B}_1$. Neither is the case when the decay is measured by repeating Hahn echoes with increasing echo time, so that the transverse relaxation time T_2 from Hahn echo measurements is generally found to be shorter than $T_{2\mathrm{eff}}$ determined with the CPMG sequence, and the variation of $1/T_2$ with varying angle θ between tendon direction and magnetic field direction is more pronounced than the variation of $1/T_{2\mathrm{eff}}$. The diffusion weight introduced to the transverse relaxation rate by the field gradient of the NMR-MOUSE is angle independent, because the gradient is in the depth direction along the rotation axis and not along the direction of the magnetic field.

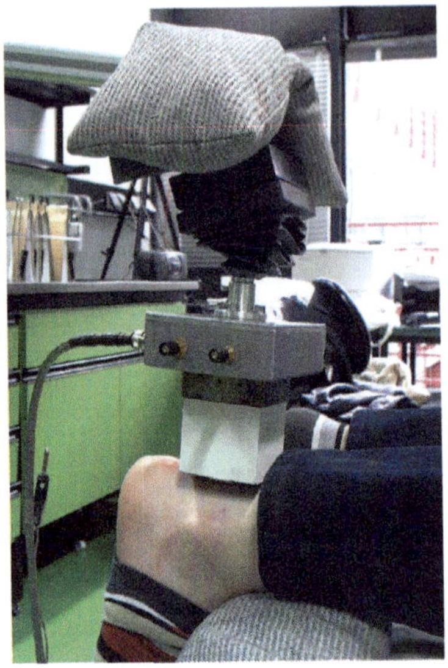

Figure 6.2.2. Photograph of a setup with an early version of the NMR-MOUSE used to measure Achilles tendon *in vivo*.

The relaxation rate is expected to vary with the square of the *second Legendre polynomial* 1/2 (3 $\cos^2\theta$ -1), which describes the *angle dependence* of the *dipole-dipole interaction*,

$$1/T_{2\mathrm{eff}}(\theta) = 1/T_{2\mathrm{eff,iso}} + 1/T_{2\mathrm{eff,aniso}}[3(\cos^2\theta - 1)/2]^2. \quad (6.2.1)$$

This equation follows from eqn. (5.1.10), which has been used to model the orientation dependence of the relaxation rate in strained rubber bands, if the angle distribution is made infinitely narrow corresponding to perfect alignment of the vectors connecting the dipolar coupled pairs of protons. Such a model is feasible for water molecules diffusing along the direction of the macroscopically well-aligned collagen fibers (Fig. 6.2.1). In addition to the mobile water protons, the protons from bound water and the collagen fibers are expected to contribute to the signal, so that the magnetization decay should show at least two components, a fast decay from bound water and collagen protons and a slow decay from water. A mono-exponential analysis is suitable when the fast decay is cut off by the dead time of the instrument.

6.2.5 Hardware

Most tissue like bone, tendon and muscle hardly ever have flat structures, and an NMR-MOUSE with a large depth range of at least 10 mm is recommended for measurements so that the entire sensitive volume can be placed inside the tendon volume. To measure relaxation anisotropy, the angle θ between magnetic field $\boldsymbol{B}_0$ and the alignment direction $\boldsymbol{n}$ of sample structures must be varied (Fig. 6.2.3a). This can conveniently be done with a stray-field sensor for which the magnetic field direction lies in the plane of the sensitive slice. Consequently, a U-shaped NMR-MOUSE (Fig. 1.2.4b) is a suitable

Table 6.2.1. Default acquisition parameters for ^{1}H stray-field NMR of tendon

Parameter	Value
Magnet, probe	NMR-MOUSE PM10
transmitter frequency ν_{rf}	18.1 MHz
transmitter attenuation of 90° pulse	−8 dB at 80 W
duration t_p of 90° pulse	10 µs
dwell time Δt	1 µs
acquisition time t_{acq}	10 µs
echo time t_E	100 µs
number n_E of echoes	1024
recycle delay t_R	1 s
number n_s of scans	256

tool, but not the bar magnet NMR-MOUSE (Fig. 1.2.4c). The anisotropy measurements reported below were done with a simple U-shaped NMR-MOUSE (Fig. 6.2.2) before the Profile NMR-MOUSE had been invented.

6.2.6 Pulse sequences and parameters

Transverse relaxation rates are measured either with several *Hahn echoes* using different echo times or with a *CPMG sequence* (Fig. 5.1.5a). The anisotropy is more pronounced if Hahn echoes are employed, because partial averaging of the dipole-dipole interaction is avoided (Fig. 6.2.3b). But Hahn echo measurements are more time consuming than CPMG measurements, so that acquiring data with the CPMG sequence is preferred for *in vivo* measurements. Because the signal from water molecules confined by the aligned tendon fibers is probed, the echo is attenuated by diffusion in addition to relaxation. Therefore, the echo time should be short, and the acquisition parameters are similar to those for soft biological tissue and rubber (Tab. 6.2.1).

6.2.7 Beginner's level measurements

Objects with macroscopic order are often elongated in one direction, for example, along the axis of the tendon, or along the fiber direction in a bundle of fibers. When measuring such objects with the NMR-MOUSE the sensitive slice should be placed completely inside the object for all orientation angles θ. Otherwise, the signal amplitude and along with it, the signal-to-noise ratio will vary with θ. For elongated, isotropic objects, which are narrower than the sensitive slice is wide, the relaxation rate $1/T_{2eff}$ has been found to show a slight angle dependence for a simple U-shaped magnet with a sensitive volume less well-defined than that of the Profile NMR-MOUSE, while for the Profile NMR-MOUSE with a constant field in the plane of the sensitive slice, this was not the case.

Using CPMG sequences, the angle-dependent relaxation rates of oriented collagen fibers have been measured with a simple U-shaped NMR-MOUSE for a *rattail*, sheep *Achilles tendon in vitro* and human Achilles tendon *in vivo* (Fig. 6.2.3). In cylinder coordinates, the angle dependence assumes the shape of a cross, similar to the angle dependence observed in strained rubber (Fig. 5.1.3e). CPMG and Hahn echo measurements are compared for the rattail in Fig. 6.2.3b. In addition to tendon, the rattail contains bone, veins and skin. The relaxation rates determined with lengthy Hahn echo measurements show a more pronounced angle dependence than those determined with the CPMG sequence. Nevertheless, *in vivo* measurements are preferably done with the CPMG sequence to keep the measurement time at reasonable values. The minimum angle range that needs to be covered is 90°. With an angle resolution of 10°, 10 measurements are required, so that the acquisition time extends beyond 10 minutes even with the CPMG sequence.

The angle dependences of the Achilles tendons from sheep *in vitro* and humans *in vivo* determined with the CPMG sequence (Figs. 6.2.3c,d) are stronger than that of the rattail (Fig. 6.2.3b) because the rattail contains a considerable amount of tissue

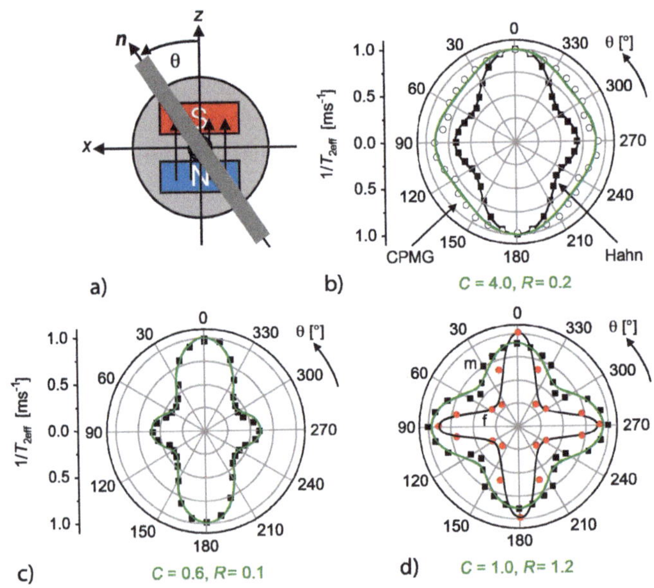

Figure 6.2.3. Anisotropy of tendon explored with a U-shaped NMR-MOUSE (adapted from [2] with permission). (a) Principle of the experiment. The direction of the tendon is $\boldsymbol{n}$, and the angle θ between tendon direction and magnetic field direction is varied from scan to scan. (b) Angle-dependent transverse relaxation rates for a rattail measured with Hahn echoes and with the CPMG sequence. (c) Angle-dependent transverse relaxation rate for a sheep Achilles tendon measured with the CPMG sequence. (d) Angle-dependent transverse relaxation rate of human Achilles tendon measured *in vivo* with the CPMG sequence for a male (m) and a female (f) volunteer. The echo time in the CPMG sequence was $t_E = 0.1$ ms. C and R are the fit parameters from eqn. (6.2.2).

Table 6.2.2. Common issues encountered when measuring tendon

– The depth of the sensitive slice is too low so that only a fraction of the sensitive volume is inside the tendon
– The noise level is too high. The object acts as an antenna and needs to be electrically shielded, e.g. by wrapping it with parachute silk
– The echo time is too long, and the signal is attenuated by diffusion

other than oriented collagen fibers. Interestingly, the cross shape of the relaxation rate dependence is more balanced for the human Achilles tendon (Fig. 6.2.3d) than for the sheep Achilles tendon (Fig. 6.2.3c).

When measuring humans and electrically conducting bodies with stray-field NMR devices, the body will act as a radio frequency *antenna* and may need to be shielded, unless the measurement is done in a shielded room. A good shield is grounded *parachute silk*, with which the limb and the NMR-MOUSE can be wrapped for measurement. Common issues encountered when measuring tendon are summarized in Tab. 6.2.2.

6.2.8 Advanced level measurements

Given short enough dead time of the instrument, it should be possible to find a fast relaxation component from the collagen fibers in addition to the slow relaxation component from the water. In either case, the motional averaging of the dipole-dipole interaction is anisotropic. While this anisotropy has been probed via the angle-dependence of the *transverse relaxation rate*, the *longitudinal relaxation rate* should show similar angle dependence. Furthermore, one could measure double-quantum filtered relaxation rates and *double-quantum build-up curves* to eliminate signal from isotropic tissue (Fig. 5.1.6a) and to enhance the angle dependence of the NMR signal. In view of long measurement times, such measurements have not been done so far with the NMR-MOUSE on biological tissue *in vivo*. But *relaxation anisotropy* is difficult to measure with a medical tomograph because the subject would need to be rotated perpendicular to the axis of the magnet. On the other hand, *diffusion anisotropy* can well be studied with pulsed gradient fields by *in vivo* MRI but not with the NMR-MOUSE, because the gradient direction and the rotation axis are parallel.

6.2.9 Data processing

The analysis of angle-dependent relaxation rates from tendon follows that for strained rubber (eqn. (5.1.10)). While the relaxation generally needs to be described by an orientation distribution function, the collagen fibers are so well aligned that for tendon the orientation distribution function can be replaced by a delta function, obtaining eqn. (6.2.1). To fit the observed anisotropy of human Achilles tendon, a second, anisotropic relaxation contribution rotated by 90° needs to be added, so that

$$1/T_{2\mathrm{eff}} = 1/T_{2\mathrm{eff},0}\left\{C + \left[\frac{1}{2}(3\cos^2(\theta) - 1)\right]^2 + R\left(C + \left[\frac{1}{2}(3\cos^2(\theta + 90^\circ) - 1)\right]^2\right)\right\}, \quad (6.2.2)$$

where, with reference to eqn. (5.1.10), $1/T_{2\mathrm{eff,aniso}} = 1/T_{2\mathrm{eff},0}$, $1/T_{2\mathrm{eff,iso}} = 1/T_{2\mathrm{eff},0}C$, and R is the relative weight of the rotated relaxation pattern.

This function has been employed to fit the outer curves in Figs. 6.2.3b,c, and d, and the values of the fit parameters are listed in the figure. C is four times larger for the *rattail* than for the *Achilles tendon* of sheep and humans indicating the presence of isotropic tissue, while R is essentially zero for the rattail and sheep Achilles tendon so that one tendon orientation dominates. In human Achilles tendon, R is about unity, indicating two equally populated fractions of highly-aligned collagen fibrils at perpendicular orientations. This is the signature of fibrils crimped or twisted like the threads in a rope (Fig. 6.2.1). It is interesting to note that this structure has neither been observed in the sheep Achilles tendon nor in the rattail, which may have to do with the fact that these studies were done ex vivo and that the tendon samples had been strained prior to measurement [3].

6.2.10 References

[1] Józsa L, Kannus P. Human Tendons: Anatomy, Physiology, and Pathology. Champaign: Human Kinetics; 1997.

[2] Blümich B, Casanova F, Perlo J. Mobile single-sided NMR. Prog Nucl Magn Reson Spectrosc. 2008; 52:197–269.

[3] Masic A, Bertinetti L, Schuetz R, Galvis L, Timofeeva N, Dunlop JWC, et al. Observation of multiscale, stress-induced change of collagen orientation in tendon by polarized Raman spectroscopy. Biomacromolecules. 2011; 12: 3989–3966.

6.3 NMR of plants and fruits

6.3.1 Introduction

Plants help to sustain life on earth. They provide energy, food, and a variety of raw materials, which are processed to products like paper, biodiesel, pharmaceuticals and dyes. The worldwide most important crop plants are wheat, rice, maize and potatoes. From a biological perspective plants are eukaryotic organisms, most always characterized by the presence of chloroplasts, which are embedded in the plant cell. They are the location for oxygenic photosynthesis, the light-driven synthesis of carbohydrates

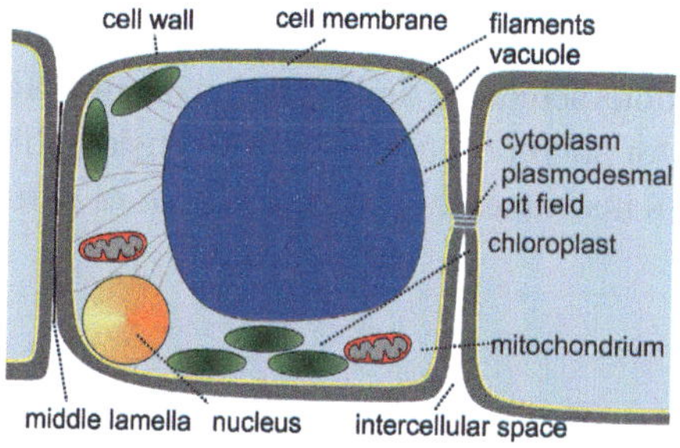

Figure 6.3.1. Drawing of a representative plant cell with some important organelles and the large water-filled vacuole.

from carbon dioxide and water. Besides the presence of chloroplasts, the typical plant cell differs from the typical animal cell by its more rigid cell wall surrounding the cell membrane and by the *vacuole*, which is filled with aqueous solution of different solutes (Fig. 6.3.1). At times the vacuole can take over nearly the entire cell volume.

Photosynthesis occurs in the leaves so that the synthesized assimilates are transported downwards to the roots through the *phloem*, one type of vascular bundle. The opposite upwards transport of mineral nutrients like potassium, phosphate, magnesium and nitrogen taken up by the roots takes place in the *xylem* (Fig. 6.3.2). In 1966, John Philip introduced the concept of *Soil-Plant-Atmosphere Continuum* (*SPAC*) to point out that the various transport processes of water and energy from soil to plant to atmosphere are coupled and must be regarded as a dynamic system. In this model water flows along the *potential gradient*, which is explained in more detail in the theory section.

6.3.2 Objective

NMR investigations of plants and fruit aim at characterizing the different tissues with respect to water status, attribution of water to compartments, oil content, diffusion, and motion or simply at imaging the internal structure of a fruit and vegetables. The motivation for such studies is basic research on the one hand, trying to understand plant growth and nutrient transport in detail, and on the other hand it is profit optimization, raising and quantifying the quality of farm products.

Investigations are performed on short and long timescales to monitor plant growth and fruit maturation in the field and in the processing plant. Compact low-field NMR can help to optimize irrigation schemes and plant growth cycles, by relating relaxation times or images with water status in different plant tissues for a range of environmental indicators like *soil moisture*, sun radiation, and temperature. Besides a characterization based on relaxation and diffusion, MRI is the method of choice if internal features are to be detected like seeds or flawed regions in fruits. Low-field

NMR relaxometry as well as imaging devices can conveniently be integrated in industrial food processing lines for continuous screening and detection of multiple internal quality attributes. The use of NMR for such purposes has had considerable impact on the development of compact NMR instrumentation, in particular of portable MRI equipment.

6.3.3 Further reading

Van As H, van Duynhoven J. MRI of plants and foods. J Magn Reson. 2013; 229: 25–34.

Slawinski C, Sobczuk H, Soil-plant-atmosphere continuum. In: Glinski J, Horabik J, Lipiec J, editors. Encyclopedia of Agrophysics. Dordrecht: Springer; 2011. pp.805–810.

Milczarek RR, McCarthy MJ, Low-field sensors for fruit inspection. In: Codd SL, Seymour JD, editors. Magnetic Resonance Microscopy. Weinheim: Wiley; 2009. pp. 289–314.

Kose K. Applications of permanent-magnet compact MRI systems. In: Codd SL, Seymour JD, editors. Magnetic Resonance Microscopy. Weinheim: Wiley; 2009. pp. 365–379.

6.3.4 Theory

The concept of a *Soil-Plant-Atmosphere Continuum* (*SPAC*) comprises all components that contribute to water flow from the soil to the atmosphere. This means that if water evaporates from the leaves into the atmosphere the local *water potential* gets more negative and causes flow from the xylem in the stem into the leaves. These processes propagate downward so that water flows from the root xylem to the xylem conduits in the stem, and eventually water is taken up from the surrounding soil by the roots (Fig. 6.3.2). Generally, water flow in saturated conduits is described following *Darcy's law*,

$$J = Q/A = -K\partial\psi/\partial z, \qquad (6.3.1)$$

where J is the flow velocity, Q the volume flow rate, A the flow conducting area, K the *hydraulic conductivity*, and $\partial\psi/\partial z$ is the water *potential gradient*. Alternatively, the water potential ψ can be expressed as hydraulic head $h = \psi/g\rho$, where g is the gravitational acceleration and ρ the density of water. In contrast to the plant's vascular bundles, soil is mostly unsaturated, and the variable saturation degree must be taken into account for the description of water flow in soil. This is modeled by the Richards eqn. (7.2.14).

In the SPAC model, different values of the *hydraulic conductivity* are distinguished. They are K_s for soil water, K_r for root water, K_x for axial transport in the xylem, and K_l

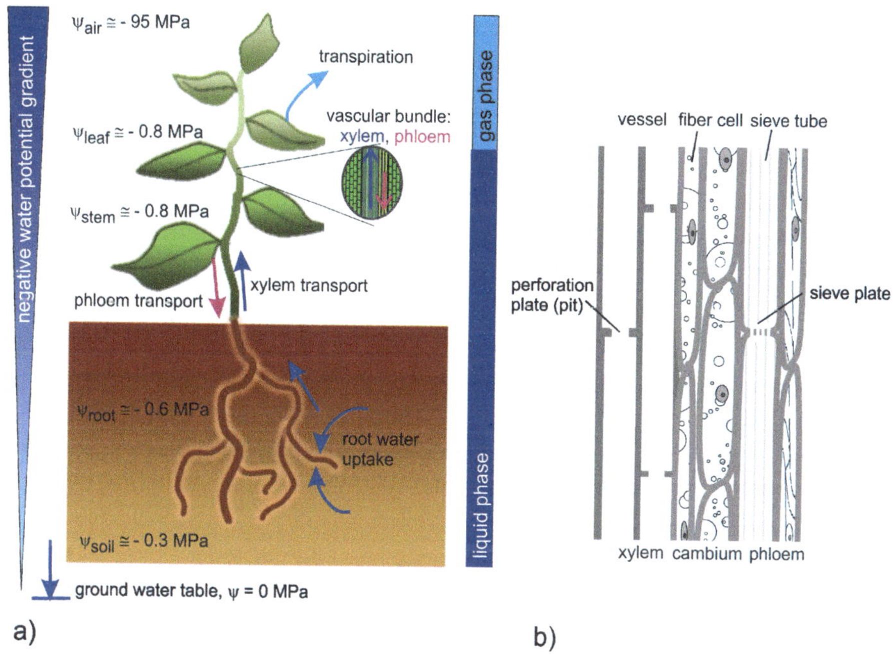

Figure 6.3.2. Water flow in plants. (a) According to the Soil-Plant-Atmosphere Continuum (SPAC) model, water flows along the potential gradient from the soil into the roots, through the roots into the part of the plant above ground, and from there into the leaves. From the leaves it transpires into the atmosphere. Representative values for the water potentials are taken from [1]. (b) Schematic longitudinal section of a vascular bundle where the upward flow in the xylem and the downward flow in the phloem take place.

for transport in the leaves. These values depend on several internal and environmental factors. K_s depends strongly on the soil water content, and K_r and K_l depend on the presence and activity of aquaporins, which are water flow controlling proteins that open and close water channels in cell membranes. K_x depends on the width of the xylem vessels, the tracheids, and on the size of the pits connecting the xylem cells which form the bottleneck for the xylem conductance.

The evaporation rate is controlled by temperature, relative humidity and wind speed, and it is regulated by the opening and closing of the stomata, which are the microscopic pores in the plant epidermis. This regulation depends also on the water stress situation, which is communicated from the root system to the stomata by hormonal signaling. Besides the *xylem* transport system for water, there is a second transport system, the *phloem* system. It serves for the transport of mainly carbohydrates as nutrients from the leaves to the root system (Fig. 6.3.2). In many cases it is also located in the vascular bundles but anatomically different from the xylem.

In addition to water flowing in the transport systems, many plant cells contain stationary water in vacuoles, cytoplasm and intercellular space, which exchanges by diffusion between the compartments. Therefore, relaxation measurements yield multi-exponentially decaying curves whose relaxation times can be assigned in the following hierarchical order: The longest T_1 and T_2 values arise for water in the vacuoles, intermediate values in the cytoplasm and short values for water in the cell wall. The *Brownstein–Tarr model* (Fig. 7.1.1) for the relaxation of fluid molecules diffusing in a pore relates the measured relaxation rate with the bulk rate in the center of a vacuole and a surface rate which is proportional to the surface to volume ratio S/V of the vacuole according to

$$1/T_2 = 1/T_{2,\mathrm{bulk}} + \rho_2\, S/V. \tag{6.3.2}$$

The proportionality constant ρ_2 correlates with the permeability of the vacuole membrane. In porous media studies it is denoted as relaxivity (eqn. (7.1.4)). In contrast to the solid-liquid interface, in soil particles containing *paramagnetic impurities*, the decisive factor for enhanced relaxation is not the membrane itself but the cytoplasm behind it. It is important to note that this model is valid only in the so-called *fast diffusion limit* (Section 7.1.4).

6.3.5 Hardware

In recent years a variety of transportable NMR setups has been developed suitable for both relaxometry and MRI of plants and fruits inside the factory and outside in the field. The great advantage of permanent magnets is design flexibility so that closed or open access magnets can be optimized for the corresponding size of the plant without being constrained by the size of the laboratory or the greenhouse equipment. However, attention must be paid to moist environment and magnetic field drift from changing temperature. The hardware should be as light as possible for ease of transportation across rugged ground. With these goals in mind, a portable MRI setup with a 0.2 T permanent magnet has been mounted on a motorized cart with a forklift to perform *in situ* measurements of fruits hanging from trees during the growth period (Fig. 6.3.3d).

An ingenious device for plant growth is the NMR-*CUFF* (Cut-open, Uniform, Force Free, Fig. 6.3.3a–c) [4], a Halbach magnet that can be opened and closed with minimal applied force. In contrast to unilateral stray-field devices, it can be mounted around thin cylindrical objects like the branch of a tree or the stems of a plant for imaging and studies of fluid transport. Larger objects like trees need devices with open access like the *tree hugger* (Fig. 6.3.3e) [3]. This portable magnet weights 55 kg and can accommodate trunks with diameters up to 100 mm between its poles for long-term outdoor imaging and relaxation measurements.

Because leaves are thin and sensitive with respect to their functionality, they are measured best with a unilateral stray-field sensor like the Profile NMR-MOUSE. But different from skin with different layers, leaves have vascular structures in the plane, so that the diameter of the rf coil limits the selectivity with which different lateral structures can be resolved.

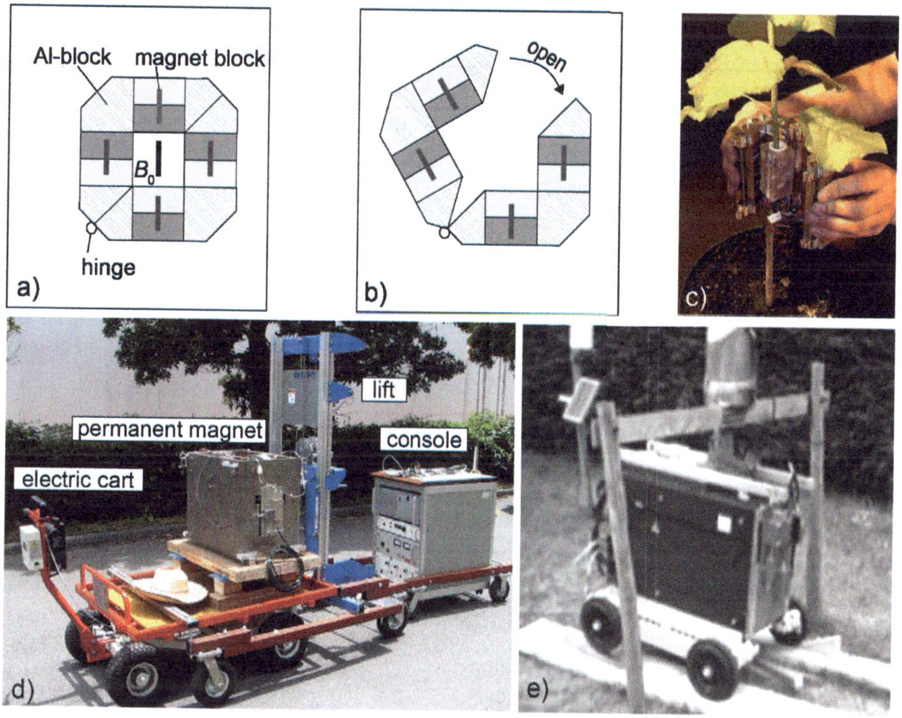

Figure 6.3.3. Magnets for investigating plants. (a–c) The NMR-CUFF, a dedicated Halbach magnet that can be opened for placing it around a plant stem (adapted from [4] with permission). (d) Motorized cart with a magnet consisting of two parallel slabs. The rf coil is wound directly around the object and the magnet is moved to the position of the object at a maximum height of 1.4 m (adapted from [2] with permission). (e) The tree hugger. It is made from two poles each comprising five concentric rings of small block magnets (adapted from [3] with permission).

6.3.6 Pulse sequences and parameters

Low-field NMR devices are frequently used for time domain measurements to determine the proton spin density, and the relaxation times T_1 and T_2. A component analysis of relaxation curves (eqn. (3.1.3), Fig. 4.3.1a) can be used to quantify water in different compartments. T_2 is typically measured with the CPMG sequence, whereas for mere water content determination a single Hahn echo is sufficient if the signal-to-noise ratio is good. The echo time should be sufficiently long, e.g. 50 μs for an NMR-MOUSE with a small rf coil, to suppress signal from solid components like cellulose, protein, and lignin (Section 4.3). To focus on vascular water the echo time can be made several milliseconds long.

T_1 can be measured by a *saturation* and *inversion recovery* sequences (Fig. 3.2.2). Imaging experiments are done with C-shaped magnets or Halbach-type magnets that partially enclose the object and provide a more homogeneous field than single-sided

Table 6.3.1. Default acquisition parameters for relaxometry of plant tissue

Parameter	Value
Magnet, probe	NMR-MOUSE PM5
transmitter frequency ν_{rf}	17.1 MHz
transmitter attenuation for 90° pulse	−8 dB at 300 W
duration t_p of 90° pulse	5 μs
dwell time Δt	1 μs
acquisition time t_{acq}	10 μs
echo time t_E	50 μs
number n_E of echoes	2048
recycle delay t_R	2 s
number n_s of scans	64

stray-field devices. To cope with remaining field inhomogeneity, spin echo sequences are preferably used (Fig. 3.3.4). Advanced sequences employ flow and diffusion filters, which require pulsed field gradients not only for space encoding but also for encoding translational motion (Fig. 3.2.5c,d). Some common issues are summarized in Tab. 6.3.2.

6.3.7 Beginner's level measurements

To characterize the ripening process of Japanese pear fruits, relaxation times were measured during the cell enlargement period of the growth process (Fig. 6.3.4a) using a portable NMR relaxometer for outdoor use (Fig. 6.3.3d). It is known that increasing cell volume goes along with the growing size of the *vacuoles* (Fig. 6.3.1) so that the relaxation time of the water protons approaches the values of the bulk. Longitudinal and transverse relaxation times determined by conventional *inversion recovery* and *CPMG* sequences in the laboratory uniformly increase with maturation time of the pear fruits. The transverse relaxation decay is bi-exponential, i.e. one can discriminate between a fast and slow relaxation process with relaxation times T_{2short} and T_{2long}, respectively. The slow relaxation rate $1/T_{2long}$ depends linearly on the inverse size of the

Table 6.3.2. Common issues encountered when measuring plants

- The recycle delay is too short so that the signal from water with long T_1 relaxation time is partially saturated
- The duration of the echo train is too short and the echo train is truncated
- The echo time is chosen too short so that solid components may contribute to the signal
- The effective echo time and the transverse relaxation times of different compartments of the object do not match so that the contrast is not optimal
- The field of view is smaller than the object so that aliasing artifacts occur in the image

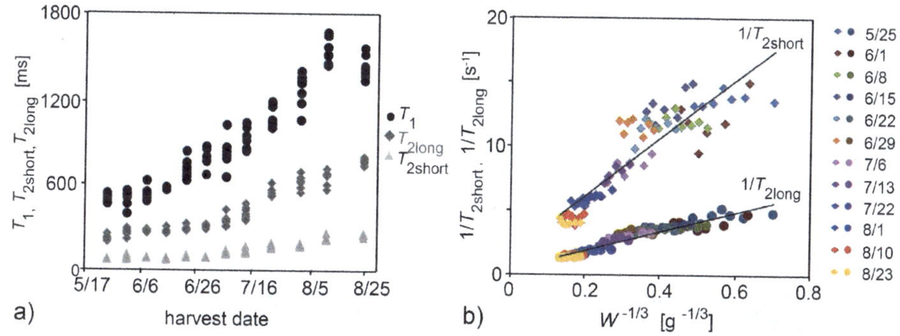

Figure 6.3.4. Relaxation measurements of Japanese pears during the ripening process. (a) Transverse and longitudinal relaxation times as a function of the harvesting date, which corresponds to the measurement day (adapted from [2] with permission). (b) Correlation of the relaxation rates with the inverse of the cubic root of the weight of the fruit, which is proportional to the inverse of the vacuole size. The linear dependence of the slow relaxation time $1/T_{2long}$ of the water protons in the vacuoles is dominated by surface relaxation (adapted from [2] with permission).

fruits (Fig. 6.3.4b). This demonstrates that *surface relaxation* at the boundary of the vacuoles is the predominant relaxation mechanism (eqn. (6.3.2)) [2].

6.3.8 Advanced level measurements: Imaging

Individual fruits can be inspected nondestructively by low-field MRI at commercially interesting throughput rates. A viable example is the counting of seeds in mandarin oranges. In the state of California, 2.5 million tons of these crops are produced annually, and the seedless variety is ranked first place in value. Different fast MRI methods have been explored with a 1 T imaging magnet having a 60 mm gap to visualize the seeds [5]. Depending on the method, i.e., *RARE*, *FLASH* (*Fast Low Angle SHot*), and *TURBO-FLASH* imaging, the contrast is different and can be further optimized with the help of additional image processing filters to identify the internal structure of the fruit (Fig. 6.3.5).

With the *RARE imaging* technique, multiple traces through $\boldsymbol{k}$-space are acquired in a single CPMG echo train (Figs. 3.3.4b and 6.3.5a). Depending on the order of the phase-encoding steps in which $\boldsymbol{k}$-space is sampled, the effective echo time, $t_{E,eff}$ describes the time when the central trace at $k_{phase} = 0$ is acquired. For example, sampling the $\boldsymbol{k}$-space in one shot with n_E echoes yields $t_{E,eff} = (n_E/2)t_E$, i.e. a strongly *T_2 weighted image* (Fig. 6.3.5d). *FLASH imaging* (Fig. 6.3.5b) is similar to Hahn echo imaging (Fig. 3.3.4a), except that the 180° pulse is omitted and a gradient echo is acquired. This allows the flip angle of the excitation pulse to be adjusted to the Ernst angle (eqn. (2.7.1)), so that the recycle delay t_R can be made short. At very short to zero recycle delay, the magnetization excited by different pulses may interfere. To avoid this,

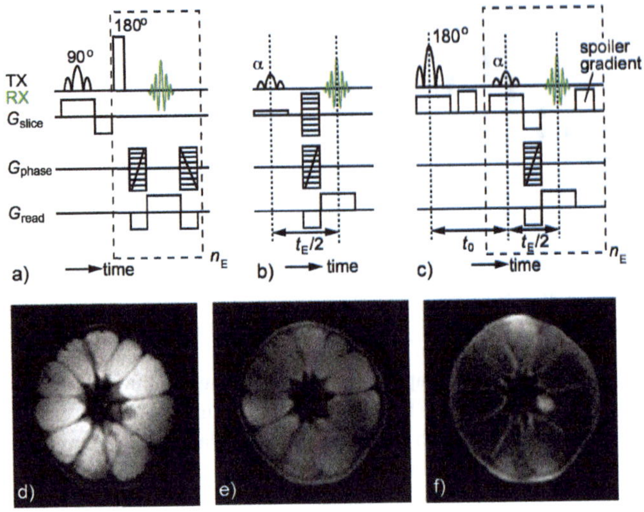

Figure 6.3.5. MRI pulse sequences and images of a mandarin orange. (a) RARE sequence. (b) 3D FLASH sequence. (c) Turbo-FLASH sequence with T_1 weighting by inversion recovery preparation. (d–f) Cross-sectional images with 128 × 128 pixels at a resolution of 0.57 mm in the plane (adapted from [5] with permission). Each image is the average of 16 central slices. (d) RARE image with $t_{E,eff}$ = 412 ms. (e) FLASH image. (f) T_1-weighted Turbo-FLASH image.

residual transverse magnetization is destroyed after each echo in the Turbo-FLASH sequence (Fig. 3.3.4c), with a spoiler gradient pulse. A T_1 filter can be added before the gradient echo is generated to introduce T_1 contrast in the image. With an *inversion recovery filter* (Fig. 3.2.2b), positive and negative amplitudes can be generated in the image. The best contrast in the images in Fig. 6.3.5 is indeed obtained with the Turbo-FLASH sequence and an inversion recovery filter (Fig. 6.3.5f).

Fungal infection is a serious disease of *fruits* and other agricultural products that can considerably impact the production yield and lower the profit. For example, the black heart disease is often found in Californian pomegranate plantations and cannot be detected by simple visual inspection because it develops only inside the fruit. Here NMR is useful in studying changes in the fruit tissue, because T_2 relaxation times are sensitive to cell compartmentalization. In healthy pomegranate arils (Fig. 6.3.6d) three modes can be differentiated in the T_2 distribution in the time range from millisecond to seconds (Fig. 6.3.6a). Water in cell walls relaxes rapidly with about 10 ms and in cytoplasm with about 200 ms, whereas water in the vacuoles relaxes with about 1 s, which is close to the bulk relaxation time. In infected fruit the fungus produces enzymes that decompose the cell wall and middle lamellar so that the internal compartmentalization is damaged and the permeability is enhanced. During this process also free radicals are formed as concluded from EPR experiments. These radicals are paramagnetic and accelerate relaxation. As a result the relaxation spectrum is shifted to shorter relaxation times and a new peak appears, which is assigned to extracellular

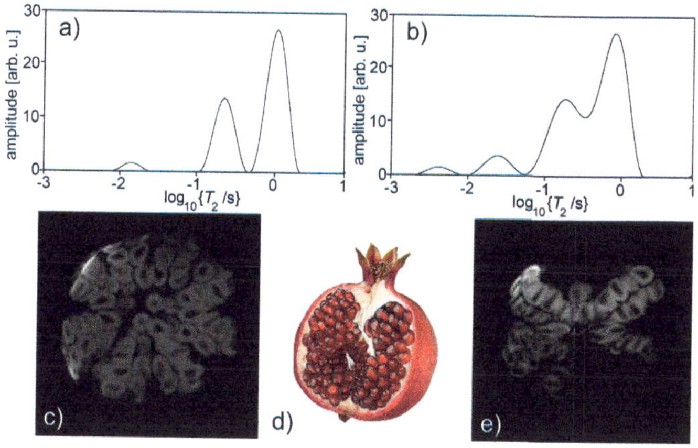

Figure 6.3.6. NMR relaxometry and imaging of a healthy (left) and infected (right) pomegranate fruit (adapted from [6] with permission). (a, b) Distributions of transverse relaxation times. (c, e) Fast spin-echo images. (d) Drawing of a pomegranate fruit with the arils encapsulated by the rind.

water (Fig. 6.3.6b). As an additional indicator T_2 weighted fast spin-echo images show the infected areas as dark regions in the fruit (Fig. 6.3.6c,e). The lack of signal is explained by the particular choice of the effective echo time $t_{E,eff}$ employed in these fast spin-echo imaging measurements. In this particular case $t_{E,eff}$ was 500 ms so that the signal from the infected areas was almost fully relaxed (Fig. 6.3.6e).

The NMR-CUFF (Fig. 6.3.3c), a lightweight Halbach magnet with a magnetic field of 0.57 T that can be opened and closed for placing it around a stem section between branches of a plant, has been demonstrated to be capable of producing an image across a castor oil plant stem and mapping the water flow in the xylem tissue. Equipped with gradient coils producing a maximum gradient strength of 630 mT/m and an rf coil wound in situ in the groves of a template holder, the space for the plant is limited to about 6 mm diameter. This is sufficient to place the magnet around the stem of a young castor oil plant and image a cross section with a conventional T_1 weighted spin echo sequence (Fig. 3.3.4a) within 26 min (Fig. 6.3.7a) [4]. The anatomically different xylem tissue surrounding the parenchyma cells in the center can be identified in the image.

The sap flow in the xylem and phloem (Fig. 6.3.2) differs not only in the flow direction but also in the flow velocities. Generally, the speed is higher in the xylem, e.g. for the transport of water and disolved nutrients from the root upwards to the leaves. The xylem flow in a 1.75 m tall intact poplar stem was measured with the NMR-CUFF by tracing the displacement of spins within a given time interval using pulsed field gradients in the same way as when measuring diffusion (Fig. 3.2.5f). By varying the gradient amplitude the distribution of displacements during an encoding period Δ = 20 ms was recorded. This distribution is known in NMR as the *average propagator*. It can be decomposed into two parts describing *coherent* and *incoherent displace-*

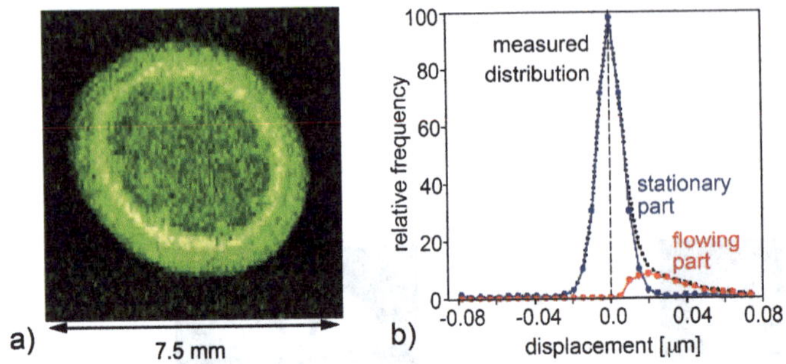

Figure 6.3.7. Measurements of a castor oil plant stem with the NMR-CUFF (adapted from [4] with permission). (a) Summation of 10 images of a young castor oil plant stem measured with a spin echo sequence and a matrix size of 64 × 128 pixels. (b) Distribution of water displacements in the xylem of a poplar stem and decomposition into two fractions, one from incoherent diffusive displacement and one from coherent flow displacement.

ments (Fig. 6.3.7b). The incoherent displacements are caused by *diffusion* and produce a peak at zero displacement. The coherent displacement is caused by directional flow and produces a peak at non-zero displacement. The low-field data reveal xylem flow velocities that are of the same order of magnitude as others measured at high field [7].

6.3.9 References

[1] Nobel PS. Physicochemical and Environmental Plant Physiology. San Diego: Academic Press; 1999.

[2] Geya Y, Kimura T, Fujisaki H, Terada Y, Kose Y, Haishi T, Gemma H, Sekozawa Y. Longitudinal NMR parameter measurements of Japanese pear fruit during growing process using a mobile magnetic resonance imaging system. J Magn Reson. 2013; 226: 45–51.

[3] Jones M, Aptaker PS, Cox J, Gardiner BA, McDonald PJ. A transportable magnetic resonance imaging system for in-situ measurements of living trees: The Tree Hugger. J Magn Reson. 2012; 218:133–140.

[4] Wind CW, Soltner H, van Dusschoten D, Blümler P. A portable Halbach magnet that can be opened and closed without force: The NMR-CUFF. J Magn Reson. 2011; 208:27–33.

[5] Milczarek RR, McCarthy MJ. Low-field sensors for fruit inspection, In: Codd SL, Seymour JD. editors. Magnetic Resonance Microscopy. Weinheim: Wiley; 2009. pp. 289–314.

[6] Zhang L, McCarthy MJ. Black heart characterization and detection in pomegranate using NMR relaxometry and MR imaging. Postharv Biol Technol. 2012; 67:96–101.

[7] Wind CW, Vergeldt FJ, de Jager PA, van As H. MRI of long-distance water-transport: A comparison of the phloem and xylem flow characteristics and dynamics in poplar, castor bean, tomato and tobacco. Plant, Cell and Environment. 2006; 29:1715–1729.

7 Porous media

A *porous medium* consists of a solid or soft-matter matrix with voids or *pores*, which, when physically connected, enable the continuous flow of fluids and gases through the material (Fig. 7.0.1). According to this definition there are numerous types of natural and synthetic porous media, for example, rock, sediments, soil, cement, concrete and filters. Whereas synthetic porous media exhibit mostly well-defined, sometimes even periodic pore structures, most natural porous media do not. They are characterized by distributions of pore sizes and shapes and connectivities of the pores. Remarkable is the range of pore sizes. It can cover several orders of magnitude, ranging from large caverns to nanoscopic capillaries, which retain liquids predominantly by adhesive forces.

Rock is formed from mineral aggregates and is the precursor of *soil*. By weathering and humification soil becomes the top cover of the earth's crust. The formation of soil is a continuous process. Structurally intermediate between rock and soil are sediments, which consist of rock fractured by weathering and erosion, i.e. the impact of water, air and temperature.

Building materials include both natural porous media like rock and stone and artificial materials like *cement*, *concrete*, and polymers. Knowledge of their pore structure is essential for the construction of buildings and roads. Moreover, the pore space of cement and mortar changes rapidly in the initial stages of the setting process.

Because most porous media are opaque, diffraction methods, such as X-ray and neutron scattering, as well as NMR are commonly employed to explore the pore space. X-ray and neutron scattering elucidate the skeleton or matrix of the medium, whereas NMR directly probes the pore space in terms of parameters like *porosity*, pore size dis-

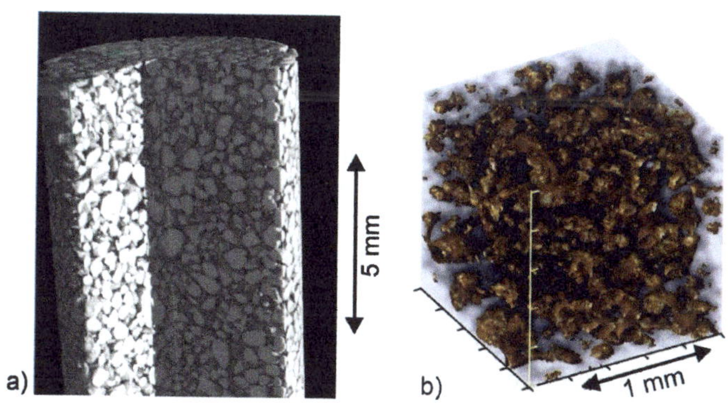

Figure 7.0.1. Sand as an example of a natural porous medium. (a) X-ray CT of the model sand FH 31, fully saturated with water. The CT image was measured with a micro-CT scanner XTec 225 HMX at a resolution of 15 µm in each direction (courtesy of U. Weller, University Tübingen). (b) Excerpt of the CT image, which illustrates the heterogeneity of the pore space.

tribution, permeability, fluid saturation and *wettability*, provided it is filled with an NMR-active fluid. Consequently, mobile NMR is applied in geophysics for *well logging* to characterize aquifers, and to analyze rock cores in the laboratory. Historically, the step from laboratory to outdoor experiments was motivated by the need of the oil industry for well logging shortly after the emergence of NMR, whereas in situ investigations of soil are still in their infancy. The reasons for this situation are manifold: First, soil studies bring a lower return on investment than well logging. Second, the pore space of soil is often not fully saturated with water, which gives rise to magnetization components with transverse relaxation times much shorter than a millisecond, while the water and oil bearing zones below are mostly saturated with fluid and provide NMR signals that decay much slower. Third, deep, fluid-saturated zones are well shielded from electrical interference, while power lines and other electrical installations degrade the signal-to-noise ratio unless noise is suppressed by suitable hardware and software. Consequently, so far only few reports exist that describe measurements of topsoil investigations outdoors.

7.1 Rock and sediments

7.1.1 Introduction

Rock is a solid, porous material composed of one or several types of minerals. It is classified into eruptive rock, for example, granite, sediments like carbonates, and metamorphic rock, for example, shale. The *porosity* of rock can vary from a few percent in granite to 40 % in sandstones. Beyond that, rock and soil are magnetic with an induced or remnant magnetization, which is predominantly determined by the content of ferri-, para- and anti-ferromagnetic minerals such as magnetite, biotite and hematite. Important for NMR investigations are the ions of the solid matrix because they impact the relaxation times by enhancing the *relaxation* of liquids in a narrow shell in contact with the pore wall. Differences in magnetic susceptibility between matrix and pore distort the applied magnetic field and give rise to field gradients inside the pores. These gradients are commonly termed *internal gradients*. When probing the pore space with fluids filling the pores, the transverse magnetization is attenuated by *diffusion* of the molecules in the internal gradients in addition to bulk and surface relaxation. These internal gradients may vary from pore to pore. Because the magnetic field distortions and with them, the internal gradients, are proportional to the strength of the applied field, the contribution of diffusion to the decay of transverse magnetization is smaller at low than at high magnetic field.

The development of NMR of porous media has been driven by the well-logging industry. This chapter does not review well-logging applications but focuses on basic measurement procedures for NMR investigations of rock samples by compact NMR in the laboratory. In a well-logging scenario, the sensor operates under extreme pressure of up to 3000 bar and temperatures up to 160° C. The total amplitude of fluid-saturated rock provides the mineralogy-independent porosity given the size of the signal-bearing volume and the hydrogen density of the pore fluid. From the combined analysis of diffusion and relaxation, the fluid type, i.e. whether gas, water or oil are present, and even information about the kind of oil can be determined [1]. In the laboratory, rock and soil samples are investigated mostly at ambient pressure and temperature and variable fluid saturation. When partially saturated with fluid, the NMR signal amplitude of rock no longer provides the total porosity but the moisture content or fluid saturation.

7.1.2 Objective

NMR studies of rock and sediments aim at specifying the *porosity*, the *pore size distribution*, and the *fluid saturation*, and at quantifying *bound fluid* and *producible fluid* fractions. In laboratory measurements, samples are typically saturated with water. Advanced measurements aim at gaining information about the connectivity of pores. In either case relaxation curves and diffusion-weighted signals from the pore-filling fluid are analyzed. Distributions of relaxation times are obtained by operations equivalent to inverse Laplace transformation of longitudinal and transverse relaxation curves. The *relaxation time* distribution maps the *pore size distribution* in the *fast diffusion limit* for fluid-saturated porous media. The amplitude of the relaxation curve or the area under the relaxation time distribution function specify the *porosity*. From the porosity and the pore size distribution the fluid *permeability* can be predicted with the help of models. Signals in the relaxation time distribution from different fluid components like oil, water and gas can be separated by 2D Laplace NMR methods. In particular a diffusion-relaxation correlation map provides valuable information for *fluid typing*.

7.1.3 Further reading

Song YQ. Magnetic Resonance of Porous Media (MRPM): A perspective. J Magn Reson. 2013; 229: 12–24.

Song YQ. Novel two-dimensional NMR of diffusion and relaxation for material characterization. In: Stapf S, Han SI, editors. NMR Imaging in Chemical Engineering. Weinheim: Wiley-VCH; 2006, pp. 163–183.

Dunn KJ, Bergman DJ, Latorraca GA. Nuclear Magnetic Resonance: Petro-physical and Logging Applications. Helbig K, Treitel S, Series editors. London: Pergamon Press; 2002.

Coates GR, Xiao L, Prammer MG. NMR Logging: Principles and Applications. Houston: Halliburton Energy Services; 1999.

7.1.4 Relaxation in porous media

The processes that give rise to relaxation of nuclear magnetization happen on the space scale of the molecule. A relaxation model must therefore consider the molecules and their environment. Fluid molecules in a pore experience two environments, the pore surface, which is often covered with paramagnetic relaxation centers, and the bulk fluid volume. A simple relaxation model for fluids in a pore was established by Brownstein and Tarr [2]. It divides the pore volume into a shell of thickness h covering the surface area S of the pore and the bulk volume V (Fig. 7.1.1). The surface layer is a few molecules thick, i.e. of the order of a few nanometers for water. For fluids in rock pores, three relaxation mechanisms have been identified. These are (i) bulk relaxation in the fluid phase, (ii) surface relaxation at the grain-fluid interface, and (iii) transverse relaxation caused by diffusion of molecules in the inhomogeneous magnetic fields inside the pores.

Bulk relaxation refers to the relaxation of the molecules that do not collide with the pore wall and are surrounded by liquid molecules during the observation time. Most molecules in the large pores of vuggy carbonates and those molecules that flow through macropores relax with the properties of the bulk fluid (Section 3.2). For example, relaxation times of water are approximately a few seconds if the concentration of dissolved paramagnetic ions is low and surface relaxation can be neglected so that bulk relaxation dominates.

Surface relaxation impacts the magnetization of those molecules that diffuse close to paramagnetic impurities at the pore wall or the grain surface. Experimental evidence indicates that relaxation at clean surface sites through the dipole-dipole interaction between surface electrons and fluid molecules rotating anisotropically near the grain surface is too weak to explain the observed relaxation of water in rock. Surface relaxation results from fluid molecules and paramagnetic ions like Fe^{3+} coordinated

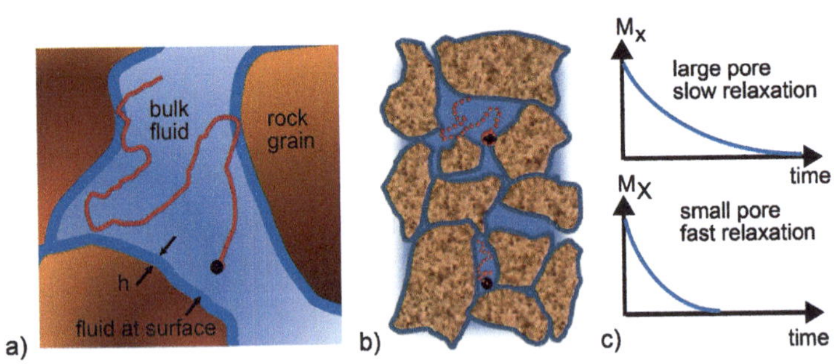

Figure 7.1.1. Illustration of the relaxation model for fluids in porous media according to Brownstein and Tarr. (a) The fluid molecules confined in a pore explore two relaxation regimes by diffusion, the surface regime (dark blue) and the bulk relaxation regime (light blue). (b) Rock contains different shapes and sizes of pores, including large and small pores. (c) Schematic relaxation curves for a large (top) and a small (bottom) pore.

at the pore surface. Most types of rock and soil contain a few percent of paramagnetic ions like Fe^{3+}, Mn^{2+}, Cu^{2+}, or Ni^{2+}, which enhance the surface relaxation relative to pure diamagnetic surfaces and therefore dominate the surface relaxation mechanism. The capacity of a surface to relax protons by T_1 and T_2 processes is expressed by the *surface relaxivities* ρ_1 and ρ_2, respectively.

In the fast diffusion limit, the relative impact of surface and bulk relaxation is specified by the *surface-to-volume ratio* S/V. This ratio is proportional to the inverse pore radius. The radius determines the frequency with which fluid molecules diffuse between bulk fluid and surface. In large pores, water molecules collide less frequently with the grain surface than in small pores, resulting in slow and fast transverse magnetization decays (Fig. 7.1.1c).

Following the *Brownstein–Tarr model*, the fluid molecules diffuse across the pore diameter a few times during the observation time in the *fast diffusion limit* so that bulk and surface relaxation are averaged. If diffusion in gradients can be neglected, the *longitudinal relaxation rate* $1/T_1$ and the *transverse relaxation rate* $1/T_2$ in a single pore can be described by

$$1/T_1 = 1/T_{1\text{bulk}} + \rho_1(S/V)_{\text{pore}} \tag{7.1.1}$$

$$1/T_2 = 1/T_{2\text{bulk}} + \rho_2(S/V)_{\text{pore}}, \tag{7.1.2}$$

where $T_{1\text{bulk}}\,T_{2\text{bulk}}$, are the bulk relaxation times, S/V the surface-to-volume ratio of a pore, and $\rho_{1,2}$ are the surface relaxivities for longitudinal and transverse relaxation. In many cases the bulk relaxation rates are small compared to the surface relaxation rates and can be neglected, so that

$$1/T_1 \approx \rho_1(S/V)_{\text{pore}}, \tag{7.1.3}$$

$$1/T_2 \approx \rho_2(S/V)_{\text{pore}}. \tag{7.1.4}$$

For spherical pores of radius r the surface-to-volume ratio is $S/V = 3/r$ and for long cylindrical pores $S/V \approx 2/r$. Then the relaxation times are proportional to the pore radius or the *pore size*, and for uniform pore size, the magnetization build-up and decay curves are mono-exponential,

$$M_z(t_0) = M_0\{1 - \exp(-\rho_1(S/V)_{\text{pore}}t_0)\}, \tag{7.1.5}$$

$$M_x(t) = M_0 \exp\{-\rho_2(S/V)_{\text{pore}}t\}, \tag{7.1.6}$$

where M_0 is proportional to the amount of fluid protons in the pore volume, t is the observation time for *transverse magnetization* M_x (eqn. (3.2.6)), and t_0 is the recovery time for *longitudinal magnetization* M_z (eqn. (3.2.5)).

Natural rock and soil exhibit a wide distribution of pore shapes and sizes, and the pores are interconnected and form a network. This network is typically approximated by a distribution of pores with different *surface-to-volume ratios*. Hence, the measured transverse magnetization derives from all pores with different sizes and is the sum of many exponential functions,

$$M(t) = \Sigma_i M_i \exp\{-\rho_2(S/V)_i t\}. \quad (7.1.7)$$

Here the index i labels pores of different surface-to-volume ratios, and the M_i measure the numbers of protons contained in each pore size class i and are proportional to the fluid volume in all pores of that size. A similar equation can be set up for longitudinal relaxation. The sum over the component amplitudes M_i is proportional to the total fluid contained in all pores. For a fluid-saturated sample it is proportional to the pore volume and can be calibrated to porosity (eqn. (7.2.2)). The component amplitudes are conveniently determined by algorithms equivalent to inverse Laplace transformation under confinement to positive amplitudes by means of a regularization method. The inverse Laplace transform of eqn. (7.1.7) is the distribution of M_i versus relaxation rates $1/T_2$, which is typically displayed on a logarithmic scale of T_2 (Figs. 1.3.1b and 3.1.6d).

Relaxation by diffusion needs to be considered when the detection field is inhomogeneous. This is the case for stray-field NMR sensors like the *NMR-MOUSE* and *logging tools* (Fig. 3.1.2) but also for fluids inside pores where the applied magnetic field is distorted as a result of differences in the magnetic susceptibility of the solid matrix and the pore fluid. An NMR-MOUSE with a penetration depth of 25 mm has a static gradient of 12 T/m. In contrast, a simple Halbach magnet exhibits an effective gradient of about 0.05 T/m across a sensitive volume of 20 mm diameter [3]. The latter is sufficiently small to excite the complete sample volume by a single pulse and to suppress the signal attenuation from diffusion in echo experiments by short echo times. The field distortions inside a pore from differences in magnetic susceptibility between solid matrix and the fluid phase are described in first order by an *internal gradient* G_{int}. This gradient is proportional to the applied magnetic field B_0 and depends on the mineralogical nature of the grains and the pore shape. Typical values are some T/m even at the low fields of compact instruments, so that the magnetic field inside a pore can vary appreciably. Fluid molecules diffuse inside the pore space, and the precession frequencies of their transverse magnetization components vary according to the random walk of the molecules, leading to an additional transverse relaxation

$$1/T_{2\text{diff}} = 1/12\, D\,(\gamma G_{\text{int}} t_{\text{E}})^2, \quad (7.1.8)$$

which depends on the internal gradient G_{int} and the echo time t_{E}.

Accordingly, the overall transverse relaxation rate is the sum of three terms (eqn. (3.2.3)),

$$1/T_2 = 1/T_{2\text{bulk}} + 1/T_{2\text{surf}} + 1/T_{2\text{diff}} = 1/T_{2\text{bulk}} + \rho_2(S/V)_{\text{pore}} + 1/12\, D\,(\gamma G_{\text{int}} t_{\text{E}})^2. \quad (7.1.9)$$

Relaxation from diffusion in internal gradients can be minimized by measuring at low magnetic fields $B_0 < 0.1$ T and choosing a short echo time t_{E}. The longitudinal relaxation rate $1/T_1$ is not noticeably affected by diffusion in magnetic field gradients so that eqn. (7.1.1) is valid also in inhomogeneous magnetic fields.

7.1.5 Hardware

Fluids in rock and soil are usually analyzed using closed magnets that accommodate the sample inside (Fig. 7.1.2a,b) or using logging tools that are inserted into the object (Fig. 7.1.2c). Dedicated *core scanners* (Fig. 7.1.2a) are marketed by different companies for the oil and gas industry to study drill cores. Laboratory experiments can be conducted with simple Halbach magnets (Fig. 7.1.2b), which need to be operated in temperature-stable environment to minimize field drift. *Well-logging tools* for the oil industry are several meters long (Fig. 7.1.2c). They include the complete NMR hardware from the magnet to the operating electronics in a steel shell, which is lowered into a borehole and withstands high temperature and pressure.

In soil and low-porosity rock with small pores, the transverse relaxation time T_2 can be very short, for example, of the order of 100 μs in fully water-saturated clay. At lower water content, T_2 may be even shorter. To measure magnetization components with such short relaxation times, short echo times are needed in CPMG experiments to avoid signal loss during the dead time. These are easier to achieve with small than with large coils. This is why the Halbach magnet in Fig. 7.1.2b has been fitted with an insert for small cylindrical samples with a diameter of 20 mm. Another benefit of Halbach magnets is that the magnetic field points transverse to the symmetry axis of the cylinder, so that solenoidal rf coils can be used for detection. These are more sensitive than the saddle coils employed with magnets where the magnetic field direction coincides with the cylinder axis, so that sensitivity can be compromised in favor of short dead time.

In principle, stray-field devices can also be employed to study porous media. However, given a gradient strength of 10 to 20 T/m for the *NMR-MOUSE*, even echo times of 20 μs are not short enough to avoid signal decay accelerated from diffusion (eqn. (7.1.8)). For small pores, the surface relaxation rate (eqn. (7.1.4)) is larger than the relaxation rate from diffusion, but for large pores it is not. As a result, relaxation time distributions measured in grossly inhomogeneous magnetic fields are distorted at long relaxation times compared to those measured in moderately inhomogeneous fields.

Figure 7.1.2. Magnets for measuring proton NMR of fluids in porous media. (a) Rock Core Analyzer from Magritek for relaxation measurements. (b) 0.22 T Halbach magnet with an inner diameter of 140 mm and with an insert to accommodate 20 mm diameter drill cores in its most homogeneous region in the center. (c) Well-logging tool with magnet and electronics.

Table 7.1.1. Acquisition parameters for ^{1}H relaxometry of water saturated rock and soil

Parameter	Value	Value
magnet, probe	NMR-MOUSE PM25 (rock)	Halbach magnet (soil)
transmitter frequency ν_{rf}	13.8 MHz	21.9 MHz
transmitter attenuation for 90° pulse	−8 dB at 700 W	−6 dB at 80 W
duration t_P of 90° pulse	15 μs	5 μs
dwell time Δt	1 μs	1 μs
acquisition time t_{acq}	10 μs	20 μs
echo time t_E	120 μs	100 μs
number n_E of echoes	2000	4000
recycle delay t_R	2.5 s	2.5 s
number n_s of scans	128	128

7.1.6 Pulse sequences and parameters

Although *longitudinal relaxation* is insensitive to diffusion in internal magnetic field gradients and therefore should be a good measure of pore properties even in inhomogeneous magnetic fields, common practice is to measure the diffusion-sensitive *transverse relaxation* instead. The main reason lies in measurement duration. Longitudinal relaxation processes can only be probed indirectly (Fig. 3.2.2) and require a rather protracted measurement series, whereas transverse relaxation can be assessed directly in a single-scan experiment with a CPMG sequence (Fig. 3.1.1b). Table 7.1.1 summarizes generic acquisition parameters for measuring CPMG echo trains of a water-saturated rock sample with the NMR-MOUSE.

7.1.7 Beginner's level measurements

Water-saturated drill cores from boreholes are investigated routinely by measuring transverse magnetization decays using standard CPMG sequences. Such measurements can be performed in a few seconds up to a few minutes time. Instead of commercial core scanners (Fig. 7.1.2a), simple Halbach magnets (Fig. 7.1.2b) can be used as long as the magnetic field is sufficiently homogeneous and as long as the magnet is kept at constant temperature. The CPMG decay (Fig. 7.1.3a) is inverted into a distribution of relaxation times (Fig. 7.1.3b) by inverse Laplace transformation. Typically this distribution is bimodal with a node separating the signal from *producible fluid* at large relaxation times from the signal of *bound fluid* at short relaxation times. For water-saturated sandstone, the so-called T_2 *cut-off*, which separates both relaxation ranges, is 33 ms. Note that the total range of relaxation times covers about four orders of magnitude, from below one millisecond to one second. If the CPMG sequence is applied in the presence of a strongly inhomogeneous field, the relaxation time distribution

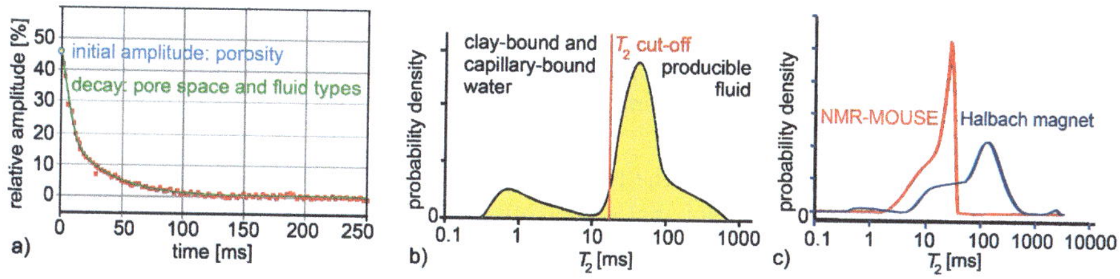

Figure 7.1.3. Core analysis by measurement of transverse relaxation from water-saturated limestone samples. (a) CPMG decay. (b) Relaxation time distribution obtained from (a) by inverse Laplace transformation. The ordinate scale is proportional to the probability density of finding a pore with a given T_2 (adapted from [4] with permission). (c) Relaxation time distributions obtained from CPMG decays measured in the inhomogeneous field of an NMR-MOUSE and in the homogeneous field of a Halbach magnet (adapted from [5] with permission).

appears compressed from the side of the long relaxation times (Fig. 7.1.3c), because the signal is attenuated not only by relaxation but also by diffusion (eqn. (7.1.9)).

Drill cores are long cylinders, and the average pore size distribution changes along the core following the changes in depth of the borehole. To map the changes the drill core is shifted through the magnet step by step. Alternatively the magnet is shifted along the drill core. The latter approach demands less total length of the device and has been realized in a prototype instrument (Fig. 7.1.4a) [3]. Along a length of just 30 cm, the relaxation time distributions vary appreciably (Fig. 7.1.4c) and so does the

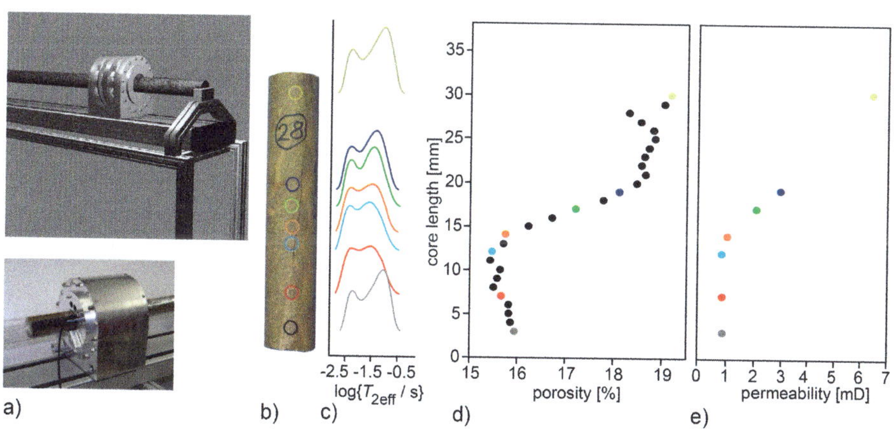

Figure 7.1.4. Geophysical properties of a drill core measured with a Halbach scanner (adapted from [3, 6] with permission). (a) The drill core rests in a plastic pipe that fits inside the Halbach magnet. CPMG echo trains are measured at different positions along the drill core by moving the magnet under control of a stepper motor. (b) Sandstone core with positions marked for the measurement. (c) Relaxation time distributions at the marked positions. (d) Change of the porosity along the core length calculated from the integrals over the relaxation distributions. (e) Fluid permeability estimated from the relaxation time distributions and the porosity.

Table 7.1.2. Common issues encountered when measuring rock and soil

– The sample is macroscopically heterogeneous or to small too be representative of the rock or soil formation under investigation
– The sample is large and covers a magnetic field region with appreciable inhomogeneity, so that the CPMG echoes are attenuated not only by relaxation but also by diffusion
– The sample is not fully fluid saturated
– The echo time is long and the CPMG echoes are attenuated not only by relaxation but also by diffusion
– The echo time is too long to detect rapidly relaxing magnetization components from liquids in small pores
– The number of echoes is high and the recycle delay short so that the rf coil heats up
– The temperature of the sample changes because the magnet temperature or the room temperature are not controlled
– Note, that the relaxation time distribution only maps the pore-size distribution in the fast diffusion limit
– Note, that the inverse Laplace transformation tends to be unstable in the presence of noise

porosity (Fig. 7.1.4d). The *porosity* corresponds to the initial amplitude of the signal decay from a fully saturated sample or to the integral of the relaxation time distribution. From the porosity and the relaxation time distribution, the fluid *permeability* can be predicted with the help of empirical models (Fig. 7.1.4e).

When studying porosity and pore size distributions, the porous material to be investigated needs to be completely fluid saturated, because only in this case the CPMG signal amplitude or the integral of the relaxation time distribution is proportional to porosity. However, full saturation is hard to achieve for a rock sample once it has dried. Drawing vacuum on the dry sample during the wetting process helps to improve the fluid uptake. In order to obtain genuine porosity values by NMR, originally wet cores should be investigated preferably immediately after sampling. When cutting small samples from large cores, regions can be selected that are void of alterations originating from the sampling process. A useful sample size is a 5 cm diameter cylinder. Common issues encountered when measuring the signals from fluids in rock are summarized in Tab. 7.1.2.

7.1.8 Advanced level measurements: 2D Laplace NMR

Most natural porous media are intrinsically heterogeneous, not only regarding the geometry of the pore network but also the fluids inside the pores. The pores of bedrock are typically filled with a mixture of water, oil and gas, the latter being mostly methane. To unravel the complexity of fluid-filled rock, the default CPMG measurements, from which distributions of relaxation times are obtained, are complemented by investigations that employ methods of *2D Laplace NMR* that lead to correlation maps of distributions of relaxation times and diffusion coefficients. Two methods have been discussed in Section 3.2.7, the T_1-T_2 correlation experiment for water-saturated sandstone and the T_2-T_2 exchange experiment for a water-saturated pack of spherical silica particles (Fig. 3.2.4). Another important 2D Laplace experiment is the D-T_2 correlation experiment.

T_1-T_2 correlation NMR

The T_1-T_2 *correlation* experiment resolves the distribution of one relaxation time with respect to the other (eqns. (7.1.1) and (7.1.9)). Apart from different bulk relaxation times and surface relaxivities, T_2 depends on the field homogeneity specified by the field gradient G at the site of the spins and the *diffusion coefficient* D, while T_1 does not. This field gradient originates from the gradient of the applied field and the magnetic field distortions inside the pore due to differences in magnetic susceptibilities at liquid-liquid and liquid-solid interfaces. The longer the *echo time* t_E and the higher the gradient, the stronger the diffusive attenuation (eqn. (7.1.8)). When *internal field gradients* play a role in samples with a pore size distribution (Figs. 3.2.4a and 7.1.5a) or the fast diffusion limit differs for T_1 and T_2, both relaxation times are not proportional to each other. Because internal gradients are stronger in smaller pores than in larger pores, the T_1-T_2 correlation ridge tilts towards a line parallel to the diagonal at short t_E (Fig. 7.1.5b). At high field, the internal field gradients are stronger than at low field, so that low fields are preferred when studying pore size distributions by distributions of transverse relaxation times.

T_2-T_2 exchange NMR

The T_2-T_2 *exchange* experiment correlates transverse relaxation time distributions via the exchange of magnetization between different relaxation sites. This exchange is mediated in fluid-filled porous media by molecules diffusing across the pore space from one relaxation site to another. The exchange proceeds not only during the mixing period but also during the evolution and detection periods (Fig. 3.2.4c) so that the

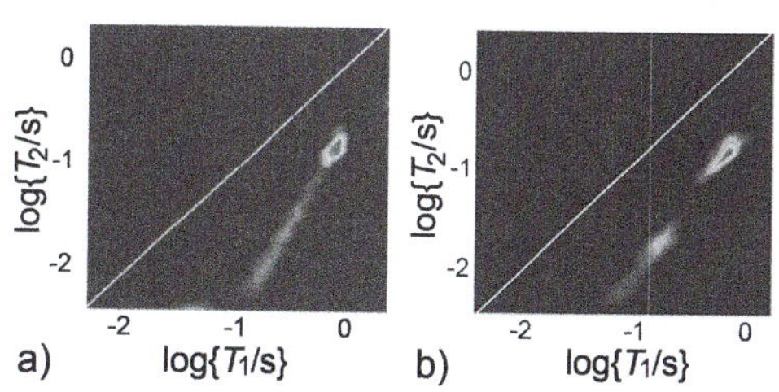

Figure 7.1.5. 2D T_1-T_2 correlation maps of saturated Allermöhe sandstone samples with a permeability of 1.85 mD and a porosity of 6% measured at 0.22 T corresponding to 9.6 MHz ^{1}H NMR frequency. (a) 0.15 ms echo time. (b) 0.06 ms echo time (adapted from [3] with permission).

resultant 2D distribution map has to be simulated to extract the exchange rates between the different relaxation sites (eqn. (3.2.8)). From these exchange rates and the diffusion coefficient, the distance between relaxation centers can be estimated to reconstruct the porespace topology once the relaxation centers have been assigned to particular sites in the pore space or to particular pore sizes.

D-T_2 correlation NMR

Diffusion-relaxation correlation experiments are of particular interest to the oil industry in *fluid typing*, that is, in quantifying the amounts of water and oil in a particular rock formation. In the 1D relaxation time distribution, the peaks from both fluids overlap, and only the amounts of producible fluid and bound fluid can be quantified in terms of peak integrals (Fig. 7.1.3b). But water and oil have very different diffusion coefficients, allowing the amounts of oil and water to be specified when the distribution of relaxation times is resolved in a second dimension by the distribution of diffusion coefficients similar to the DOSY experiment (Fig. 4.1.5), with which the distribution of frequencies is resolved by the distribution of diffusion coefficients.

The D-$T_{2\mathrm{eff}}$ map acquired with a logging tool from a water- and oil-bearing rock formation with the pulse sequence shown in Fig. 3.2.7a shows separate peaks for oil and water (Fig. 7.1.6). Average diffusion coefficients and relaxation times for most oils lie on a straight line at an oblique angle in the D-$T_{2\mathrm{eff}}$ map. The peak on this line identifies the oil peak. The other peak shows the bulk diffusion coefficient of water. The integrals under each peak provide the *water saturation* and the *oil saturation* of the rock. While in the projection of the 2D map onto the relaxation axis the signals from water and oil cannot be separated, they are identified as separate peaks in the projection onto the diffusion axis.

Alternatively, the D-$T_{2\mathrm{eff}}$ map can be derived from a set of *CPMG echo trains* which have been acquired with different echo times t_E [8]. The signal from a single pore de-

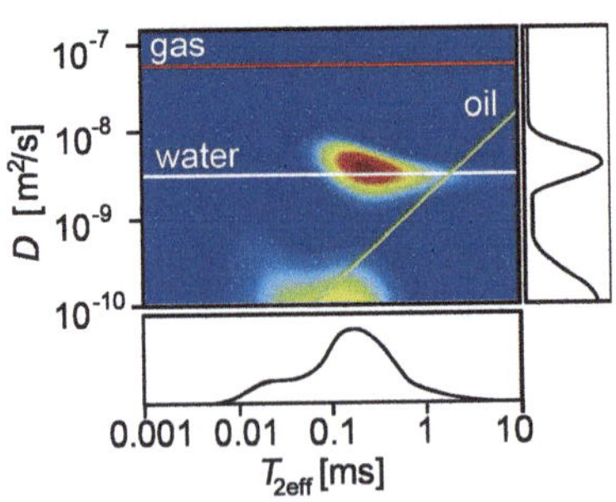

Figure 7.1.6. D-$T_{2\mathrm{eff}}$ map of a water and oil bearing formation acquired with a well-logging tool (adapted from [7] with permission). The horizontal lines mark the diffusion coefficients of bulk water and methane gas. The tilted yellow line identifies the most common D and $T_{2\mathrm{eff}}$ values of oils.

cays according to

$$s(t) = s(0)\exp\{-1/12\,(\gamma G t_E)^2 D t\}\exp\{-t/T_{2eff}\}. \tag{7.1.10}$$

By storing the CPMG decays in a 2D array with axes labeled $t_E^2 t$ and t, a 2D Laplace inversion produces a 2D correlation map $S(1/12(\gamma G)^2 D, 1/T_{2eff})$, which shows $1/12(\gamma G)^2 D$ on one logarithmic axis and $1/T_{2eff}$ on the other logarithmic axis. The gradient G can refer to the internal gradient, the gradient of the applied magnetic field or to both. When neglecting internal gradients so that G is the gradient of the NMR-MOUSE, then S is a map correlating the distributions of D and T_{2eff}. With this procedure, T_{2eff} distributions unaffected by diffusion (Fig. 7.1.3c) can be measured with the NMR-MOUSE. This technique has been applied in the Villa of the Papyrus in Herculaneum to one of the moist walls (Fig. 8.1.6b).

7.1.9 Data processing

Distributions of relaxation times and 2D correlation maps of T_1, T_2, and D are computed from the experimental data by 1D and 2D inverse Laplace transformation [7]. To interpret a *relaxation time distribution* in terms of a *pore size distribution* requires the validity of the *fast diffusion limit*, which is usually fulfilled for water in pores smaller than 10 μm. The *surface relaxivity* parameters ρ_1 or ρ_2 in eqns. (7.1.3) and (7.1.4) are typically determined by matching relaxation time distributions with pore size distributions obtained by *nitrogen adsorption porosimetry* or *mercury intrusion porosimetry* experiments (Fig. 7.1.7). Perfect agreement cannot be achieved for several reasons. Both fluids, nitrogen gas and liquid mercury, can only ingress into pores that are connected to the pore network, while NMR can detect water trapped in dead volumes. Moreover, nitrogen quickly saturates small pores up to the point of condensation and mercury is subject to capillary forces, both of which falsify porosimetric results.

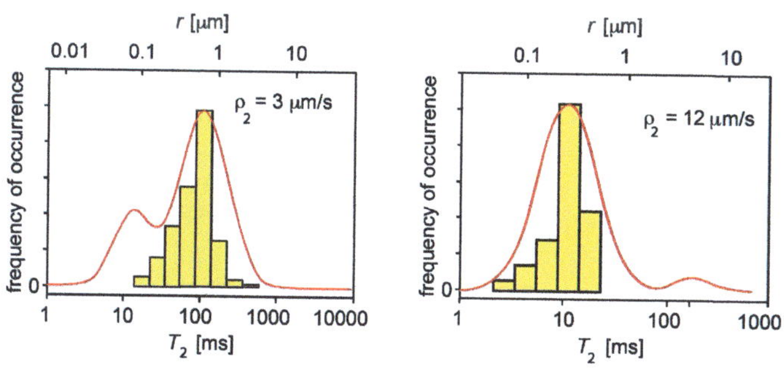

Figure 7.1.7. Calibration of pore size distributions on limestone samples from the Great Bahama Bank (adapted from [5] with permission). Continuous curves: Relaxation time distributions measured with a 0.3 T Halbach magnet. Bars: Distributions of pore sizes from mercury intrusion porosimetry.

The total integral of the relaxation time distribution specifies the total amount of hydrogen in the porous medium. Given the hydrogen density and the size of the sensitive volume, the *fluid saturation* and the *porosity* can be derived. The same applies to integrals of particular peaks in the distribution. For example, in a bimodal distribution, the integral of the peak at long relaxation times measures the amount of producible fluid, and the integral of the peak at short relaxation times, the amount of bound fluid. The *porosity* Φ is the volume of the fluid divided by the sensitive volume for fully fluid-saturated samples. From it the fluid *permeability* κ can be predicted following models for different types of rock. For carbonates the *Kenyon model* is widely accepted,

$$\kappa = aT_{2\mathrm{lm}}^2 \Phi^4, \tag{7.1.11}$$

where $T_{2\mathrm{lm}}$ is the logarithmic mean value of the T_2 distribution. The constant a depends on the surface relaxivity ρ_2 and is determined empirically. Usually, for sandstones a = 4 mD/(ms)^2 where 1 mD = 9.87 10^{-16} m^2.

7.1.10 References

[1] Hirasaki GJ. NMR applications in petroleum reservoir studies. In: Stapf S, Han SI, editors. NMR Imaging in Chemical Engineering. Weinheim: Wiley-VCH; 2006. pp. 321–340.

[2] Brownstein KR, Tarr CE. Importance of classical diffusion in NMR studies of water in biological cells. Phys Rev. 1979; 19:2446–2453.

[3] Anferova S, Anferov V, Arnold J, Talnisnikh E, Voda MA, Kupferschläger K, Blümler P, Clauser C, Blümich B. Improved Halbach sensor for NMR scanning of drill cores. Magn Reson Imaging. 2007; 25:474–480.

[4] Anferova S, Anferov V, Rata DG, Blümich B, Arnold J, Clauser C, Blümler P, Raich H. Mobile NMR device for measurements of porosity and pore size distributions of drilled core samples. Conc Magn Res. 2004; 23B:26–32.

[5] Arnold J, Clauser C, Pechnig R, Anferova S, Anferov V, Blümich B. Porosity and permeability from mobile NMR core-scanning. Petrophysics. 2006; 47:306–314.

[6] Blümich B, Casanova F, Perlo J. Mobile single-sided NMR. Prog Nucl Magn Reson Spectrosc. 2008; 52:197–269.

[7] Song YQ. Novel two-dimensional NMR of diffusion and relaxation for material characterization. In: Stapf S, Han SI, editors. NMR Imaging in Chemical Engineering. Weinheim: Wiley-VCH; 2006, pp. 163–183.

[8] Mitchell J, Chandrasekera TC, Gladden LF. Obtaining true transverse relaxation time distributions in high-field NMR measurements of saturated porous media: Removing the influence of internal gradients. J Chem Phys. 2010; 132:244705.

7.2 Soil

7.2.1 Introduction

Most of the continental earth surface is covered by *soil*. Soil is at the interface between atmosphere and earth. In its top layer it hosts the root system of plants, while its bottom layer is in contact with ground water. In contrast to the bottom layer, the top layer is exposed to the direct impact of climate in terms of rain, temperature and wind. As a consequence, soil carries fluxes of mostly water, carbon dioxide, many salts, air, and heat energy, and its composition changes continuously through the impact of these fluxes and the biological activity on the surface. The soil above the ground water table is only partially saturated with water and is termed the *vadose zone* from the Latin word 'vadosus' meaning 'shallow'.

7.2.2 Objective

The main objective of studying soil by NMR is to characterize the fluxes of matter and energy from the pore scale upward. In this regard water fluid saturation, diffusion, and the geometry of the pore space as a function of water supply and drainage are prime targets with and without the impact of the biosphere and erosion. Whereas NMR spectroscopy of soil materials is a domain of high-field NMR for reasons of sensitivity, low-field NMR instruments are employed for measuring *relaxation* and *diffusion* in soil employing essentially the same NMR methods as those used for investigating rock.

Investigations of soil are simplified by the fact that the liquid typically is water and that the gas phase usually is air. Although the gas phase is often saturated with water vapor, its signal is negligible compared to that of liquid water. Similar to rock, the signal amplitude provides the level of *fluid saturation*, but the relaxation decay from which the *relaxation time distribution* is calculated, is difficult to interpret in unsaturated porous media. Nevertheless, relaxation exchange experiments and imaging experiments provide information about fluid transport on timescales close to the diffusion limit. Few experiments are carried out on natural soil and most studies refer to soil models, because natural soil is difficult to obtain with sufficiently well-reproducible properties, and it often contains paramagnetic compounds, which can seriously shorten the relaxation times to the limit where most of the signal decays within the dead time of the instrument. *Soil models* are constructed from cleaned and sieved compounds, which are carefully repacked to obtain reproducible bulk density.

7.2.3 Further reading

Bayer JV, Jaeger F, Schaumann GE. Proton Nuclear Magnetic Resonance (NMR) relaxometry in soil science applications. Open Magn Res J. 2010; 3: 15–26.

Jury WA, Horton R. Soil Physics. Hoboken: Wiley; 2004.

Hemminga MA, Burman P. Editorial: NMR in soil science. Geoderma. 1997; 80: 221–224.

Marshall TJ, Holmes J, Rose CW, Soil Physics. 3rd edition. Cambridge: Cambridge University Press; 1996.

7.2.4 Soil Physics

Types of soil

Soil is a heterogeneous material composed of solids, liquids, and gases. It is subject to continuous transformation processes of inorganic and organic materials by supply of energy and matter, and by biological activity. The *liquid phase* consists of mostly water with dissolved salts and biomass. The *gas phase* is water-saturated air often mixed with CO_2 and other gases. The *solid phase* has been produced mainly by weathering of larger sediments and consists largely of minerals and smaller amounts of organic compounds such as humic acids. The mineral fraction is classified in terms of its *grain size distribution*. Soil with grain sizes smaller than 2 μm contains reactive clay minerals. Soil particles larger than 2 μm are divided into *silt*, *sand* and *gravel*. Silt has grain sizes between 2 μm and 63 μm, sand between 63 μm and 2 mm, and gravel consists of stones with sizes between 2 mm and 63 mm in diameter. Gravel is usually separated from the soil in pore and transport studies of soil. Different grain size distributions give different soil textures. The most common soil textures are clay, silty loam and sand in increasing order of their average grain size, where *loam* contains more humus than sand. The smaller the average grain size, the broader the *grain size distribution*.

Pore space and structures

On the microscale soil consists of individual particles and the corresponding pore space. However, many particles may cluster and form aggregates, which define the mesoscale. On the macroscale soil is organized in horizontal layers and the pore space includes wormholes and fractures. Accordingly, water flow is driven by different mechanisms on the different scales. On the micro- and mesoscale water moves either by diffusion in the pore space or flows along potential gradients. Beyond the pore scale, water movement in unsaturated soil is described by the *Richards equation*. On the macroscale additional preferential flow phenomena can occur, e.g. fast downward flow in macropores and wormholes following strong precipitation.

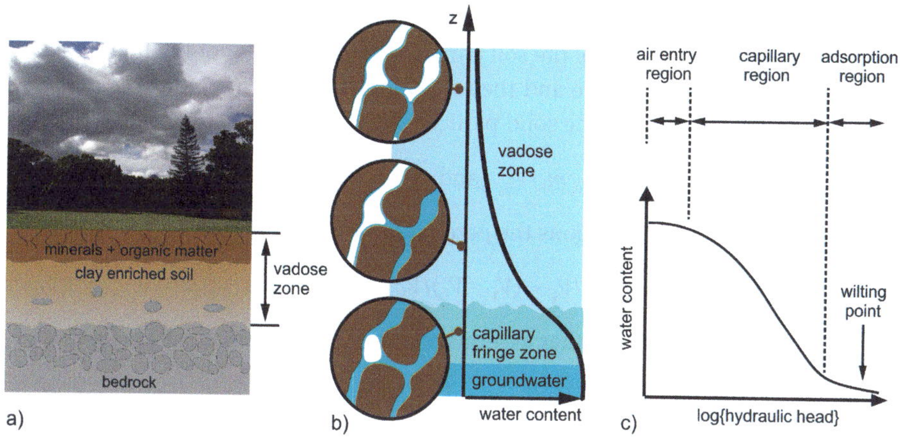

Figure 7.2.1. Soil layers. (a) Soil is classified into three major horizons differing in their stage of development. The lowest horizon is the bedrock from which the soil develops with time by weathering. The horizon above consists of finer grade material from bedrock enriched with clay particles. The top horizon is strongly influenced by biotic activity including roots and exhibits a high content of organic matter resulting from degradation of plant material and animals. (b) Water saturation regimes in an idealized soil profile (adapted from [1]): In the uppermost soil layer the water content is low and the water phase is discontinuous. Water flow and airflow are decoupled. In the layer underneath, the water phase becomes continuous. The water pressure is determined by the capillary forces and by gravity. Water and airflow are strongly coupled. Further below, the air phase is discontinuous. (c) Water retention curve of a homogeneous soil material showing the characteristic features: air entry region, capillary region and adsorption region. The wilting point denotes the water content below which water can no longer be taken up by plants.

The horizontal soil layers are called *horizons* (Fig. 7.2.1a). Each horizon from the surface to the *bedrock* exhibits characteristic physical and chemical properties, color, and soil biota. The top horizon is strongly influenced by biotic activity including the formation and decay of roots and exhibits a high content of organic matter resulting from degradation of plant material and animals. The horizon underneath consists of finer grade material from bedrock and is enriched with clay particles. The lowest horizon is the bedrock from which the soil develops with time by weathering. Similar to rock, soil can be highly heterogeneous. Pore sizes can range from nanometers in clay aggregates to several millimeters and more in fractures and wormholes.

Parameters of the soil structure

The soil texture and the settling of particles that form the soil define the details of the pore space, which is not delineated by a rigid matrix as in rock, but can swell and shrink with water uptake and drainage depending on the wettability of the different fractions. In this regard, soil exhibits properties similar to those of pastes, but needs to be approximated in terms of a three-phase system from solid particles of different size and wettability, water, and gas. Accordingly, four volumes are employed to describe soil. These are the volume V_s of the soil particles, the volume V_w of the liquid phase, the volume V_g of the gas phase and the total volume V_t. Given the masses m of the components, the densities of the solid particles and the bulk soil are given by

$$\rho_s = m_s/V_s \quad \text{and} \quad \rho_t = m_t/V_t, \tag{7.2.1}$$

respectively. With these definitions the *porosity* Φ can be written as

$$\Phi = (V_w + V_g)/V_t = (V_t - V_s)/V_t = (\rho_s - \rho_t m_s/m_t)/\rho_s. \tag{7.2.2}$$

The *water content* is defined either in terms of a volume fraction or a mass fraction,

$$\theta_v = V_w/V_t \quad \text{and} \quad \theta_m = m_w/m_t. \tag{7.2.3}$$

When fully saturated with water, the volume of the gas phase in the soil is zero, and the *volumetric water content* θ_v is identical to the porosity Φ. The *water saturation*, on the other hand is defined by

$$\Theta_s = V_w/(V_w + V_g). \tag{7.2.4}$$

Often only the normalized water saturation with producible water is of interest, so that the saturation of bound water needs to be subtracted, giving

$$\Theta = (\theta_v - \theta_r)/(\Phi - \theta_r), \tag{7.2.5}$$

where θ_r is the bound water content, which cannot be removed.

Water distribution

The water content changes continuously depending on rainfall, drainage, evaporation and water uptake by the plant roots. In the fully water-saturated *groundwater* zone water moves largely in horizontal direction, while in the unsaturated *vadose zone* above, the water motion has a vertical component due to the effects of gravity-driven downward flow and evaporation-driven upward flow. With increasing depth, the water content θ_V in the vadose zone generally increases and the gas saturation decreases (Fig. 7.2.1b). Near the surface, the air phase is continuous while the water phase is not, and the motion of the water phase is largely decoupled from that of the air phase. In a narrow region further down, both air and water phases are continuous. In this continuous multiphase regime, the air and water flows are strongly coupled. Even further down, the air phase becomes discontinuous. In this capillary fringe zone, the water is retained in the soil by capillary forces. The water conductivity is governed by the pore geometry and molecular self-diffusion while the air conductivity is governed the interdiffusion of air through water.

To assess the impact of soil processes on vegetation and atmosphere, macroscopic hydraulic properties like water retention and hydraulic conductivity of soil are defined, which are related to soil properties such as texture, porosity and chemistry. To specify the state of water in soil the sum over all forces acting on the water molecules is decisive. However, it is easier to measure the work necessary to bring a portion of water from a reference point to a certain level than to measure forces. This is conveniently formulated in terms of potential energy, in brief potential, which does not include kinetic and thermal energy. The *total potential* of water in soil is defined as a sum of following partial potentials

$$\psi = \psi_m + \psi_z + \psi_o + \psi_\Omega + \psi_p. \tag{7.2.6}$$

The *matrix potential* ψ_m is caused by all interactions of water molecules with the surface of the solid phase and ψ_z is the *gravitational potential*. The *osmotic potential* ψ_o is due to hydration of dissolved ions, the *overburden potential* ψ_Ω is caused by a load pressing on the top surface and ψ_p is an additional air pressure potential. Experimentally the total potential in unsaturated soil is often approximated by the *hydraulic potential* ψ_H, neglecting the other terms,

$$\psi_H = \psi_m + \psi_z, \tag{7.2.7}$$

which is the sum over matrix and gravitational potentials. The relation between the matrix potential and water content θ is expressed by the *water retention curve* or water characteristic function. It describes the increasing matrix potential ψ_m with decreasing water content. Notice, that the matrix potential is a negative quantity. It is zero in fully saturated matrix (e.g. at the groundwater table) and becomes more negative with progressive moisture loss, e.g. with increasing elevation. It is also customary to plot the *hydraulic head* $h = \psi/(\rho g)$, i.e. the length of a hanging water column instead of the matrix potential against the water content (Fig. 7.2.1c). Generally this function can be divided into three major regions: in the air-entry region the water content is constant but the matrix potential changes its magnitude. To remove water from large pores of saturated soil, a minimum suction, the so-called air-entry suction, must be used. For coarse sand this value is approximately 5 to 10 cm and changes to larger values for soils with finer texture. In the intermediate part of the water retention curve, the suction increases and the smaller pores in the capillary region are drained. The third part of the curve is called adsorption region. It is also relatively flat and large changes in the matrix potential correspond to small changes in the water content, because the water molecules are tightly bound to the solid soil particles.

An important point on the water retention curve is the *wilting point*. Below the wilting point the water content is too low to sustain plants growth. The *permanent wilting point* is defined as the water saturation at a matrix potential of -1.5 MPa. While water retention curves provide useful information on water storage in soil, measuring natural saturation and drainage processes may take weeks to months. Measurements with forced dynamic processes, particularly with applied pressure, are faster, but may perturb the system under investigation.

The water retention curve $\theta_v(h)$ (Fig. 7.2.1c) can be approximated with the *van Genuchten model* [2], which is fitted to experimental data by taking θ_r, Φ, α, and n

as fit parameters

$$\theta_v(h) = \{\Phi - \theta_r\}/[1 + (\alpha h)^n]^{(1-1/n)} + \theta_r, \tag{7.2.8}$$

where θ_v is the actual water content (eqn. (7.2.3)) at a given hydraulic head h or matric potential ψ_m, and Φ and θ_r are the water contents when fully saturated and fully drained, respectively. The parameter $\alpha > 0$ is related to the inverse of the air entry suction. The parameter $n > 1$ describes the slope of the curve and is related to the *pore size distribution*. The narrower the pore size distribution, the steeper the retention curve and the larger n. Some typical values are $\alpha = 0.03\ \text{cm}^{-1}$ and $n = 4$ for sandy soil and $\alpha = 0.01\ \text{cm}^{-1}$ and $n = 1.6$ for a silty loam.

Water transport

To characterize the water transport properties for a one-dimensional laminar flow in the pore space under stationary conditions, the *hydraulic conductivity* K is introduced to specify the flux density corresponding to the flow velocity J following *Darcy's law* (eqn. (6.3.1)),

$$J = Q/A = -K(\Delta h/\Delta z), \tag{7.2.9}$$

where Q is the flow rate in units of m^3/day, A is the cross-sectional area of the water flow, and the quotient relates the difference of the hydraulic head Δh across the flow distance Δz. The hydraulic conductivity K depends only on the soil properties. It is proportional to the *permeability* κ according to

$$K = \kappa g \rho/\eta, \tag{7.2.10}$$

where η is the dynamic viscosity of the fluid.

Assuming stationary three-dimensional flow, mass balance requires

$$\partial[K \cdot \partial h/\partial x]/\partial x + \partial[K \cdot \partial h/\partial y]/\partial y + \partial[K \cdot \partial h/\partial z]/\partial z = 0. \tag{7.2.11}$$

Following division by K the Laplace equation for the hydraulic head is obtained,

$$\partial^2 h/\partial x^2 + \partial^2 h/\partial y^2 + \partial^2 h/\partial z^2 = 0, \tag{7.2.12}$$

which relates the hydraulic head h to the space coordinates for calculation of the stationary flow field.

To describe the water flux through unsaturated soil under transient conditions, e.g. by infiltration or evaporation, an advanced model must be used which takes into account that the net water flow summed over all directions is no longer zero but equal to the changes of water content with time, so that

$$\partial J/\partial x + \partial J/\partial y + \partial J/\partial z + \partial \theta_v/\partial t = 0. \tag{7.2.13}$$

This equation relates changes in water flux and water storage. If necessary, sources or sinks of water can be added for example for root-water uptake. Additionally, the hydraulic conductivity K is no longer constant but a function of water content or matrix potential. By inserting Darcy's law for J, one obtains an equation to predict water content or matrix potential for transient flow. For the most important case of vertical flow one obtains

$$\partial\theta_v/\partial t = \partial/\partial z[K(h)(\partial h/\partial z + 1)]. \tag{7.2.14}$$

This equation contains two unknown variables θ_v and h, and cannot be solved directly. Following Richards it can be rewritten in two ways as

$$\partial\theta_v/\partial t = \partial/\partial z[D(\theta_v)\partial\theta_v/\partial z] + \partial K(\theta_v)/\partial z, \tag{7.2.15}$$

$$C_w(h)\partial h/\partial t = \partial/\partial z[K(h)(\partial h/\partial z + 1)], \tag{7.2.16}$$

where $D(\theta_v) = K(\theta_v)\,\partial h/\partial\theta_v$ is the soil-water diffusivity, and $C_w(h) = \partial\theta_v/\partial h$ is the *water capacity* function. Both quantities are known from the water retention curve and the hydraulic conductivity curve. These equations are known as the water content form and the matrix potential form of the *Richards equation*, respectively. They can be solved by introducing parameterized models for the water retention curve and hydraulic conductivity curves. A popular model is the *van Genuchten equation* (7.2.8) and its extension by *Mualem* for the relative hydraulic conductivity, which is defined as the ratio of hydraulic conductivities at partial and at full saturation [3],

$$K(\Theta)/K_s = \Theta^{1/2}[1 - (1 - \Theta^{1/m})^m]^2. \tag{7.2.17}$$

Here the water saturation Θ is defined in eqn. (7.2.5), and $m = 1 - 1/n$.

7.2.5 Hardware

Water saturation is determined from the signal amplitude, that is, from the number of protons in the sensitive volume of the sensor. The first *in situ measurements* of soil moisture were reported by Paetzold and coworkers 1985 [4] using a single-sided U-shaped electromagnet, which was mounted on a sledge and pulled by a tractor across the field while acquiring proton signal (Fig. 7.2.2a). Today commercially available unilateral NMR sensors, e.g. the NMR-MOUSE (Fig. 7.2.2b), oil-well logging tools (Fig. 7.1.2c), *slim-line logging tools* (Fig. 7.2.2c), and closed magnets with moderate field homogeneity are employed to study soil (Fig. 7.1.2b). To investigate the water content of topsoil *in situ*, depth profiles can be acquired with the NMR-MOUSE down to a maximum depth of 25 mm. Moisture at larger depth can be measured by two different approaches. One is *surface NMR*, or *NMR sounding*, which employs surface coils with diameters up to 200 m that measure the NMR signal in the earth's magnetic field from water-bearing layers up to 200 m deep [6]. The other method employs *logging tools* like those employed for oil-well logging (Fig. 7.1.2c) or *slim-line logging tools* (Fig. 7.2.2c) that are built following the same concept as oil-well logging tools, but are narrower in diameter, operated in armed holes in the ground and typically comprise the magnet and

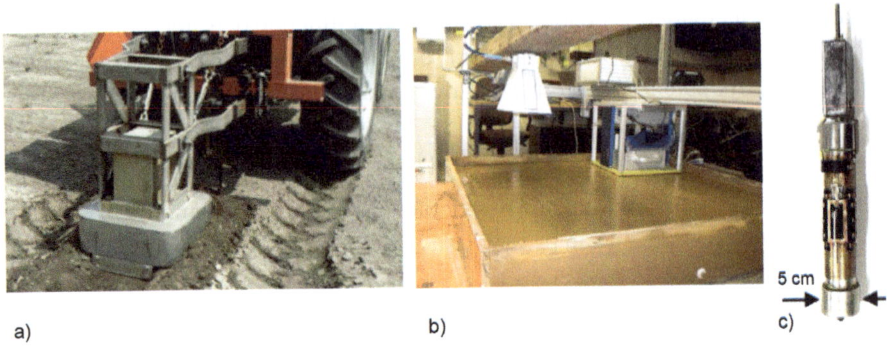

Figure 7.2.2. NMR devices for measuring soil in the field and laboratory. (a) Electromagnet mounted on a sledge and pulled by a tractor across the field while NMR signals are recorded continuously to monitor the moisture [4]. (b) NMR-MOUSE hanging from a frame facing downwards together with a ground penetrating radar antenna above a lysimeter for monitoring evaporation processes from the top soil surface (photograph courtesy of S. Merz). (c) Slim-line logging tool for ^{1}H NMR at 3.3 MHz with slice selection at 18 mm distance from the borehole wall to measure moisture content in natural soil. The tool does not include the spectrometer electronics. It consists only of the magnet and the coil with the associated rf circuitry (adapted from [5] with permission).

the coil circuit only while they are connected to the spectrometer outside the hole via a long cable [7, 5, 8]. However, today soil is still mostly investigated by measuring relaxation and diffusion of water in soil models in laboratory environments using closed magnets.

7.2.6 Pulse sequences and parameters

Although *soil* is often investigated at partial water saturation, the pulse sequences are the same as those for studying *rock* (Section 7.1.6). The standard measurement applies the CPMG sequences (Fig. 3.1.1b). Occasionally also longitudinal magnetization build-up curves are measured (Fig. 3.2.2). In soil, *relaxation time distributions* are often broad ranging from clay-bound water to water in macropores, so that transverse magnetization decay curves and longitudinal magnetization build-up curves need to be measured with many data points, requiring long *CPMG echo trains* at short echo time and measurements for many different recovery times t_0, respectively. Moreover, field inhomogeneities from magnetic components as well as interfaces between solid, liquid and gas phases lead to internal gradients, which impact the apparent transverse relaxation decay according to eqn. (7.1.9) even when measured in a homogeneous magnetic field. This effect can be minimized when measuring at low field and by using short echo times t_E. Consequently CPMG echo trains are required to be long and may contain up to 32,000 echoes. The rf energy dissipated in the rf coil from so many pulses will raise the temperature of the coil, so that recycle delays much longer that 5 $T_{1\max}$

are recommended to let the setup cool down, where $T_{1\,max}$ is the longitudinal relaxation time of the water in the macropores. Typical CPMG pulse sequence parameters for measuring soil with a Halbach magnet are collected in Tab. 7.1.1.

7.2.7 Beginner's level measurements

In most cases, model soil is investigated in the laboratory, which has been sieved, repacked and saturated with water. For fully saturated samples the amplitude of the CPMG echo train provides the *porosity*, but because soil may swell with the uptake of water such porosity measurements have to be interpreted with due caution. The echo envelope is the Laplace transform of the relaxation time distribution function, which only relates to the *pore size distribution* for fully saturated soil samples. The interpretation of the relaxation time distributions for partially saturated soil is a matter of ongoing study, because the distribution depends on the wettability and the stage in the wetting and drying cycle, which may show a hysteresis. The overall observation is, that upon drying, the signals from peaks at large relaxation times vanish first while peaks at short relaxation times are largely unaffected (Fig. 7.2.5). This behavior is the same as in fluid-saturated rock (Fig. 7.1.3b), in which *producible fluid* gives signal at large relaxation times and *bound fluid* gives signal at short relaxation times.

The T_2 distribution functions of different types of water-saturated soil can differ considerably (Fig. 7.2.3a). Depending on soil texture, they cover a relaxation time range of about four decades, from 0.1 ms to 1 s. The lower limit is determined by the dead time of the instrument and thus by the minimum echo time of the CPMG sequence but not the sample. In fact, signals from moisture in small pores and bound

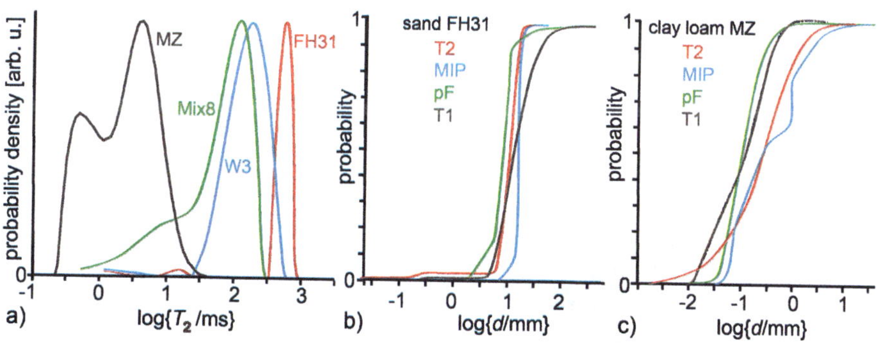

Figure 7.2.3. Distribution functions of relaxation times and pore sizes $d = 2r$ for water-saturated soil samples measured with a Halbach magnet at $B_0 = 0.15$ T. (a) Probability densities of T_2 of sand (FH31), silt (W3), a mixture of 92% sand and 8% clay (Mix8), and a natural clay-loam soil (MZ) (adapted from [9] with permission). (b) Integral pore size distributions of the sand FH31 derived from water retention curves (pF$\equiv -\log\{h\}$), mercury intrusion porosimetry (MIP), as well as from T_1 and T_2 relaxation time distributions. (c) Integral pore size distributions of the soil MZ.

fluid are typically not detected. The dead time can be reduced by employing smaller rf coils and higher magnetic field. But then sample size is reduced accordingly and the measured distribution may be less representative of the soil than that determined from a larger sample. The T_2 *distribution functions* of Fig. 7.2.3a have been measured in a Halbach magnet with $B_0 = 0.15$ T from saturated soil samples in 4 cm diameter glass tubes. The echo time was $t_E = 0.15$ ms, and up to 15,000 echoes were recorded in one CPMG scan. Common issues to consider when measuring soil are similar to those of measuring rock. They are summarized in Tab. 7.1.2. Different from most rock is the swelling and shrinking of soil samples so that the soil volume may depend on the water saturation and even the particular wetting and drying procedures.

7.2.8 Advanced level measurements

Laboratory studies

Similar to rock, soil is studied by *2D Laplace NMR* methods to gain insight into the *pore space geometry* and the *fluid transport* in fully and partially saturated soil. The most important experiments measure T_1-T_2 correlations, which probe the heterogeneity of the relaxivity and the effects of internal gradients on the pore scale, and T_2-T_2 exchange maps, which identify diffusive pathways that are explored by water molecules.

The T_1-T_2 *correlation* map of water-saturated sandy loam (Fig. 7.2.4) shows that T_1 and T_2 are proportional to each other and so are the relaxivities ρ_1 and ρ_2, with a T_1/T_2 ratio slightly higher than the value of 1.6 commonly found in sandstone. Furthermore, it can be concluded that the impact of diffusion in the internal field gradients at the field strength of $B_0 = 0.52$ T and the echo time $t_E = 0.3$ ms of the experiment are negligible. Note from Fig. 7.2.4a that part of the signal is already lost during the first encoding period due to fast longitudinal relaxation.

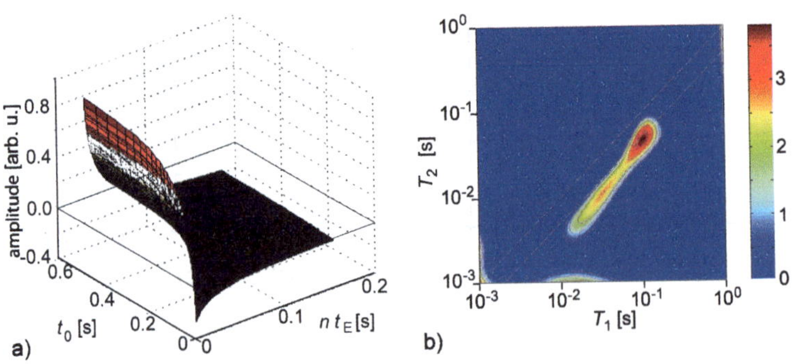

Figure 7.2.4. T_1-T_2 correlation NMR of a sandy loam soil sample acquired with the inversion recovery sequence. (a) 2D relaxation map where the amplitudes of CPMG decays are plotted as a function of the recovery time t_0 and multiples of the echo time $t_E = 0.3$ ms. (b) T_1-T_2 correlation map obtained by inverse 2D Laplace transformation of the data shown in (a).

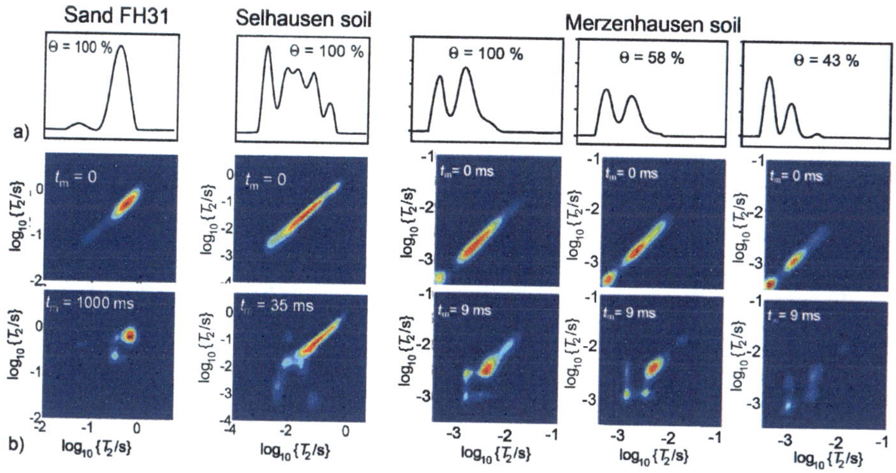

Figure 7.2.5. Relaxation time distributions and exchange maps of different soil samples at different saturation levels Θ measured in a Halbach magnet at 0.5 T. (a) 1D T_2 distributions with arbitrarily scaled amplitudes. (b) 2D T_2-T_2 exchange maps of different soils and for different exchange times t_m.

T_2-T_2 *exchange* maps provide information on fluid transport between relaxation centers. At zero and short exchange time t_m, the T_2-T_2 exchange maps are essentially diagonal, revealing only the T_2 distribution at a particular saturation level (Fig. 7.2.5) in which the signals at large relaxation times vanish first upon drying. With increasing exchange time, cross peaks appear. These mark the exchange of protons between different relaxation centers. Depending on the connectivity of the relaxation centers, these exchange maps are usually asymmetric when more than two relaxation centers are involved in the exchange process. At full saturation, T_2-T_2 exchange maps can be simulated to extract the exchange rates that provide boundaries to the diffusive distance between relaxation centers, from which the pore geometry can be estimated so long as the relaxation centers can be related to geometric features of the pore space (eqn. (3.2.8)). Typically, relaxation centers are assigned to pore sizes. If this assumption is valid, the existence of exchange peaks in unsaturated soil hints at diffusive transport between small pores of different sizes while at full saturation exchange can be observed between all pores.

Field studies

A *logging tool* for soil moisture is basically designed in the same way as a logging tool for oil wells, but it operates at far milder pressures and temperatures. The primary objective for its use is to map the *moisture content* at low depths by moving the tool from one position to another inside a pipe and collecting signal from soil regions outside near the pipe.

At shallow depth, that is, in the *vadose zone*, soil is not completely saturated with water so that the sensor is little shielded from environmental electromagnetic noise. Furthermore, the sensor has a smaller diameter than logging tools, resulting in a sensitive volume smaller than that of a well-logging tool. Both factors lead to a low signal-to-noise ratio.

In field measurements, such a tool should be driven automatically, because measurement times are long due to extensive signal averaging for signal-to-noise improvement (Fig. 7.2.6a). Even across a depth range of just one meter, the CPMG decays and the relaxation time distributions vary considerably (Fig. 7.2.6b). This may be attributed in part to the impact of drilling the hole for inserting the pipe. Accordingly, also the initial amplitudes vary and indicate varying moisture content (Fig. 7.2.6c). The data were acquired with a 5 cm diameter prototype logging tool (Fig. 7.2.2c) in a pipe that was shielded around half of its circumference to attenuate external noise (Fig. 7.2.6d). The sensitive slice was 1.8 mm outside the pipe wall. Routine field experiments are becoming possible with sensors of this sort that exhibit higher penetration depth, larger sensitive volume and lower sensitivity to electromagnetic noise [7, 8].

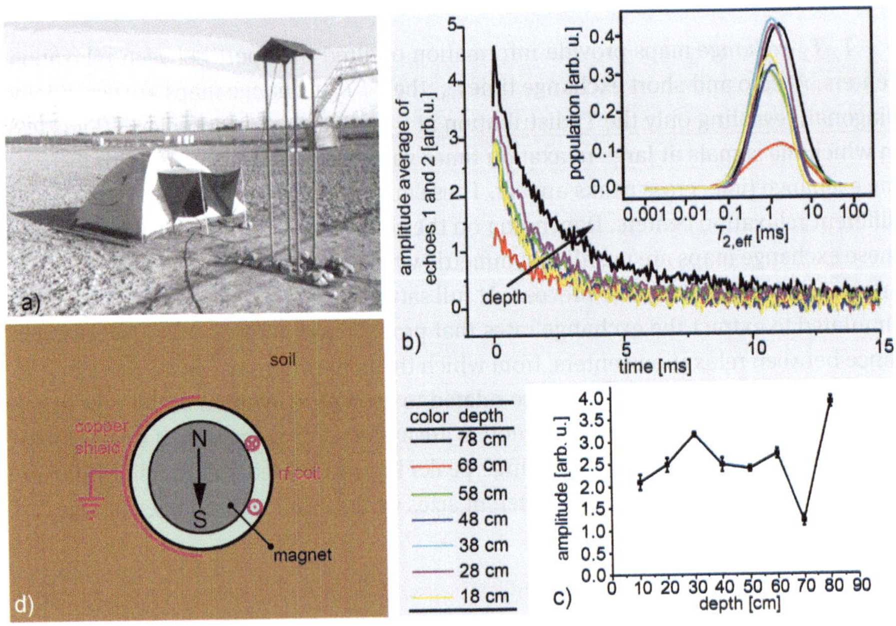

Figure 7.2.6. Field measurements of the water content at Selhausen in the west of Germany with a slim-line logging tool (adapted from [10] with permission). (a) Measurement site showing a scaffold above the borehole in the ground to assist automated probe positioning in a vertical pipe. (b) CPMG echo train envelopes and corresponding relaxation time distributions from different depths. (c) Variation of the signal amplitude calculated as the sum of the first 8 CPMG echo amplitudes versus depth. (d) Schematic drawing of the borehole with the pipe and a grounded copper shield on one side of the pipe to reduce external electromagnetic noise.

7.2.9 Data processing

The CPMG experiment provides the initial amplitude of the echo train envelope that corresponds to the integral under the relaxation time distribution function (Fig. 7.2.3a) and is proportional to the volumetric *water content*. The water content of a fully saturated, rigid porous medium provides the *porosity*, given the spin density of water or the pore-saturating fluid at the sample temperature and the size of the sensitive volume. If these values are unknown, the signal amplitude from the water-saturated soil sample can be normalized to the signal amplitude of pure water measured from a sample of the same size in order to determine porosity. Note that soil samples can swell in contrast to most rock samples.

Given the surface relaxivity ρ_2 of the soil, the *relaxation time distribution* can be rescaled to a *pore size distribution* as long as the pores are small enough so that the fast diffusion limit is fulfilled. If the surface relaxivity is not known, it can be determined by matching the relaxation time distribution with the pore size distribution or parameters thereof measured in a different way, e.g. by porosimetry. While the differential distribution functions, which are proportional to probability density, provide considerable detail on the fluid in different environments of the pore network, the details of the pore space are probed differently by different methods. To determine the relaxivities ρ_1 (eqn. (7.1.3)) and ρ_2 (eqn. (7.1.4)) such details may be omitted. Then, instead of matching probability densities (Fig. 7.1.7) integral distribution functions can be compared (Fig. 7.2.3b,c), which provide the probability of finding pores with diameters smaller than or equal to a particular diameter. Integral distribution functions are obtained from differential ones by integration. They are commonly measured by *mercury intrusion porosimetry* (Hg) and can be derived from *water retention curves* (pF). Mercury intrusion porosimetry is difficult to perform on soil, because high pressures are applied, which disturb the soil structure, and to measure water retention curves takes days to weeks. A faster way of determining ρ_2 is by matching the average T_2 value with the average *surface-to-volume ratio* determined from the nitrogen absorption isotherm following the *BET* method according to Brunauer, Emmett and Teller. Once the surface relaxivities are known for a particular type of soil, pore size distributions can rapidly be determined by NMR relaxation measurements.

7.2.10 References

[1] Roth K. Soil Physics Lecture Notes, Chapter 6. Heidelberg: Ruprecht-Karls-Universität; 2012. (http://www.iup.uni-heidelberg.de/institut/forschung/groups/ts/soil_physics/students/lecture_notes05).

[2] van Genuchten MT. A closed-form equation for predicting the hydraulic conductivity of unsaturated soils. Soil Sci Soc Am J. 1980; 44:892–898.

[3] Mualem Y. A new Model for predicting the hydraulic conductivity of unsaturated porous media. Wat Res Res. 1976; 12:513–522.

[4] Paetzold RF, Matzkanin GA, De los Santos A. Surface soil water content measurements using pulsed nuclear magnetic resonance techniques. Soil Sci Soc Am J. 1985; 49:537–540.
[5] Sucre O, Pohlmeier A, Minière A, Blümich B. Low-field NMR logging sensor for measuring hydraulic parameters of model soils. J Hydrol. 2011; 406:30–38.
[6] Hertrich M. Imaging of groundwater with nuclear magnetic resonance. Progr Nucl Magn Reson Spectr. 2008; 53:227–248.
[7] Perlo J, Danieli E, Perlo J, Blümich B, Casanova F. Optimized slim-line logging NMR tool to measure soil moisture in situ. J Magn Reson. 2013; 233: 74–79.
[8] Walsh D, Turner P, Grunewald E, Zhang H, Butler Jr. JJ, Reboulet E, Knobbe S, Christy T, Lane Jr. JW, Johnson CD, Munday T, Fitzpatrick A. A small-diameter NMR logging tool for groundwater investigations. Ground Water. 2013; doi: 10.1111/gwat.12024.
[9] Stingaciu LR, Weihermüller L, Haber-Pohlmeier S, Stapf S, Vereecken H, Pohlmeier A. Determination of pore size distribution and hydraulic properties using nuclear magnetic resonance relaxometry: A comparative study of laboratory methods. Wat Res Res. 2010; 46:W11510.
[10] Blümich B, Casanova F, Dabrowski M, Danieli E, Evertz L, Haber A, Van Landeghem M, Haber-Pohlmeier S, Olaru A, Perlo J, Sucre O. Small-scale instrumentation for nuclear magnetic resonance of porous media, New J Phys. 2011; 13:015003.

7.3 Cement and concrete

7.3.1 Introduction

Besides natural building materials, such as clay, stone and wood, many man-made products are in use. Among them, *cement* and *concrete* are the most important ones in addition to brick. For example in 2007 over 2.8 billion tons of cement were produced worldwide, most of it for making concrete. The consumption of concrete is only superceded by the human consumption of water. Cement is an inorganic and non-metallic building material, which is mostly employed as binder between stone materials. It is produced in a cement plant from limestone, clay, sand and iron ore. The mixture is fired at temperatures of around 1400° C, and after cooling down it is finely ground with the admixture of gypsum. In a hydration reaction cement hardens and becomes a durable material. The main products of the hydration reaction are insoluble, stable *calcium silicate hydrates* (C-S-H), which form fine needle-shaped crystals (Fig. 7.3.1b). The final high strength of cement and concrete is provided by the tight interlocking of the individual crystallites with each other and with additives such as sand and stones (Fig. 7.3.2).

Concrete is a composite building material from cement and a mixture of gravel or crushed rocks, sand, and water (Fig. 7.3.1a). After casting the fresh liquid concrete into a mold, it is degased with a vibrator to remove most of the air bubbles, which would otherwise weaken the material strength. In the modern age, concrete has become the most important building material for high-rise buildings due to its long life cycle, formability, pressure resistance, and heat resistance. These and other extreme characteristics are achieved with the addition of special additives.

Figure 7.3.1. Concrete and cement. (a) Photo of a cross section through a concrete block showing the main constituents, coarse stones and cement. The field of view is 5 mm × 5 mm. (b) Scanning electron micrograph of a cement particle after three hours of hydration. Visible are young C-S-H needles and $Ca(OH)_2$ platelets grown on the cement particle surface (adapted from VDZ, Zementtaschenbuch 2003, Verband der Zementwerke, Düsseldorf, with permission).

7.3.2 Objective

The main objective in studying building materials like cement, concrete and ceramics is to characterize their *microstructure*, which defines many physical properties like stiffness, hardness, and durability. These properties need to be well understood and optimized for the construction of buildings, roads, and bridges and critically depend on the hydration process and the free and bound water. ^1H NMR can probe the hydrates and the water pools directly and characterize the microstructure in terms of *porosity*, *pore size* distribution, and *pore connectivity* with 1D and 2D relaxation and diffusion methods. When measuring building materials in their natural state, the primary goal is to determine the moisture content (Section 8.1.9). When measuring water-saturated building materials, the objective is to determine the pore size distribution (Fig. 7.1.7) and to study the pore connectivity, e.g. by T_2-T_2 exchange NMR (Fig. 7.2.5).

7.3.3 Further reading

Valori A, McDonald PJ, Scrivener KL. The morphology of C-S-H: Lessons from ^1H nuclear magnetic resonance relaxometry. Cem Concr Res. 2013; 49: 65–81.

McDonald PJ, Mitchell J, Mulheron M. Cement products: Characterization by NMR and MRI. In: Buschow KHJ, Cahn RW, Flemings MC, Ilschner B, Kramer EJ, Mahajan S, Veyssière P, editors. Encyclopedia of Materials: Science and Technology. Amsterdam: Elsevier; 2005.

Mehta PK, Monteiro PJM, editors. Concrete: Structure, Properties and Materials. Englewood Cliffs: Prentice-Hall; 1993.

7.3.4 Theory

Cement hydration

When curing cement, anhydrous cement powder is hydrated by water in a reaction that proceeds in three different periods lasting from minutes up to months (Fig. 7.3.2e). Immediately after mixing cement and water, hydration products are formed, i.e. *Calcium-Silicate-Hydrate* (C-S-H), calciumhydroxide and ettringite start to surround the anhydrous cement particles (clinker) in a thin layer, but their amount is too low to fill the voids between the cement particles so that the mixture is still malleable (Fig. 7.3.2a). This hydration product is referred to as *cement gel*. The average size in cement gel is in the nanometer range. The second period starts after one hour when needle shaped C-S-H crystallites grow on the surface of the clinker particles into the free volume so that the hardness increases and the porosity start to decrease (Fig. 7.3.2b, e). The cement gel is composed of randomly stacked sheets and intercalated water, which is sometimes differentiated as intra- and inter-C-S-H gel pore water. But often the entire water in the cement gel is summarized as gel water. The curing process terminates after some months when a rigid structure has ultimately been formed in the third period (Fig. 7.3.2c,d).

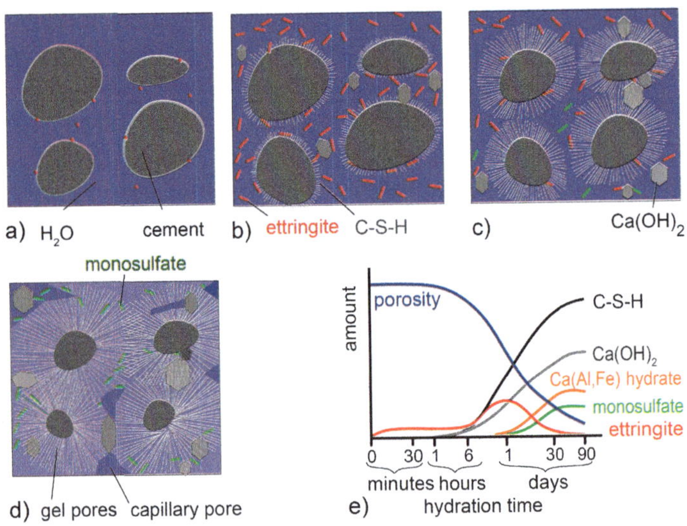

Figure 7.3.2. Schematic drawing of the hydration process of cement. (a) The anhydrous cement particles react with water to Calcium-Silicate-Hydrate (C-S-H), calciumhydroxide and ettringite surrounding the cement particles in thin layers. (b) C-S-H crystallites grow rapidly on the surface of the cement particles into the voids between the particles. (c) The growth rate of the C-S-H particles slows down as the gel layers thicken, and the mixture starts to harden. (d) Final state of the hydration process where the voids are nearly filled with C-S-H, Ca(Al,Fe) hydrate, solid $Ca(OH)_2$ and monosulfate, transformed from ettringite. (e) Temporal development of the hydration products. It can be divided into three different periods.

Depending on the original water-to-cement mass ratio (w/c) water can exist in different phases. Most of the water is bound in the solid matrix as hydration water or crystal water. When this water is removed at high temperature, the cement matrix is destroyed. Other water is bound inside *gel pores* with nanometer diameters. When the w/c ratio exceeds a value of about 0.4, *capillary pores* with diameters larger than some ten micrometers form, that can take up and release water. If the capillary pore fraction exceeds 25 %, the pores become connected and the fluid permeability increases strongly. Furthermore, concrete and mortar exhibit coarse pores filled with air or water. The fraction of these pores highly depends on the processing conditions.

Relaxation theory

The protons of water in porous materials are probed by NMR relaxation and diffusion measurements. An analysis of NMR relaxation data (Sections 3.2.4 and 7.1.4) can provide the concentration of each water population in the porous system and the pore size distribution. Compared to rocks and soils a larger fraction of chemically bound, i.e. hydration water is part of the microstructure of the solid cement matrix. Hence the corresponding transverse relaxation time T_2 is controlled by dipole-dipole interactions and therefore short while T_1 may be much longer. For water in the gel and capillary pores, the fast diffusion limit is valid, which means that the water molecules exchange rapidly between the surface and the bulk pore, and the measured relaxation rates are an average of the bulk and surface relaxation rates (eqns. (7.1.1) and (7.1.2)). Specifically for cement, the local dynamics at the solid-liquid interphase in the pores can lead to a frequency dependence of T_1, which is explained in a model developed by Korb [1].

7.3.5 Hardware

The hardware required for studying cement and concrete is the same as that for rock. In closed magnets of the Halbach type small samples can be investigated in rather homogeneous magnetic field so that the impact of diffusion in external field gradients on the transverse magnetization decay is minimized. Experiments on larger specimens and outdoor measurements are performed with unilateral devices like the Profile NMR-MOUSE® (Figs. 3.1.2b, 3.2.1b, and 7.3.3a) or the *Surface GARField magnet* (Fig. 7.3.3b) [2]. The latter device consists of an array of three magnet blocks with alternating polarization forming a planar Halbach magnet. The rf magnetic field matches the profile of the static magnetic field so that both fields are perpendicular to each other in a large depth region. The depth range of the 20 kg sensor extends to 50 mm above the magnet surface, where the resonance frequency in an extended planar sensitive slice is 3.2 MHz for protons and the constant gradient is only 3.25 T/m. When dry samples of grey cement are studied, the relaxation times can be very short, so that sensors with

Table 7.3.1. Acquisition parameters for measuring transverse relaxation decays of water in grey cement and porous ceramic

Parameter	Value (cement)	Value (ceramic)
NMR-MOUSE	PM25	PM10
transmitter frequency ν_{rf}	13.8 MHz	18.1 MHz
transmitter attenuation for 90° pulse	−6 dB at 700 W	−8dB at 300 W
duration t_p of 90° pulse	8 μs	7 μs
dwell time Δt	1 μs	0.5 μs
acquisition time t_{acq}	5 μs	5 μs
echo time t_E	140 μs	60 μs
number n_E of echoes	256	3250
recycle delay t_R	1 s	1 s
number n_s of scans	256	512

short dead times need to be employed. Such sensors typically have high field and small coils. In this case, desktop instruments with closed magnets are preferably employed (Fig. 7.1.2a,b).

7.3.6 Pulse sequences and parameters

Grey cement is contaminated with relaxation centers that make it hard to detect signal at all from material with natural moisture content, because the transverse magnetization largely relaxes within the dead time of the sensor. This is why usually only the tail of the free water signal can be detected either with the very first few echoes of a CPMG echo train measured with a stray-field sensor or with an *FID* in a closed magnet with moderate field inhomogeneity. At high magnetic field, *bound water* can be measured as well due to shorter dead time. To detect signal from *gel water* and crystal water in addition to signal from *free water* with compact NMR sensors at low field, most studies are conducted on white cement, where paramagnetic relaxation centers are far more dilute and the transverse relaxation times are longer.

The decision on the particular choice of the signal detection method depends on the available hardware and the specific aim of the investigation (Section 7.2.6). Short relaxation components are acquired with an FID or with solid echoes (Fig. 3.1.4, Fig. 5.1.5b), whereas long relaxation components are acquired with the CPMG sequence (Fig. 5.1.5a) to determine amplitude and effective transverse relaxation time, or with a saturation or inversion recovery sequence (Fig. 3.2.2) appended with a CPMG sequence (Fig. 3.2.4a) to determine the longitudinal relaxation time. Typical parameters for measuring relaxation decays in cement are summarized in Tab. 7.3.1.

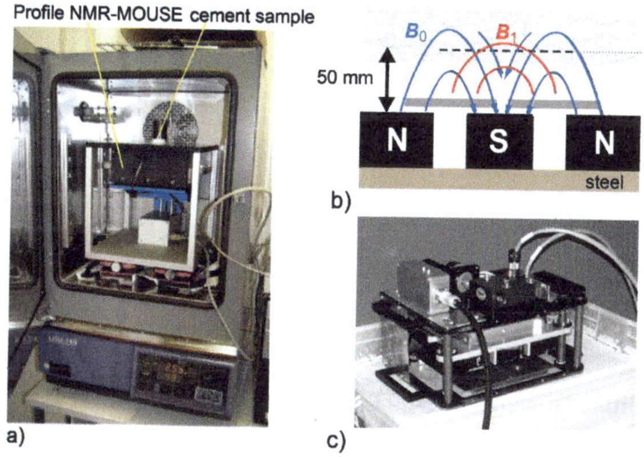

Figure 7.3.3. Hardware for measuring cement. (a) Profile NMR-MOUSE PM25 on a lift inside a climate chamber. (b) Schematic drawing of the surface GARField magnet. The lines of the permanent magnetic field B_0 and the lines of the rf field B_1 are perpendicular to each other in a wide depth range above the magnet. (b) Photo of the surface GARField magnet positioned on a large concrete block (adapted from [2] with permission).

7.3.7 Beginner's level measurements

Hardening of cement paste

^{1}H NMR relaxation has been studied extensively to learn about the water dynamics and the cement-water interactions, which are fundamental to the hydration process. Even today they are not fully understood. Based on the three-phase relaxation model of Schreiner [3] longitudinal and transverse relaxation develop as follows: Directly after mixing dry cement with water the single bulk components of T_1 and T_2 split into several components, which cover a time range from microseconds up to tens of milliseconds. The shortest T_2 component belongs to water bound in solid phases, mainly in calcium hydroxide. This water is totally immobile and exhibits T_1 values longer than 100 ms. Water in gel pores is more mobile and its relaxation times are in the range of 0.5 – 1.0 ms due to enhanced surface interaction. As the gel shrinks, capillaries may develop, which collect water with the longest T_2 of several milliseconds.

For example, while curing a cement paste prepared with a w/c ratio of 0.5, after 1 h two relaxation components are still visible (Fig. 7.3.4a). They can be assigned to hydration water and mobile water. Already after one day, a third component develops. With the hydration reaction proceeding, the relaxation peaks shift to shorter times. After 28 days three peaks are observed at 9 μs, 80 μs, and 350 μs. They are assigned to chemically bound water, water in the gel pores, and capillary water, respectively [4]. The volume fractions of the different constituents of fully-hardened Portland cement produced from paste with a water-to-cement ratio of 50% in a sealed container are 40% cement, 52.5% total water, i.e. bound water, gel water, and capillary water, and

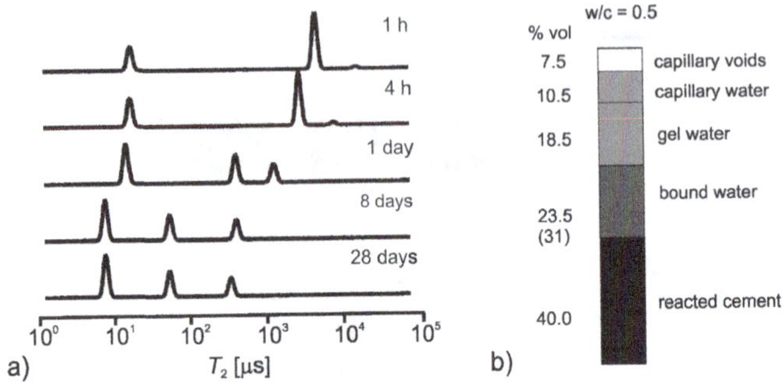

Figure 7.3.4. Hardening of Portland cement paste. (a) Distribution functions of the transverse relaxation time from a cement paste with a w/c ratio of 0.5 at different hydration times. (b) Calculated volume fractions of fully hardened cement. The value in brackets refers to the original volume of bound water before shrinking (adapted from [4] with permission).

7.5% capillary voids (Fig. 7.3.4b). The seeming increase in water content results from shrinkage of the sample upon hardening [4].

Moisture content in concrete

The signal amplitude is proportional to the proton density and thus the *moisture content* (Section 8.1.9) in cement. But ordinary grey Portland cement is highly contaminated with *paramagnetic impurities*, so that the transverse magnetization relaxes extremely fast, and quantitative measurements are highly demanding. When attempting to determine the moisture content in concrete, the measurements are further aggravated by the presence of stones in the sensitive volume, which take up less moisture than the cement between them (Fig. 7.3.1a). Then signal amplitude is no longer a representative measure of the overall moisture content of the material. Nevertheless, water signal could be detected with the PM5 NMR-MOUSE up to 25 mm depth in the concrete structures of a traffic tunnel and a lock (Fig. 3.1.2b) [5].

Reference samples from these sites were subsequently investigated at different depths in the laboratory in a wetting and drying study (Fig. 7.3.5). When plotting the moisture content derived from the signal extrapolated to zero time versus effective relaxation time for the two types of concrete, the data scatter in only one diagonal half of the graphs. This scatter arises from different signal amplitudes for each relaxation time due to the presence of a varying rock fraction in the sensitive slice of the sensor. The limiting line at maximum amplitude reveals a relationship between effective relaxation time $T_{2\text{eff}}$ and signal amplitude in pure cement, correlating $T_{2\text{eff}}$ and moisture content. The relationship is linear at total water concentrations higher than that of bound water. The slopes of the lines and the amplitude offsets are characteristic for the type of cement. This result is in agreement with the concept that mobile water molecules explore the pore space in the fast exchange limit [6], and it suggests that

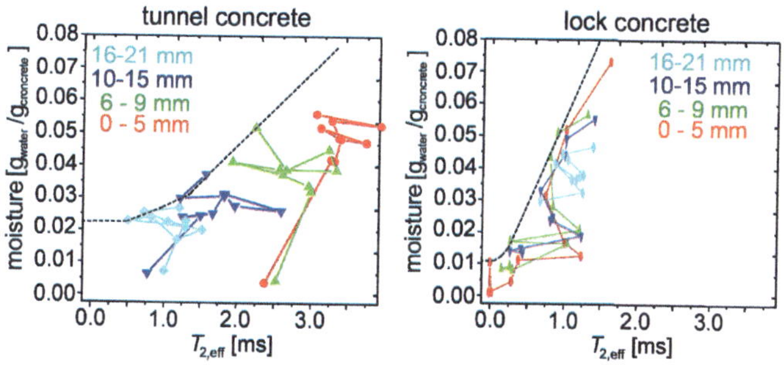

Figure 7.3.5. Plots of signal amplitude calibrated to moisture content versus effective relaxation time for two types of concrete from grey cement (adapted from [5] with permission). The dotted lines are drawn as guides for the eye.

moisture content of cement in concrete should be determined from the effective relaxation time and not the signal amplitude.

Porosity of a ceramic filter

Diesel particulate filters were introduced in the year 2000 when the first automobile manufacturer equipped the standard diesel-engine cars with such a filter. It reduces particles from soot and unburned hydrocarbons in the exhaust of diesel engines. The filter consists of a block of parallel rectangular channels with porous ceramic walls of silicon carbide through which the exhausts gases are forced to pass (Fig. 7.3.6a). The pore size distribution of the ceramic wall is one of the most important factors for the function of the filter. It is routinely determined by *mercury intrusion porosimetry* (MIP). This method is destructive, expensive, time consuming, and generates toxic waste. The same and even more information can be obtained from transverse relaxation decays and depth profiles measured with the NMR-MOUSE. The channels and the pore

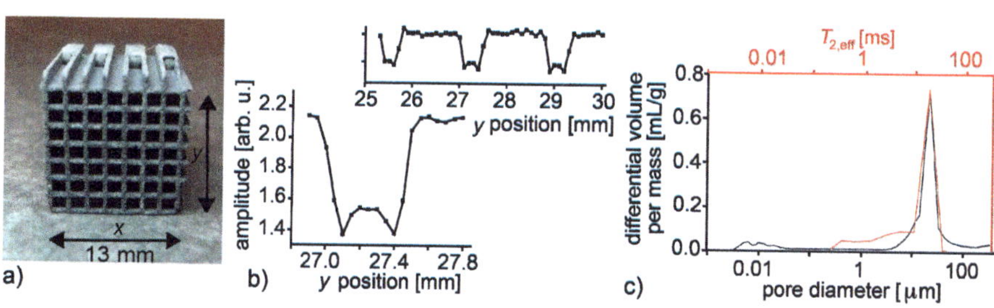

Figure 7.3.6. Mapping porosity in the walls of a water-soaked diesel particle filter. (a) Photo of a section of the filter showing the channels for the exhaust gas and the porous walls. (b) Depth profiles across one wall (bottom) and three walls embedding two channels (top). (c) Comparison of the $T_{2\text{eff}}$ relaxation time distribution and the differential volume curve from mercury intrusion porosity (adapted from [5] with permission).

Table 7.3.2. Common issues encountered when measuring cement-based materials

- The object is macroscopically heterogeneous and the sample is to small too be representative for the object
- The sensitive volume is too small for the signal to give a representative average of the heterogeneous material
- The signal decays rapidly within the echo time due to paramagnetic impurities in the material
- When measuring natural moisture content, room temperature and humidity need to be carefully controlled and the object equilibrated to the climatic conditions
- When measuring porosity, the pore space needs to be fully liquid saturated, preferably by pumping vacuum on the sample
- Note, that the relaxation time distribution only maps the pore-size distribution in the fast diffusion limit
- Note, that the inverse Laplace transformation becomes unstable at low signal-to-noise ratio

space need to be fully saturated with water. A depth profile produces high signal in the channels, where the porosity is 100%, and low signal across the walls (Fig. 7.3.6b). The *depth profile* further reveals that the porosity is lower at the wall surfaces than in the interior. This difference is hard to measure with MIP. On the other hand, the distribution of $T_{2\mathrm{eff}}$ determined by Laplace inversion of a CPMG decay matches the pore-throat distribution determined by MIP well (Fig. 7.3.6c). Consequently the time-consuming and hazardous MIP measurements are not needed for quality control of diesel particulate filters.

Following the hydration kinetics of cement, analyzing the moisture content of concrete and determining porosity and pore size distributions in porous objects with the NMR-MOUSE require only the use of multi-echo sequences. The major difficulty encountered when studying standard building materials arises from the impact of impurities in the material, which enhance the transverse relaxation rate and can give rise to very short relaxation decays, so that only mobile but not bound water is detected. Common issues encountered when measuring cement-based materials are summarized in Tab. 7.3.2.

7.3.8 Advanced level measurements

Hardening of cement paste

The hardening of cement paste strongly depends on temperature and humidity. Compact stray-field NMR is well suited for tracking changes in the water content profile upon drying at different temperatures and humidity, because the sensor can be put right into the climate chamber together with the object (Fig. 7.3.3a). Depth profiles were measured for a white cement paste at a w/c ratio of 0.4 with a PM25 NMR-MOUSE across a depth range of 22 mm in 1 mm steps at exponentially spaced hydration time intervals from 1 to 32 hours. With temperature increasing from 5° C to 35° C the water

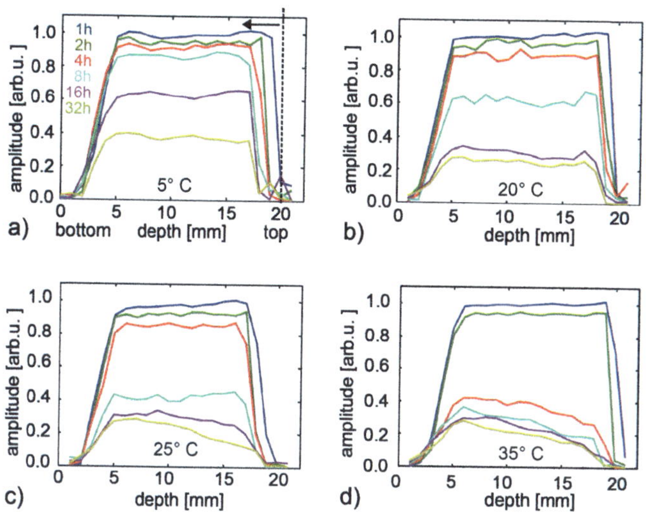

Figure 7.3.7. Hardening of white cement paste with a w/c ratio of 0.4 studied with the Profile NMR-MOUSE in terms of depth profiles inside climate chamber. Depth profiles were measured at 1 h, 2 h, 4 h, 8 h, 16 h and 32 h hydration time and temperatures of 5 ° C (a), 20 ° C (b), 25 ° C (c) and 35 ° C (d). Experimental parameters: CPMG echo trains, $t_E = 140\,\mu s$, $n_E = 128$, $t_R = 1\,s$, $n_s = 48$ for hydration times up to drying time of 4 h. For longer times, $n_E = 64$, $t_R = 0.7\,s$. The amplitudes were determined from fits of the CPMG echo trains with mono-exponential functions. The arrow in (b) points at the receding evaporation front on the top of the sample (adapted from [7] with permission).

content in the interior of the sample decreases and the profiles remain mostly flat in this region at $T < 25°C$ (Fig. 7.3.7a,b). Given the minimum echo time of $t_E = 140\,\mu s$, only the *capillary pore water* with T_{2eff} larger than 0.5 ms is detected and defines the profile amplitude. A flat profile thus means that the capillary water is homogeneously distributed and can evenly be transformed into gel pore and bound water during the hydration process. Furthermore, it was observed that the top layer of the sample dries with increasing time (arrow in Fig. 7.3.7a). The homogeneous curing conditions change at temperatures above 25° C. The flat region in the profiles is maintained only for a few hours. At later times the profile amplitude decreases towards the surface indicating a loss of capillary water (Fig. 7.3.7c,d) [7].

Another parameter to characterize the kinetics of the *hydration process* is the inflection point of signal amplitude versus the curing time (Fig. 7.3.8a). In the temperature interval between 5° C and 35° C it shifts towards shorter curing times with increasing temperature. The relaxation time T_{2eff} decreases along with it (Fig. 7.3.8b). At each temperature, amplitude and relaxation time closely follow the same relationship, which is described by a power law with an exponent of three (Fig. 7.3.8c). This is consistent with the fact that the volume of the water-filled capillary pores scale with the third power of the pore radius r and T_2 scales with the surface-to-volume ratio, i.e. with the first power of r in the fast diffusion limit according to eqn. (7.1.4).

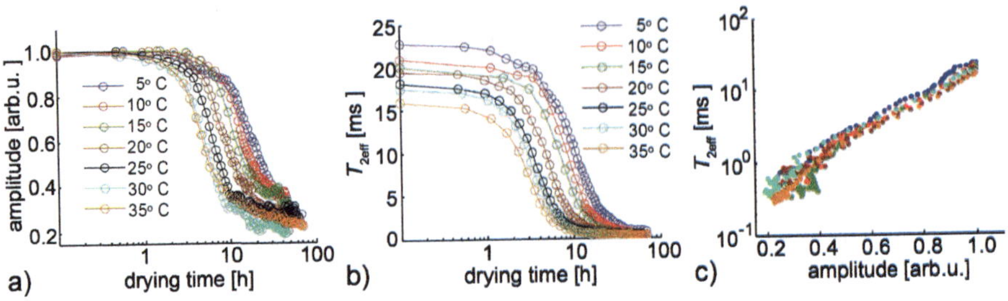

Figure 7.3.8. Drying kinetics of white cement with c/w = 0.4 at 10 mm depth from the surface. Amplitudes and relaxation times extracted by fitting mono-exponential functions to CPMG echo trains acquired with 128 echoes up to 4 h drying time and 64 echoes at longer time. (a) Amplitude versus drying time at different temperatures. (b) Effective transverse relaxation time $T_{2\mathrm{eff}}$ versus drying time. (c) The correlation of $T_{2\mathrm{eff}}$ with amplitude reveals a universal relationship (adapted from [7] with permission).

2D Laplace NMR

Most of *two-dimensional relaxation correlation* studies on cement hydration have so far been performed in homogeneous magnetic field with bench-top magnets [9]. But in view of the growing interest in on-site measurements in a built environment 2D T_1-T_2 correlation experiments (Fig. 3.2.4a) were investigated to monitor the hydration reaction of white cement with a w/c ratio of 0.4 in a feasibility study with the unilateral surface GARField magnet (Fig. 7.3.3c). After two days of hydration, the correlation maps show a single peak with $T_1 = 2.25\, T_2$ as predicted by Korb [1]. Such short relaxation times are typical for water in gel pore structures (Fig. 7.3.9 left). With increasing curing time the peak splits into two components (middle) and after 7 days three distinct peaks are observed (right). The peak with the longest relaxation time $T_1 \approx 40$ ms can be assigned to capillary pores formed by shrinking of the cement paste. Although these measurements done in an inhomogeneous field do not show the details measur-

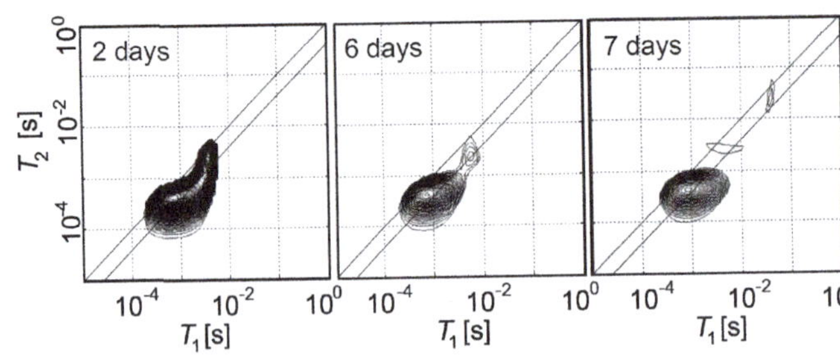

Figure 7.3.9. Hydration study of white cement paste with w/c ratio of 0.4 using the surface GARField sensor. The time series of T_1-T_2 relaxation-correlation maps reports the evolution of the relaxation time distributions during the hydration process. The water signal from the gel pores splits after two days into two signals and after 6 days into three signals. The two diagonal lines indicate $T_1 = T_2$ and $T_1 = 2.25\, T_2$ (adapted from [9] with permission).

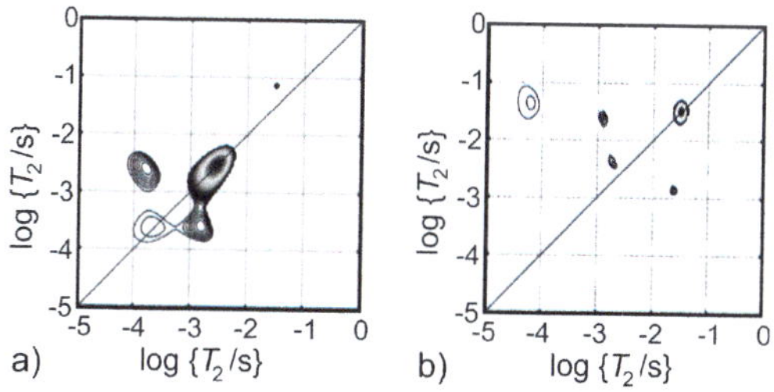

Figure 7.3.10. T_2-T_2 relaxation exchange maps of measured with a bench-top spectrometer at 20 MHz. (a) White Portland cement. The peaks on the diagonal correspond to relaxation in narrow pores and show an exchange of water indicated by the off-diagonal peaks. (b) Pure C-S-H. The synthesized sample shows the same exchange pattern but shifted to longer relaxation times due to higher sample purity (adapted from [8] with permission).

able in a homogeneous field, essential features of the hydration process are observed so that *in situ* measurements of cement hydration on model walls with realistic moisture dissipation become a viable option.

Pore connectivity is probed by monitoring water exchange between different environments in 2D T_2-T_2 exchange experiments (Fig. 3.2.4c). White Portland cement has been studied by T_2-T_2 exchange NMR after hardening for one day at 20° C [8]. The relaxation exchange map measured with a mixing time of 3 ms reveals two peaks on the diagonal at T_2 values of 0.2 ms and 3 ms (Fig. 7.3.10a). They are attributed to water in different pores in the cement gel without a clear identification on the particle and pore size scale. It is speculated that the two types of pores refer to C-S-H interlayer and gel pore water. Clearly visible are off-diagonal peaks due to exchange between the two types of pores. The same experiment performed on a highly pure, synthetic C-S-H sample shows that the exchange pattern remains constant, but due to lower impurity concentration it is shifted to longer relaxation times (Fig. 7.3.10b).

7.3.9 Data processing

The data from cement, concrete and other porous materials are processed similar to those from rock (Section 7.1.9) and soil (Section 7.2.9). *Relaxation curves* are analyzed by either 1D or 2D inverse Laplace transformation and plotted as distribution functions or correlation and exchange maps for T_1-T_2 and T_2-T_2 experiments, respectively. They can be scaled to pore size distribution functions if the surface relaxivities ρ_1 and ρ_2 are known and constant for different pore sizes and environments. For quantification, the studies should be done with devices that are capable of measuring the short transverse relaxation components of the solid phases and bound water.

7.3.10 References

[1] Korb JP. Nuclear magnetic relaxation of liquids in porous media. New J Phy. 2011; 13:035016.

[2] McDonald PJ, Aptaker PS, Mitchell J, Mulheron M. A unilateral NMR magnet for sub-structure analysis in the built environement: The Surface GARField. J Magn Res. 2007; 185:1–11.

[3] Schreiner LJ, Mactavish JC, Miljkovic L, Pintar MM, Blinc R, Lahajnar G, Lasic D, Reeves LW. NMR line–shape spin–lattice relaxation correlation study of Portland cement hydration. J Am Ceram Soc. 1985; 68:10–16.

[4] Bohris AJ, Goerke U, McDonald PJ, Mulheron M, Newling B, Le Pagea B. Broad Line NMR and MRI study of water and water transport in Portland cement pastes. J Magn Reson Imag. 1998; 16:461–544.

[5] Blümich B, Casanova F, Dabrowski M, Danieli E, Evertz L, Haber A, Van Landeghem M, Haber-Pohlmeier S, Olaru A, Perlo J, Sucre O. Small-scale instrumentation for nuclear magnetic resonance of porous media, New J Phys. 2011; 13:015003.

[6] Halperin WP, Jehng JY, Song YQ. Application of spin-spin relaxation to measurement of surface area and pore size distributions in a hydrating cement paste. Magn Reson Imag. 1994; 12:169–73.

[7] Van Landeghem M, d'Espinose de Lacaillerie JB, Blümich B, Korb JP, Besson B. The roles of hydration and evaporation during the drying of a cement paste by localized NMR. Cem Concr Res. 2013; 48:86–96.

[8] Valori A, Rodin V, McDonald PJ. On the interpretation of ^{1}H 2-dimensional NMR relaxation exchange spectra in cements: Is there exchange between pores with two characteristic sizes or Fe^{3+} concentrations? Cem Concr Res. 2010; 40:1375–1377.

[9] McDonald PJ, Mitchell J, Mulheron M, Montheilhet L, Korb JP. Two-dimensional correlation relaxation study of cement pastes. Magn Res Imag. 2007; 25:470–473.

8 Cultural heritage

The study of objects of *art* and *cultural heritage* requires nondestructive tools of analysis fulfilling demands similar to those posed by medical diagnostics. *Magnetic resonance imaging* (*MRI*) and *X-ray computer tomography* (*CT*) are the most popular imaging methods in medicine. In *art diagnostics* X-ray analysis is well established but not MRI. This is largely due to the fact that MRI requires bulky magnets, which can neither be moved from the laboratory to the measurement site nor can MRI generally analyze objects of arbitrary size. The use of the NMR-MOUSE is not constrained by these limitations. The sensor is small, *portable*, and provides imaging information similar to that of MRI, albeit on a pixel-to-pixel basis, where each *pixel* requires a separate measurement. Compact NMR is a recent addition to the growing number of *portable* instruments for *nondestructive analysis* of art and cultural heritage objects. It competes with other techniques like *FTIR* and *Raman spectroscopy*, *UV imaging*, *X-ray-diffraction*, *-fluorescence*, and *-radiography* analyses, *multi-spectral imaging*, *optical coherence tomography*, and *terahertz spectroscopy*. Each of these techniques uses *electromagnetic waves* at different frequencies and consequently with different absorption properties. NMR is the least sensitive one, because the frequency is lowest. Consequently the focus volume or sensitive volume is large to enhance the signal and the spatial resolution is low. But opposed to most other techniques, *NMR* works well in the presence of moisture and provides an abundance of contrast features like *concentrations*, *relaxation times* and *diffusion coefficients*.

Depending on the type of sensor, an average signal is derived from a flat slice as thin 10 μm and with a lateral extension of about 1 cm^2 to 20 cm^2. Thus the sensor needs to be properly aligned so that the sensitive slice resides inside the region of interest if amplitude information is required and parallel to the layer structure if the *stratigraphy* is to be resolved. When the layers curve across the *sensitive slice* the depth resolution will be reduced. In some cases different signal contributions may be disentangled when decomposing the *CPMG decay* into components with different relaxation times.

8.1 Painted walls and stone

8.1.1 Introduction

Walls are painted using pigments with and without binder. When painted with binder, the technique is called *secco*, meaning dry, and the painting is drawn on the dry wall. When painted without binder, the technique is called *fresco*, which derives from the Italian word 'affresco', meaning 'fresh', and the painting is drawn on fresh mortar. The art of fresco painting goes far back, and there are several different fresco-making methods developed by different cultures from east to west. China and Egypt were among the first civilizations to use finely finished wet clay walls to paint. Minoans adapted some techniques of the Egyptians to make buon frescoes or true frescoes. Greeks and Romans in Herculaneum and Pompeii refined the art of fresco making, and the Pompeii style is closely related to the Italian renaissance fresco technique. Pompeii style *wall paintings* are made from several layers of lime mortar with grain sizes typically becoming finer towards the painted surface layer as described by Marcus Vitruvius Pollio in De Architectura, L. VII, c. III 7, 33–20 BC. This structure helps to maintain the delicate moisture balance required for a day's work of painting, can produce a shiny finish of the surface, and results in a durable wall surface.

8.1.2 Objective

The interest in measuring painted walls and stone by stray-field NMR is to learn about their state of conservation and moisture content, and to unravel the stratigraphy of the wall. The *stratigraphy* may have been established in the making of the object, be the result of aging and deterioration by weathering and biological attack, or a consequence of restoration measures. For example, with the NMR-MOUSE, the variation in the *pore structure* can be assessed as a function of depth to analyze the effect of consolidation treatment [1], and the mortar layers supporting a fresco can be fingerprinted.

8.1.3 Further reading

Capitani D, Di Tullio V, Proietti N. Nuclear Magnetic Resonance to characterize and monitor Cultural Heritage. Prog Nuc Magn Reson Spectr. 2012; 64:29–69.

Blümich B, Casanova F, Perlo J. Mobile single-sided NMR. Prog Nucl Magn Reson Spectrosc. 2008; 52:197–269.

8.1.4 Theory

When painting a *fresco*, the color pigments are directly applied to the freshly laid lime plaster where the pigments are embedded in the calcium hydroxide forming from the reaction of lime with water. Subsequently the calcium hydroxide reacts with carbon dioxide to form calcium carbonate. To maintain a moist wall for painting for an extended time period, several mortar layers are applied with different consistencies. The first mortar layer is allowed to partially dry, and then further mortar layers of the same or finer aggregate are applied. Each layer may vary not only in thickness and composition but also in the pressure treatment of the wet mortar. Both, composition and pressure, determine the pore space of the particular layer, and the detailed layer structure is a fingerprint of the craftsmen's school. The NMR-MOUSE gives unique information

Figure 8.1.1. NMR-MOUSE PM25 with 25 mm depth range set up on a scaffold to measure diffusion in a moist wall in Herculaneum, Italy, which shows salt deposits on the surface (access to the fresco was provided by the Herculaneum Conservation Project).

about the layer structure of frescoes by probing the *moisture distribution* and *mobility* across several layers. Moreover the sensor may help to find *preservation treatments* from the past, which are rarely documented and often are of organic nature. To measure the mortar layer structure, a sufficient amount of water must be present in the wall to provide the signal. Similarly, stone can only be measured when it is sufficiently wet.

8.1.5 Hardware

The natural *moisture content* in walls is usually low. This is why an *NMR-MOUSE* PM25 with 25 mm maximum depth access is recommended, because it has a large sensitive volume (Figs. 3.2.1 and 8.1.1). Spacers between the magnet and the rf coil should be inserted to reduce the maximum depth access to the minimum value needed, because this increases the sensitivity and shortens the measurement time. However, the depth resolution of such a modified sensor is lower than that of smaller sensors with lower maximum depth ranges. To scan depth profiles into vertical walls, the NMR-MOUSE is mounted on a step-motor-controlled precision sled (Fig. 8.2.1), which is a variant of the conventional lift (Fig. 1.2.2b).

8.1.6 Pulse sequence and parameters

Wall paintings, *stone*, and other porous materials are investigated nondestructively with the NMR-MOUSE employing the CPMG sequence (Fig. 3.1.1b). Dry walls contain only bound water, which gives rise to weak and rapidly decaying CPMG signals with only a few echoes at short echo time. Water confined in large pores leads to slow decays with thousands of echoes. The signal decay of water in brick and concrete is often enhanced by fluid contact with paramagnetic relaxation centers in the material, and the transverse relaxation decays are shorter. *Relaxation time distributions* require complete CPMG decays to be recorded, and many echo pulses need to be applied at short echo time (Tab. 8.1.1). The energy dissipated in the coil is likely to heat up the coil, so that the recycle delay needs to be extended for the coil to cool down between scans. *Concentrations*, on the other hand, can be determined from the first few echoes of a CPMG signal. Such measurements are quick, because the recycle delay only serves for the longitudinal magnetization to recover to its thermodynamic equilibrium value. The parameters needed to measure signal proportional to spin density are the same as those provided in Tab. 8.1.1, except that the number n_E of echoes may be reduced to 10 and the recycle delay t_R to 300 ms, depending on T_1.

8.1.7 Beginner's level measurements

A relaxation analysis provides information about material properties in terms *component amplitudes*, *relaxation times* and *diffusion coefficients*. In most cases, it provides information about content and state of water and organic constituents. *Organic constituents* may exist as a result of *conservation treatments* with waxes and varnishes. *Spin density* may be converted to moisture or wax content by extrapolating to the amplitude of the CPMG signal from the volume of the sensitive slice to zero time (Fig. 3.1.4). Variations of the profile amplitudes, relaxation times and diffusion coefficients reveal different layers and provide benchmarks to determine the thickness of different layers. To this end, the wall section often needs to be moisturized with deionized water. Repetitive measurements at different times and measurements of the molecular self-diffusion coefficient can provide information not just on moisture content but also on *moisture mobility* (Section 3.2.8). Barrier layers at interfaces between layers can be detected in this way. Such barriers may arise from hidden layers of paint or multiple cement layers, which can be detected by *depth profiling* of moisture content and moisture migration. Original and restored sections of wall paintings will also give rise to differences in depth profiles, be it from consistency of the mortars, moisture contents, or adhesive used to glue fractured parts back into place.

Moisture distribution maps

One of the simplest NMR measurements is to measure signal amplitude. If the signal decay is long compared to the minimum echo time, the signal amplitude is well approximated by the amplitude of the *Hahn echo* (Fig. 3.2.3a) or the sum of the first few echoes of a *CPMG echo train* (Fig. 3.1.1b). Otherwise the signal amplitude needs to be determined by extrapolation of the CPMG signal to time zero (Fig. 3.1.4b).

Capillary water transport through painted walls damages the painting because soluble salts are carried along, which crystallize at the surface as the water evaporates (Fig. 8.1.1). Because the moisture flux is proportional to the moisture content, the

Table 8.1.1. Acquisition parameters for measuring signal amplitudes of CPMG echo trains with the NMR-MOUSE from partially dry walls

NMR-MOUSE®	PM 5	PM 10	PM 25
transmitter frequency ν_{rf}	17.1 MHz	18.1 MHz	13.8 MHz
transmitter attenuation for 90° pulse	−8 dB at 300 W	−8 dB at 80 W	−6 dB at 300 W
duration t_p of 90° pulse	5 μs	7 μs	6–22 μs (spacer dependent)
dwell time Δt	0.5 μs	0.5 μs	0.5 μs
echo time t_E	40 μs	70 μs	50 120 μs (spacer dependent)
number n_E of echoes	64	32	128
recycle delay t_R	1 s	1 s	1 s
number n_s of scans	512	1024	512

moisture content is usually taken as a first indicator for potential damage. The moisture distribution of the wall painting ‘St. Clement at mass and the legend of Sisinnius’ located at the second hypogeous level in St. Clement Basilica in Rome six meters below road level has been mapped with *Hahn echoes* acquired at short echo time [2]. It is strongly deteriorated due to water rising by capillary forces from water flowing under the foundation of the Basilica. The moisture distribution was mapped quantitatively following calibration of the Hahn echo integral to *gravimetric moisture content*. The moisture content at a depth of about 1 mm rose up to 13 % and was affected also by the climatic conditions in the room and the presence of salt efflorescence (Fig. 8.1.2a). The distribution at a depth of 5 mm showed maximum moisture content of 8 % closer to the floor marking the path of the rising water (Fig. 8.1.2b). Note that low moisture content may indicate detached mortar layers where moisture transport from the wall to the surface is hindered by an air cushion [3].

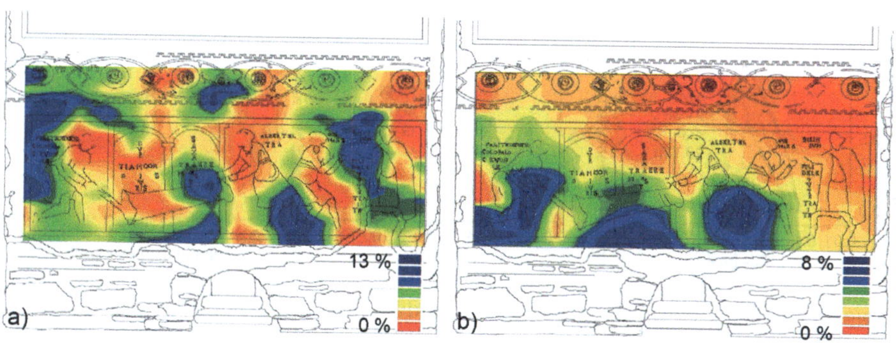

Figure 8.1.2. Moisture distribution maps of the wall painting ‘St. Clement at mass and the legend of Sisinnius’ in St. Clement Basilica in Rome. (a) At 1 mm depth. The moisture content scales from 0 to 13%. (b) At 5 mm depth. The moisture content scales from 0 to 8% (adapted from [2]).

Relaxation time distributions

The relaxation time distribution from a liquid in a porous medium is obtained by *Laplace transformation* of CPMG decay curves. When saturated with liquid, the relaxation time distribution reports about the pore size distribution of the medium and is a fingerprint of the state of the rock, the brick or mortar material. It changes with the manufacturing process, degradation and applied conservation measures. If the relaxation time distribution is measured in the absence of a gradient and if the fast diffusion limit applies, the relaxation time distribution and the pore size distribution are proportional to each other. Following eqn. (3.2.9), water molecules diffuse about 10 μm in 1 s in bulk water near room temperature, so that as a rule of thumb, the fast diffusion limit applies to pores smaller than 10 μm. This is often fulfilled for natural rock and many building materials. Although the gradient of the NMR-MOUSE cannot be turned off, the relaxation time distribution measured with the NMR-MOUSE maps the *pore size distribution* despite signal attenuation by diffusion, but the relationship between pore size and *effective relaxation* time is nonlinear. The signal attenuation from diffusion in the field gradient of the sensor is larger, the bigger the pore is, so that the relaxation time distributions measured in field gradients are squeezed at large relaxation times with reference to those measured in homogeneous field (Fig. 7.1.3c).

The standard way to characterize the pore space of solid porous matrices is *mercury-intrusion porosimetry* (*MIP*). This method, however, does not produce the pore size distribution but the *pore-throat distribution*. Both methods, NMR and MIP, have been compared for a selection of five ancient Roman bricks [4]. Each brick shows a different distribution. The peaks of the NMR *relaxation time distributions* measured with an NMR-MOUSE (Fig. 8.1.3a) follow the same order as the peaks of the *MIP pore-throat distributions* (Fig. 8.1.3c) except for brick number 2, which had ferromagnetic impurities. Interestingly, the agreement between the shapes of the distributions from the NMR-MOUSE and MIP is better than that between the distributions from the NMR-MOUSE (Fig. 8.1.3a) and the one measured in homogeneous field (Fig. 8.1.3b).

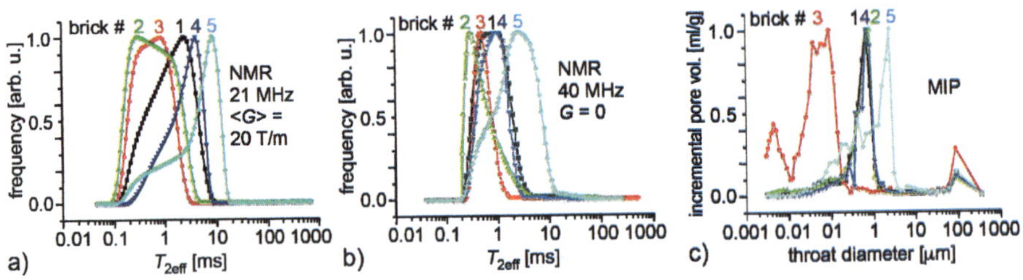

Figure 8.1.3. Characterization of the pore-size distributions of five Roman brick samples (adapted from [4] with permission). (a) $T_{2\mathrm{eff}}$ distributions obtained by Laplace transformation of CPMG signals measured with an NMR-MOUSE at 21 MHz with an average gradient of 20 T/m from water inside the pores. (b) $T_{2\mathrm{eff}}$ distributions measured in a largely homogeneous magnetic field at 40 MHz. (c) Differential pressure curves from conventional mercury intrusion porosimetry (MIP).

Table 8.1.2. Common issues encountered when measuring wall paintings

- The wall may not be flat resulting in lower depth resolution. The sensor needs to be adjusted so that it conforms to the irregularities of the wall surface for maximum resolution
- The sensor presses against the wall. As the lift moves back and the scaffold moves forward, the slice essentially does not move inside the wall
- The echo time is too long to detect rapidly relaxing magnetization components from liquids in small pores
- The number of echoes is high and the recycle delay short so that the rf coil heats up
- The temperature of the sample changes because the magnet temperature or the room temperature are not controlled
- T_1 is estimated inaccurately, and the repetition time is too short
- The number of echoes in saturated, partially saturated, and almost dry walls may vary by 2 to 3 orders of magnitude

The apparent agreement between NMR-MOUSE and MIP distributions, however, is believed to be accidental, because none of them correctly maps the pore size distribution. Nevertheless, the relaxation time distributions measured with the NMR-MOUSE are excellent fingerprints of these materials and can be used to study their origin, their degradation pathways, and their conservation states. Note that fluid-saturated porous media provide long relaxation decays, so that many echoes need to be acquired, and heating of the rf coil may be a problem (Tab. 8.1.2).

Depth profiles

NMR depth profiles can be measured when mobile protons are present, which occur in binder, wax and fluids. When the wall or the building material is dry, it is often acceptable to spray it with deionized water or, if a fragment is available, to soak the fragment in water. Then, for example, the application depth of *stone strengthener* (Fig. 8.1.4a) and *hydrophobic treatments* [1] can be probed. In the treated depth region near the surface, the relaxation times are shorter than further below at depths that the consolidant has not reached. The modification of the pore space also shows up in a change of the relaxation time distribution of moisturized stone when comparing treated and untreated regions. This has been demonstrated in a study of treated and untreated sandstone window frames of Paffendorf Castle in Germany (Fig. 8.1.4b,c).

The stratigraphy of frescoes is more challenging to study than the effects of stone strengthener, because most frescoes have undergone several restoration cycles and have often been treated with wax. While wax at the surface and the adhesive in detached frescoes can readily be detected with the NMR-MOUSE, *wax* on intact frescoes hinders the uptake of water sprayed on the surface, which is needed to map the stratigraphy of the mortar layers. The *stratigraphy* of untreated frescoed walls has been studied at the Villa of the Papyrus in Herculaneum [6], and fresco fragments

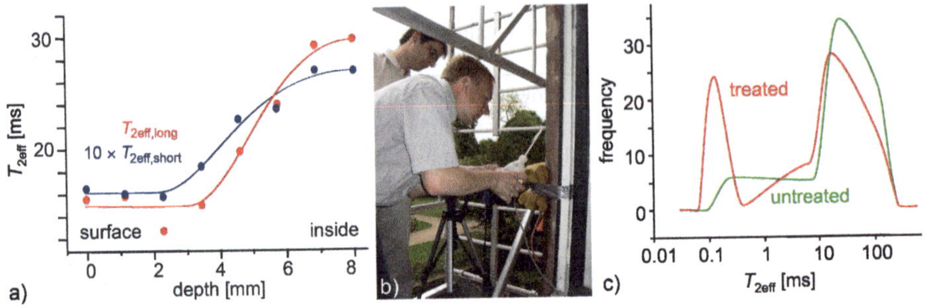

Figure 8.1.4. Probing stone conservation treatment with the NMR-MOUSE. (a) Depth profiles of the relaxation times extracted from bi-exponential fits of CPMG decays from a water-soaked sandstone sample treated with stone strengthener from one side (adapted from [4] with permission). (b) Wetting of a sandstone window frame at Paffendorf Castle before measurement. (c) Relaxation time distributions of partially wet sandstone window frames from Paffendorf castle treated with stone strengthener and untreated (adapted from [5] with permission).

from Herculaneum and nearby sites have been analyzed at the Herculaneum excavation site (Fig. 8.1.5a). Good signal could be obtained after carefully moisturizing the fresco fragments by applying deionized water with a soft brush to paper tissue covering the painted side. The *depth profiles* of the signal average of the first few echoes from a CPMG train reveal layer structures that are not readily visible to the human eye (Fig. 8.1.5b). The peaks in the profiles show layers with high water content. The differences in water content across the profiles arise from differences in the particular formulation of the lime mortars constituting individual layers and from the way of compacting the wet layers before painting. Both features are particular to the school of trade and the craftsmen who manufactured the layers. Different ways of preparing walls for fresco painting can thus be classified by NMR depth profiling (Fig. 8.1.5c) and eventually be put into an art historical context.

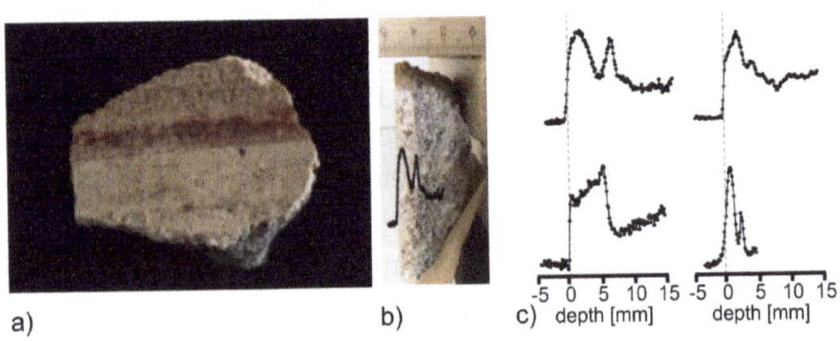

Figure 8.1.5. Depth profiles from fresco fragments (the fresco fragments were provided by the Herculaneum Conservation Project). (a) Photo of a fresco fragment. (b) Side view of a fragment with the NMR depth profile overlaid. (c) NMR depth profiles through different fresco fragments.

8.1.8 Advanced level measurements

In addition to *transverse relaxation*, *longitudinal relaxation* and *diffusion* can be measured from *moisture* in walls. NMR is unique in the sense that it can measure depth profiles of diffusion coefficients nondestructively in intact walls. Diffusion is particularly interesting, because it provides information about translational mobility, which, together with moisture content or signal amplitude, relates to moisture flux. *Moisture flux* is the quantity that is responsible for the transport of salts to the surface of the wall. The salts damage the surface of the wall when they crystallize and form the observed white deposits on the wall surface (Fig. 8.1.1). Even with moderate moisture content, the diffusion coefficient of water can be measured. The standard sequence is the CPMG detected stimulated echo sequence (Fig. 3.2.7a), because both echo time and diffusion time can be varied step by step to interrogate the signal decay (eqn. (3.2.12)) under the impact of diffusion. At low moisture content, however, transverse relaxation is fast, and the *Hahn echo sequence* (Fig. 3.2.5a) should be used.

For example, the diffusion coefficient of water in a naturally moist, painted wall from a building in Herculaneum has been measured with the stimulated echo sequence by varying the diffusion time (Fig. 8.1.6a). The observed diffusion coefficient of $D = (0.33 \pm 0.05) \times 10^{-9}$ m^2/s is lower than that of bulk water. It varies with the saturation level, and in some cases may even be higher than the bulk water diffusion coefficient of 2.3×10^{-9} m^2/s. This is the case when the diffusion length in the pore is enhanced by gas-phase diffusion enabled by exchange of molecules between the liquid and the gas phases.

D-T_2 *correlation* maps provide further insight into water dynamics. They can be measured in different ways by stray-field NMR, either by means of a Hahn or stimulated echo sequence with CPMG detection (Fig. 3.2.7a) or with a set of CPMG measurements employing a range of *echo times* (eqn. (7.1.10)). The latter scheme is simpler, but diffusion and relaxation encoding are entangled and not separated in successive evolution and detection time periods as with the other schemes. The CPMG scheme

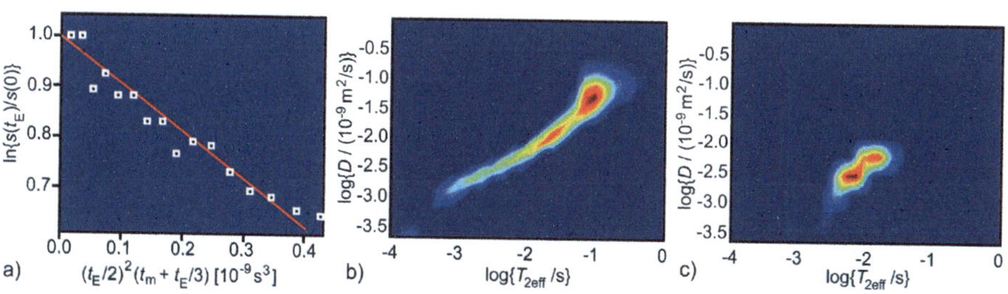

Figure 8.1.6. Diffusion in moist walls (the Herculaneum Conservation Project is acknowledged for permission to do these measurements). (a) Diffusion data from the wall depicted in Fig. 8.1.1. (b) *D*-T_2 correlation map from a wall inside the Villa of the Papyrus in Herculaneum in May 2011 at high air humidity. (c) Same as (b) but measured in March 2012 at lower air humidity.

has been tested at two occasions measuring the moisture in a wall in the Villa of the Papyrus in Herculaneum once at high air humidity (Fig. 8.1.6b) and once at lower air humidity (Fig. 8.1.6c). At lower humidity signal is observed only at low diffusion coefficients and relaxation time, indicating that the large pores are void of bulk water.

8.1.9 Data processing

Moisture content can be determined quantitatively with the NMR-MOUSE, because the sensitive volume V_t is well defined. For the PM25 NMR-MOUSE this volume is about $45 \times 45 \times 0.\,2\,\text{mm}^3$. To quantify the water content, the NMR signal s_{ref} of a fully saturated reference sample is needed. If the NMR signal amplitude s_0 at time zero of the echo envelope is from water in the wall, then the *water saturation* is given by (eqn. (7.2.4))

$$\Theta_s = s_0/s_{\text{ref}}. \tag{8.1.1}$$

If the dry wall also gives signal amplitude s_{dry}, for example, from bound water, then this signal needs to be subtracted, and the water saturation is calculated as (eqn. (7.2.5))

$$\Theta_s = (s_0 - s_{\text{dry}})/s_{\text{ref}}. \tag{8.1.2}$$

In civil engineering, *water content* is measured in units of mass, i.e. by the ratio of the weight m_w of water to the weight m_t of the dry sample (eqn. (7.2.3)),

$$\theta_m = m_w/m_t. \tag{8.1.3}$$

If maximum *gravimetric water content* $\theta_{m,\max}$ of a fully water-saturated sample is known, then the gravimetric water content is calculated from the *NMR signal amplitude* as

$$\theta_m = ((s_0 - s_{\text{dry}})/(s_{\text{wet}} - s_{\text{dry}}))\theta_{m,\max}. \tag{8.1.4}$$

This procedure was used to quantify the moisture maps in Fig. 8.1.2 [2].

8.1.10 References

[1] Di Tullio V, Proietti N, Capitani D, Nicolini I, Mecchi AM. NMR depth profiling as a non-invasive analytical tool to probe the penetration depth of hydrophobic treatments and inhomogeneities in treated porous stones. Anal Bioanal Chem. 2011; 400:3151–3146.

[2] Di Tullio V, Proietti N, Gobbino M, Capitani D, Olmi R, Priori S et al. Non-destructive mapping of dampness and salts in degraded wall paintings in hypogeous buildings: The case of St. Clement at mass fresco in St. Clement Basilica, Rome. Anal Bioanal Chem. 2010; 396:1885–1896.

[3] Proietti N, Capitani D, Lamanna R, Presciutti F, Rossi E, Segre AL. Fresco paintings studied by unilateral NMR. J Magn Reson. 2005; 177:111–117.

[4] Sharma S, Casanova F, Wache W, Segre AL, Blümich B. Analysis of historical porous building materials by the NMR-MOUSE. Magn Reson Imaging. 2003; 21:249–255.

[5] Blümich B, Casanova F, Perlo J, Anferova S, Anferov V, Kremer K, et al. Advances of unilateral, mobile NMR in nondestructive materials testing. Magn Reson Imag. 2005; 23:197–201.
[6] Haber A, Blümich B, Souvorova D, Del Federico E. Ancient Roman wall paintings mapped nondestructively by portable NMR. Anal Bioanal Chem. 2011; 401:1441–1452.

8.2 Easel paintings

8.2.1 Introduction

An *easel* is a portable, wooden, and adjustable stand to hold a piece of art for a painter. The word derives from the Germanic word 'ezel' or 'Esel' meaning donkey. In the context of paintings this means painter's donkey and denotes a wooden frame that holds the painting. Most paintings are produced on wood or *canvas* supports. Before painting, these supports are prepared by further layers of materials to produce mechanical strength and a smooth surface. Then paint layers consisting of pigments and binder are drawn and subsequently a varnish is applied to give the painting a shiny finish and protect the paint layers.

8.2.2 Objectives

Easel paintings are analyzed nondestructively by stray-field NMR relaxation measurements with variable aims. One goal is to study the *softness* or *brittleness* of the *binder* in terms of relaxation rates, which change with *aging*, *solvent exposure*, *storage conditions* and conservation impact. Another goal is to learn about the *stratigraphy* of the wooden panel or canvas. Different paint layers can only be identified in fortunate cases when the layers are sufficiently flat across the diameter of the sensitive slice. This is rarely the case with canvas paintings, because many paintings show *craquelure*, a network of fine cracks, which forms when the layers of painting contract upon *aging* and the resulting flakes bend.

From a restorer's point of view, the *solvent-paint interaction* is of interest, in particular, the *ingress of solvent* when cleaning the painting and repairing damaged sections. On the other hand, a forger might be interested in artificially aging the paint layers to the same brittleness as the original. Exposure to solvents, natural and artificial aging change the molecular mobility in the paint layers, and the mobility can be studied by NMR depth profiling and relaxation measurements. Compared to other nondestructive techniques for the analysis of thin layers such as optical coherence tomography, multispectral imaging, and terahertz spectroscopy, the *depth resolution* of the NMR-MOUSE is limited to about 10 μm, the *sensitive spot* is a wide slice of about 10 mm diameter, and the technique demands hydrogen nuclei to be present. On the one hand these are limitations, but on the other hand depth ranges of several millime-

Figure 8.2.1. Depth profiling of a painting in the National Gallery of Umbria in Perugia. A PM5 NMR-MOUSE is mounted on a horizontal displacement stage. It is retracted from the painting step by step at each depth decrement. The spectrometer is seen on the table behind the motorized sled with the NMR-MOUSE.

ters can be probed. With the exception of X-ray radiography, these are large compared to the ranges that other techniques like IR reflectography, optical coherence tomography and terahertz imaging can investigate. Furthermore, NMR relaxation is sensitive to *molecular mobility*, so that *solvent effects*, *paint drying* and the rigidity of paint layers can be investigated rather conveniently with simple CMPG measurements even when the painting is mounted in a frame and on display in a museum (Fig. 8.2.1).

8.2.3 Further reading

Alfeld M, Broekaert JAC. Mobile depth profiling and sub-surface imaging techniques for historical paintings – A review. Spectrochim Acta B. 2013; 88:211–230.

Capitani D, Di Tullio V, Proietti N. Nuclear Magnetic Resonance to characterize and monitor Cultural Heritage. Prog Nuc Magn Reson Spectr. 2012; 64:29–69.

Stoner JH, Rushfield R, editors. The Conservation of Easel Paintings, London: Routledge; 2012.

Blümich B, Casanova F, Perlo J, Presciutti F, Anselmi C, Doherty B. Noninvasive testing of art and cultural heritage by mobile NMR. Acc Chem Res. 2010; 43:761–770.

Pinna D, Galeotti M, Mazzeo R, editors. Scientific Examination for the Investigation of Paintings. A Handbook for Conservator-Restorers. Florence: Centro Di; 2009.

Blümich B, Casanova F, Perlo J. Mobile single-sided NMR. Prog Nucl Magn Reson Spectrosc. 2008; 52:197–269.

Casanova F, Perlo J, Blümich B. Depth profiling by single-sided NMR. In: Stapf S, Han SI, editors. NMR Imaging in Chemical Engineering. Weinheim: Wiley-VCH; 2006. pp.107–123.

8.2.4 Theory

Easel paintings are works of art typically done on *wood* or *canvas*, which are prepared with a ground layer and then painted and varnished. Valuable paintings have often been retouched and undergone several restoration cycles in intervals of a few decades. At each restoration stage the techniques of the time are applied, which by their very nature are invasive and at best partially reversible. *Solvents* are used for cleaning and removing the old varnish layer. But these solvents also affect the paint layers underneath by diffusion and migration of solvent through cracks and the solvent may wash out the softer, lipid components of the *binder*. Thus each painting carries its own history that NMR depth profiles may help to unravel by reporting the *stratigraphy* of the painting including its support. Moreover, NMR depth profiling can assist in developing soft conservation strategies by monitoring the effects of solvents when cleaning and retouching paintings.

The preparation of canvas and wood panels for historical paintings is different. *Wood* panels were sometimes covered with layers of canvas to prevent cracking. Then a primer or the 'ground' is applied to smooth the surface and to make it less absorbent for the paint. An often-used primer is *gesso*, a mixture of calcium carbonate and animal glue. *Canvas* is frequently primed with *mestica*. This is a mixture of oil and fine-grained calcium carbonate or earth pigments, which stays more flexible than gesso. Afterwards paint is applied in different layers, depending on the technique of the artists and the techniques of the restorers. Finally the painting may be finished with layers of varnish.

Paint consists of inorganic or organic pigments, which are mixed with oil or tempera as binders. *Tempera* is a water-oil emulsion with the addition of glue often made from egg yoke and linseed oil. There are many formulations for pigments and binders, which have been developed over the centuries. Depending on the formulation, the drying and aging kinetics are different. Upon drying the volatile components evaporate. However, the loss of small molecules from the paint does not stop once the paint is dry, but continues with *aging* and may be accelerated when the painting comes in contact with *solvents*, for example during *restoration procedures*. Such procedures mostly concern the renewal of the varnish layer, the repair of damaged areas, and the application of a second canvas behind the original one to provide new mechanical strength to an aging painting support.

8.2.5 Hardware

Depth profiles of paintings require high resolution and a short depth range, for example a PM5 NMR-MOUSE with a 5 m depth range. The sensor needs to be mounted on a motorized horizontal stage or sled so that it can be moved step by step from the painting when acquiring a depth profile (Fig. 8.2.1). To avoid damaging the painting, the measurement starts at the largest depth and the sensor is retracted in small steps.

8.2.6 Pulse sequence and parameters

The standard pulse sequence to measure paintings is the CPMG sequence (Fig. 3.1.1b). A first depth scan is conducted with large depth increments of 250 μm and a few echoes only to locate the paint layers. Subsequently a depth profile is measured at higher resolution with a step size of 15 μm to 50 μm in the depth range of interest. Complete CPMG echo trains are acquired in the second run to calculate the w-parameter (eqn. (3.1.5)) and extract signal amplitudes and relaxation times from the echo trains. At selected positions the longitudinal magnetization build-up curves can also be measured with the saturation recovery sequence (Fig. 3.2.2a).

The whole procedure can be automatized. In automatic mode a step size of 50 μm may be chosen, and at each depth the presence of signal is checked by comparing the sum of the first few echoes of a CPMG echo train with a threshold value. If below the threshold value, the sensor is advanced one more step. If above, then a complete *CPMG echo train* and a *saturation recovery* curve are acquired. The acquisition time for a spin density depth profile from a painting in automatic mode depends on the step size and the depth range, but an order-of-magnitude estimate is half an hour depending on the signal-to-noise ratio. Default parameters applicable to investigating a painting with a CPMG sequence are listed in Tab. 8.2.1. For future reference it is important to document the measurements from paintings, relevant data about the object, and photographs

Table 8.2.1. Default acquisition parameters for the PM5 NMR-MOUSE to measure paintings

Parameter	CPMG sequence
transmitter frequency ν_{rf}	17.1 MHz
transmitter attenuation for 90° pulse	−8 dB at 300 W
duration t_p of the 90° pulse	5 μs
number n_E of echoes	64
dwell time Δt	1–2 μs
acquisition time t_{acq}	10 μs
echo time t_E	40 μs
number n_s of scans	512
recycle delay t_R	0.5 s

of the measurement positions. These also include the parameters of the measurement and the type of sensor used.

8.2.7 Beginner's level measurements

Before starting any measurement with the NMR-MOUSE on a painting, the painting should be checked with a weak magnet for any magnetic components such as *nails*, which are often found near the frame and sometimes in the wood panels. In many cases the position of nails are known from X-ray radiography analyses and documented in the archives of the museum. When measuring depth profiles, the distance between the painting and the sensor has to be stable with micrometer accuracy during the measurement time. For large canvas paintings, this is a challenge, because the painting will deform from the draft of visitors passing by or doors opening and closing. Common issues encountered when measuring paintings are summarized in Tab. 8.2.2.

Depth profiling

In depth profiles through *easel paintings* on wood, the *stratigraphy* from wood, canvas, ground and paint can readily be identified (Fig. 8.2.2a), and the layer thickness can be quantified from the NMR measurements (Fig. 8.2.2b). In parts of a wood panel, where two boards have been glued together, it is not unusual to find more than one canvas layer to provide extra protection against crack formation. In such a position, the signal from the textile layer is correspondingly broader than in other places (Fig. 8.2.2c).

In paintings on canvas it is often difficult to find contrast between layers at all, so that different types of contrast like different settings of the w-parameter (eqn. (3.1.5)), the signal amplitude or the effective relaxation time $T_{2\mathrm{eff}}$ should be explored to maxize contrast. Features visible in one parameter may not be visible in another. For example, in the *Rembrandt self-portrait* at the Wallraf-Richartz Museum in Cologne (Fig. 8.2.2d),

Table 8.2.2. Common issues encountered when measuring easel paintings

- The painting is magnetic because of nails in the frame or the wooden panel. In case of a small nail, the distance between nail and sensor should be at least 10 cm
- The NMR-MOUSE is not well aligned with the painting
- The painting is large and moves from air draft in the museum
- The recycle delay is too short, and the signal cannot build up in between scans
- The measurement spot is too close to the frame not allowing access for the sensor. A minimum of 5 cm from one edge is required or the frame needs to be removed
- The paint layers are thin and the step size is so large that the paint layers are not visible in the depth profile
- The rf noise is high, and the NMR setup needs to be covered with grounded parachute silk

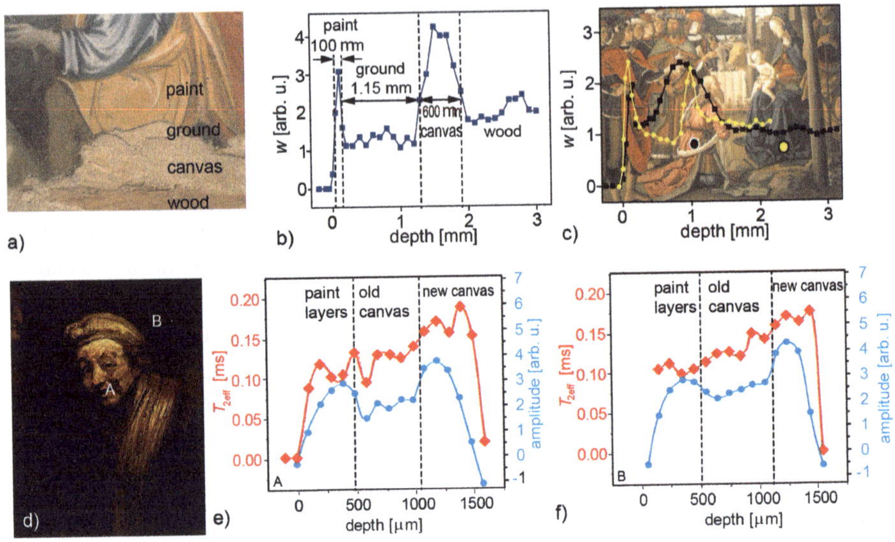

Figure 8.2.2. Depth profiles of easel paintings. (a) Close-up of a damaged part of an altar frontal by Maestro del Farneto from 1290 AD in the National Gallery of Umbria in Perugia showing the stratigraphy of a painting on wood. (b) Depth profile identifying the thickness of each layer, paint, ground, and canvas. (c) Depth profiles at the two marked positions through Adoration of the Magi by Perugino from 1470 AD revealing differences in the thickness of the textile layer (adapted from [1] with permission). (d) Self-portrait of Rembrandt at the Wallraf-Richartz Museum in Cologne. (e) Amplitude and T_{2eff} parameter depth profiles at position A. (f) Amplitude and T_{2eff} parameter depth profiles at position B.

three layers are clearly identified in the amplitude profiles (Fig. 8.2.2e,f). They are assigned to the paint layer, the original canvas and the backing with a new canvas. But different paint layers or even the varnish are hard to discriminate. This may have to do with the way the artist has applied the paint, with interdiffusion of different paint layers as a result of past conservation work, and with the unevenness of the paint layers.

Only in rare cases is it possible to clearly see different paint layers. One example is the Cologne painting 'Die hl. Elisabeth kleidet Arme, die hl. Elisabeth pflegt Kranke' on wood at the Walraff-Richartz Museum in Cologne from 1330, where the faces of Jesus and Elisabeth had been painted over in 1430 (Fig. 8.2.3a). The newer paint layer was later removed from the face of Jesus and in parts from Elisabeth. Because the wooden panel is very flat, these layers could be resolved in NMR depth profiles (Fig. 8.2.3b). It is speculated that they are so well visible, because the paint layer from 1430 had been applied on top of a varnish layer which gives little to no signal in the NMR depth profiles.

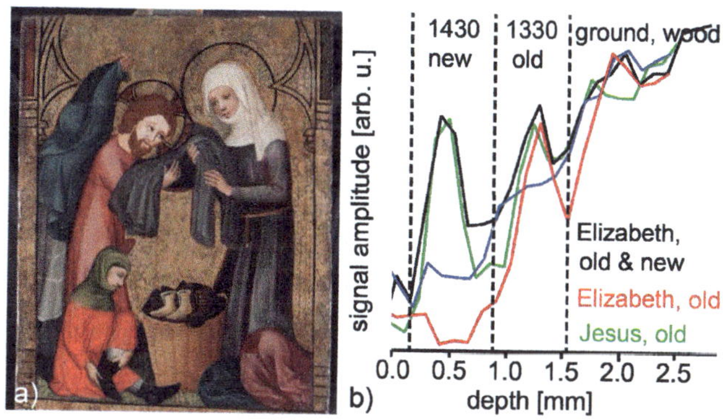

Figure 8.2.3. 'Die hl. Elisabeth kleidet Arme, die hl. Elisabeth pflegt Kranke' on wood, Walraff-Richartz Museum, Cologne, 1330. (a) The painting was painted over in 1430, and parts of the newer paint layer were removed from the face of Jesus and in some area from the face of Elizabeth. (b) The different paint layers show up clearly in NMR depth profiles.

Aging

Relaxation times are sensitive to molecular motion and therefore to the flexibility or elasticity of the paint binder. The elasticity changes with the drying of the paint and with the aging as a result of slow chemical reactions and evaporative loss of low molecular weight compounds from the binder. Real-time aging is usually difficult to reproduce in the laboratory with acceleration by elevated temperature or light exposure. *Photo aging* of tempera paints affects primarily T_2 while natural aging over a few centuries affects both T_1 and T_2 (Fig. 8.2.4). But dating paintings based on the relaxation times of the binder is hardly possible, because the type of pigment impacts the relaxation times as well as most conservation procedures. On the other hand, a comparison of relaxation times between different paintings may in well-documented cases assist the identification of *forged* and *artificially aged paintings*.

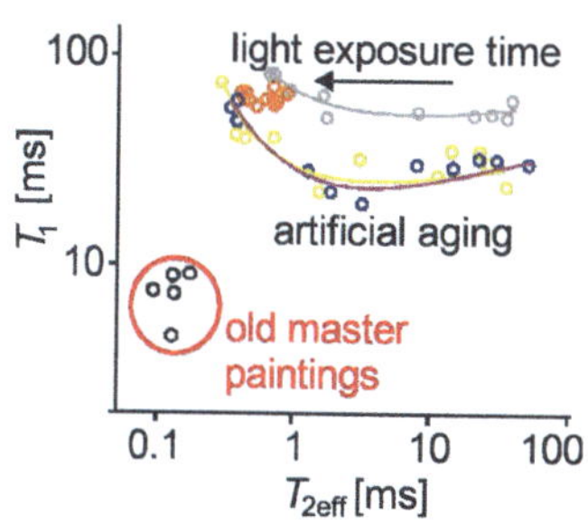

Figure 8.2.4. Artificial photo aging of tempera paints with different pigments on a T_1-T_2 map in comparison with relaxation times from old master paintings (adapted from [1] with permission).

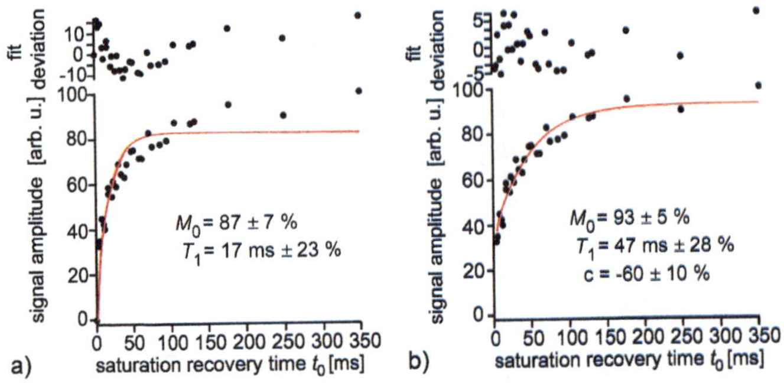

Figure 8.2.5. Measurement of longitudinal relaxation from binder in a painting using the saturation-recovery method. The acquired data (dots) were fitted with an exponential function (red) in two ways, and the fit deviations are plotted above the fitted data. (a) Fit without offset. The deviations appear correlated. (b) Fit with offset. The deviations appear random.

8.2.8 Advanced level measurements

Paintings are made from rigid materials with mostly short transverse relaxation times. This is why the measurement methods are simple and largely limited to multi-echo sequences like the CPMG sequence (Fig. 5.1.5a) and the multi-solid echo sequence (Fig. 5.1.5b). Already longitudinal relaxation is usually hard to measure (Fig. 3.2.2), because the signal-to-noise ratio is low when the total measuring time is limited to the order of one hour per position and depth profile with selected T_1 measurements included (Fig. 8.2.5).

8.2.9 Data processing

Depth profiles typically are processed with the operating software of the spectrometer by setting the parameters of the weight function w (eqn. (3.1.5)), or amplitude and T_2 relaxation time are extracted from a usually mono-exponential fit of the multi-echo decay of the transverse magnetization following eqn. (3.2.6). But the signal-to-noise ratio of data from paintings is often low, and the experimental data may be offset from zero by a constant, which may arise from inferior spectrometer settings in demanding on-site measurements. An example are the longitudinal relaxation data in Fig. 8.2.5 from the paint layer of an old master painting which were acquired with the *saturation recovery sequence* (Fig. 3.2.2a). When fitted with eqn. (3.2.5) without an offset, the errors appear to be scattered in a systematic way, and a relaxation time $T_1 = 17$ ms is extracted from the data (Fig. 8.2.5a). When a constant offset c is added to the fit function (eqn. (3.2.5)),

$$M_z = M_0(1 - f\exp\{-t_0/T_1\}) + c, \tag{8.2.1}$$

the extracted relaxation time is T_1 = 47 ms, and the errors appear to be scattered in a more random way (Fig. 8.2.5b). In either case, the relative error is high, but the longer relaxation time is the more credible one because of the random scatter of the errors.

8.2.10 References

[1] Presciutti F, Perlo J, Casanova F, Glöggler S, Miliani C, Blümich B, Brunetti BG, Sgamellotti A. Noninvasive nuclear magnetic resonance profiling of painting layers. Appl Phys Lett. 2008; 93:033505.

8.3 Wood

8.3.1 Introduction

Wood is the oldest material used by mankind as a resource for energy, construction and art. Dry wood is hygroscopic and swells when it absorbs water. Water resides in the *cell walls*, in the protoplasm of the living cells and in the *lumen* of dead cells as well as in voids. Depending on the growth period, spring or fall, *earlywood* and *latewood* form, which have low and high density, respectively, and give rise to the tree rings. Due to their different densities, earlywood and latewood swell and shrink differently upon variation of moisture content, so that wood deforms with drying and swelling. This is why air humidity needs to be controlled during transport and storage of wooden products. The *residual moisture content* of air-dried wood is between 8 % and 16 %. Objects from wood of interest in the context of cultural heritage are the backs of paintings, statues, beams and columns, furniture, and musical instruments.

8.3.2 Objective

The signal from wood most readily detected by mobile NMR is that from free and *bound water*, because the signals from cellulose and lignin have short relaxation times and require solid-state NMR techniques for their analysis. The nondestructive characterization of wood by single-sided NMR, therefore, concerns primarily studies of moisture. Nevertheless, with short dead times of 30 μs and less, signal from amorphous cellulose is detected as well. The primary aim of analyzing wooden objects relevant to cultural heritage is to learn about wood density and moisture content from signal amplitudes. *Depth profiles* can provide information about wood treatment from manufacturing or restoration, for example, in terms of varnish layers. Maps of signal amplitude over an extended region can be rescaled to gravimetric moisture content correlated with variations in air humidity and temperature [1, 2].

8.3.3 Further reading

Capitani D, Di Tullio V, Proietti N, Nuclear Magnetic Resonance to characterize and monitor Cultural Heritage. Prog Nuc Magn Reson Spectr. 2012; 64:29–69.

Blümich B, Casanova F, Perlo J. Mobile single-sided NMR. Prog Nucl Magn Reson Spectrosc. 2008; 52:197–269.

Rowell RM, editor. Handbook of Wood Chemistry and Wood Composites. Boca Raton: CRC-Press; 2005.

Maunu SL. NMR studies of wood and wood products. Prog Nucl Magn Reson Spectr. 2002; 40:151–174.

8.3.4 Theory

Wood is the most versatile natural composite material known to mankind. It provides strength, flexibility, and durability. From prehistoric times to the modern era wood has been used for making a wide variety of goods, from tools to musical instruments, altars, and buildings. It is also used as insulation material, can be reused, and finally used as fuel.

Wood is a composite material of cellulose and hemicellulose microfibrils embedded in a matrix of lignin. *Lignin* provides compressive strength and *cellulose* provides tensional strength. *Hemicellulose* is mostly present in the secondary wall of the wood cell (Fig. 8.3.1a). It is the most soluble part of cellulose, and it is the first component of the cell wall that starts to degrade.

One distinguishes living sapwood with growing cells and essentially dead heartwood, which provides stability to a tree. The lumen of wood cells is surrounded by an inner layer S1, a middle layer S2, and an outer layer S3 consisting of cellulose, hemicellulose, and lignin (Fig. 8.3.1a). Cellulose and hemicellulose are aligned forming microfibrils. The orientation of the microfibrils changes from layer to layer as well as the amounts of the different components (Fig. 8.3.1b). The most outer layer of a wood cell is made from pure lignin. Adjacent cells are held together by the middle lamella, which is composed of mostly lignin.

Wood is saturated with water when it is cut. Depending on the humidity, the free water content varies along with swelling and shrinking of the wood. Both bound and free water can be detected in NMR relaxation measurements and correlated with the type and state of wood. Upon drying, the *Fiber Saturation Point FSP* is reached when the lumen has no free water but the cell walls are still saturated with water. On average the FSP is at about 30% moisture content. Depending on the type of wood, the wood strength increases, the electric conductivity decreases, and resistance to biological attack increases upon drying [5].

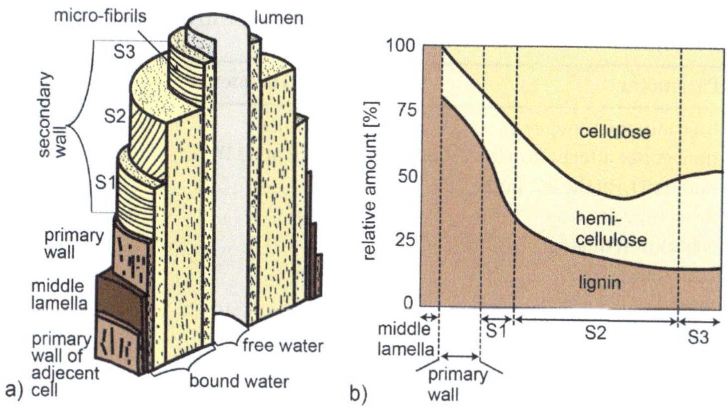

Figure 8.3.1. Wood cell and its composition. (a) Schematic of a wood cell and the wall of the adjacent cell with middle lamella (adapted from [3]). Notice the directions of fibrils in the subdivisions of the secondary wall. (b) Quantities of lignin, hemicellulose, and cellulose in the cell wall sections (adapted from [4]). The middle lamella is composed of lignin only. The water inside the lumen is free while the bound water refers to the water inside the cell walls, that is, to bound water or water in micro pores.

8.3.5 Hardware

Studies of wooden objects by single-sided NMR require an NMR-MOUSE with a short dead time in order to detect the signal from bound water and amorphous cellulose in addition to the signal from free water. Sensors with high field strength have shorter dead times than sensors that operate at lower field strength, but their depth range is shorter and the gradient is higher. For studies of *musical instruments* (Fig. 8.3.2a) and *wooden panels*, sensors with short depth ranges are acceptable, because degradation and treatments of wood in most cases begin at the surface. Consequently, a Profile NMR-MOUSE with 2 mm to 5 mm depth access is recommended for depth profiling (Fig. 8.3.2b), while a bar magnet NMR-MOUSE (Fig. 1.2.1, top right) with even shorter

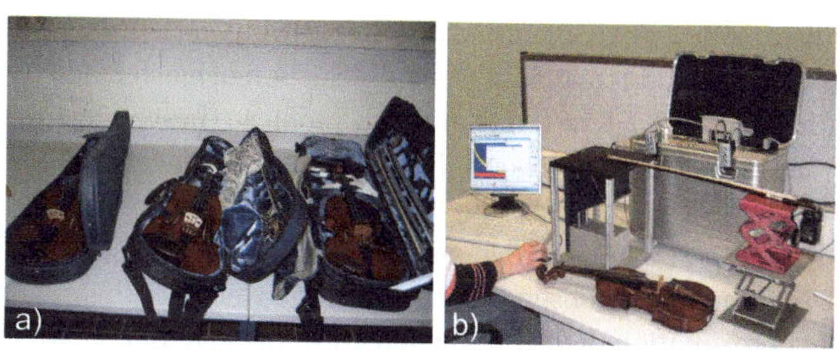

Figure 8.3.2. Measurements of wood from master violins and bows. (a) Collection of instruments. (b) Setup of a PM2 NMR-MOUSE on a lift to measure bows.

Table 8.3.1. Default acquisition parameters for a PM2 NMR-MOUSE to measure wood

Parameter	CPMG sequence
transmitter frequency ν_{rf}	29.3 MHz
transmitter attenuation for 90° pulse	−8 dB at 80 W
duration t_p of the 90° pulse	5 μs
dwell time Δt	0.5 μs
acquisition time t_{acq}	5 μs
echo time t_E	50 μs
number n_E of echoes	128
recycle delay t_R	0.5 s
number n_s of scans	512

dead time will allow for a more accurate separation of different signal contributions, but does not offer good depth resolution.

8.3.6 Pulse sequences and parameters

Depending on the type of information needed, either CPMG echo trains (Fig. 5.1.5a) and multi-solid echo trains (Fig. 5.1.5b) are measured, or advanced 2D Laplace measurements such as T_1-T_2 correlation maps and T_2-T_2 exchange maps (Fig. 3.2.4). Complete multi-echo decays need not be acquired if only spin density is needed, i.e. the initial amplitudes of multi-echo trains. Complete relaxation decays can be decomposed into the components from *bound water*, *free water*, and possibly cellulose based on differences in the relaxation times. Signal overlap in 1D relaxation time distribution can be unraveled in T_1-T_2 correlation maps, and T_2-T_2 exchange maps reveal the connectivity of the different water reservoirs at different moisture levels. Parameters for CPMG measurements on wood are summarized in Tab. 8.3.1.

8.3.7 Beginner's level measurements

Part of the exquisite sound of good master *violins* is attributed to the selection of the wood and another to the wood treatment. A small selection of master violins has been analyzed with the NMR-MOUSE in terms of depth profiles. These reveal the varnish layer and provide a signal from the wood (Fig. 8.3.3a). In some cases there are indications of two or more varnish layers, and considerable differences are found in the signal amplitudes of the wood, which denotes here the sum of the first two echoes. Because a total of 512 echoes were observed, the relaxation weight of this sum is low, so that it is a good approximation of the proton density of the material. This signal is essentially from bound water with possibly a small contribution from the wood polymer. Interestingly, the depth profiles from *bows* are less steep in the beginning (Fig. 8.3.3b), which is likely to be caused by the curvature of the wood so that only part of the bow

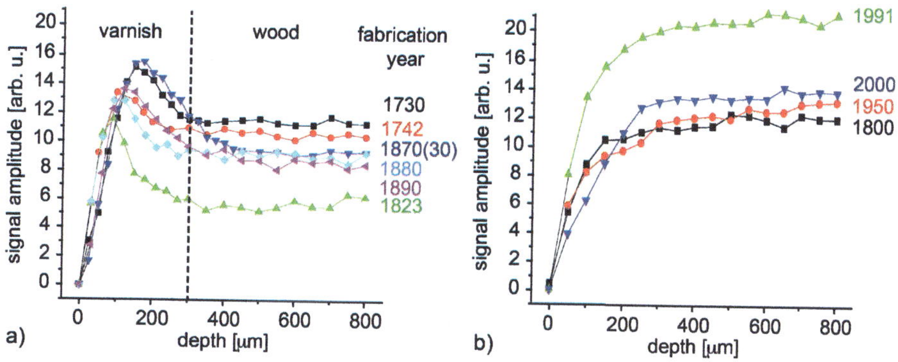

Figure 8.3.3. Amplitude depth profiles recorded from master violins and bows manufactured in different years. The signal amplitude is given by the sum of the first two echoes in a CPMG echo train. (a) Rear panels of violins. (b) Lower ends of bows.

resides within the sensitive slice of the sensor at low depths. But the signal amplitudes of the bows are higher than those of the violins, because bows are made from harder wood. It is therefore not unreasonable to expect the signal amplitude to scale with the wood density.

When plotting the signal amplitude at 0.7 mm depth over the reported fabrication year of the violins, an increase of the *proton density* with the age of the violin was found for most master *violins* investigated (Fig. 8.3.4). In contrast, the *wood density* of the master bows investigated decreased with age. This observation can be explained by use of harder wood for making bows in more recent times. The data of both, the violins and the bows, give strong evidence that the wood density is an indication for the age of the instrument and that it plays a role in the quality of master violins and bows. This information can be used in the selection of wood for producing modern

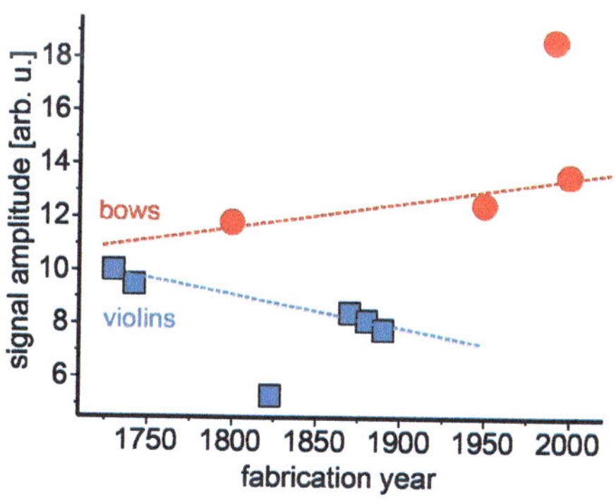

Figure 8.3.4. Proton density of the wood from the rear panels of violins at 0.7 mm depth versus the year of fabrication.

Table 8.3.2. Common issues encountered when measuring wood

– The wood may contain iron nails which attract the sensor
– The moisture content varies due to temperature gradients or varying air humidity
– The NMR-MOUSE is not well aligned with the object
– The sensitive slice is not completely inside the object. Then moisture content cannot be quantified
– The recycle delay is too short, and the signal cannot fully recover in between scans

master instruments and as evidence in the authentication of master violins. Outliers may identify suspicious instruments that need further investigation.

When preparing to measure objects from wood with the NMR-MOUSE, one needs to make sure that the object is free from nails and other magnetic components in the wider region of the position of the sensor. Apart from that, the measurement is not complicated. If the moisture content is to be quantified, then the humidity in the room needs to be controlled. Common issues that may be encountered when analyzing wood are summarized in Tab. 8.3.2.

8.3.8 Advanced level measurements

Water in wood is located in different pools, some of which have similar relaxation rates. To unravel signal from different sites in sapwood and heartwood from spruce and to quantify the *gravimetric moisture content* θ_m in the different sites, T_1-T_2 correlation experiments have been performed [6]. Four peaks can be identified in T_1-T_2 *correlation* maps (Fig. 8.3.5a,b). They are assigned to free water in the lumen (A), water in the cell walls (B, C), and wood polymer (D). Because peaks B and C overlap on the T_2 scale and peaks B and D overlap on the T_1 scale only three different pools would be identified in 1D distributions of T_1 and T_2, whereas four peaks can be identified in the T_1-T_2 correlation map. From the changes of the peak integrals the moisture content of each site can be determined as a function of the total moisture content. Peak A drains first so that it is readily assigned to the *free water* in the cell *lumen*. Peak D is sampled incompletely, because most of the signal has already relaxed before acquisition of the first data point. It is assigned to the signal from the *polymer matrix*. The assignment of peaks B and C to the cell morphology is more complicated but points at two distinct sites for bound water. T_2-T_2 *exchange maps* at a mixing time of $t_m = 12$ ms (Fig. 8.3.5c) reveal water exchange between peaks α and β, γ corresponding to free water from reservoir A and bound water from reservoirs B and C, identifying the bound water from at least one of the reservoirs B and C as weakly bound [6].

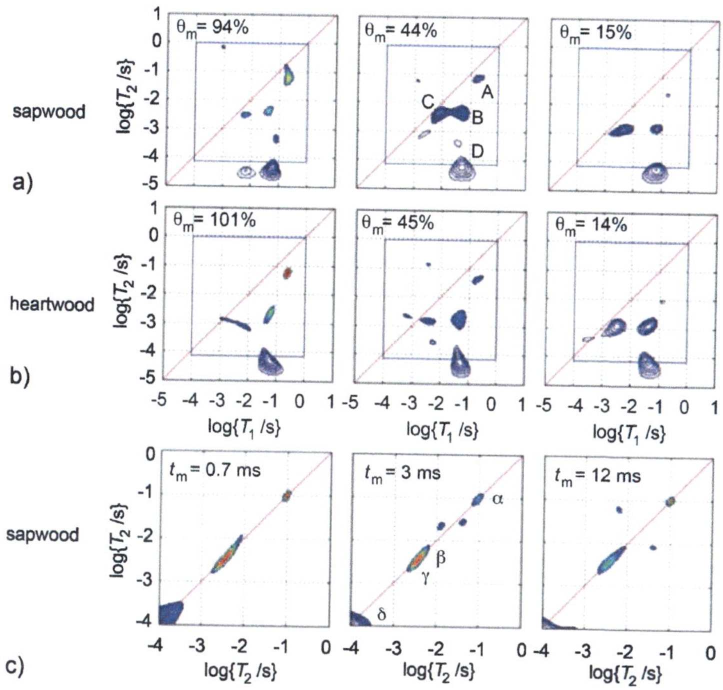

Figure 8.3.5. 2D correlation NMR of sapwood and heartwood from spruce (adapted from [6]). (a) T_1-T_2 correlation maps at different moisture content. Peaks B and C overlapping on the 1D T_2 scale and peaks B, D overlapping on the 1D T_1 scale are separated in the 2D correlation maps. The content of free water in the lumen (peak A) decreases with decreasing water content. The square marks the limits of the sampled time window. (b) T_2-T_2 exchange maps at 45% moisture content for different mixing times. Water molecules exchange between reservoirs α and β,γ.

8.3.9 Data processing

Gravimetric moisture content θ_m is calculated from the NMR signal amplitude s_0 following eqn. (8.1.4), given that the moisture content $\theta_{m,max}$ of the fully water-saturated sample and the signal amplitudes s_{wet} and s_{dry} of the fully wet and completely dry samples are known. Wood density can be estimated from the NMR signal amplitude with the help of an experimental calibration curve relating the NMR signal amplitude s_0 determined with a particular NMR-MOUSE to the density of a number of different wood samples. The object and all samples need to have been equilibrated within the same air humidity and temperature, because the NMR signal of wood derives mostly from water.

8.3.10 References

[1] Casieri C, Senni L, Romagnoli M, Sanramaria U, De Luca F. Determination of moisture fraction in wood by a mobile NMR device. J Magn Reson. 2004; 171:364–372.
[2] Senni L, Casieri C, Bovino A, Gaetani MC, De Luca F. A portable NMR sensor for moisture monitoring of wooden works of art, particularly paintings on wood. Wood Sci Tech. 2011; 43:167–180.
[3] Siau JF. Transport Processes in Wood. Berlin: Springer-Verlag; 1984.
[4] Panshin AJ, de Zeeuw C, Brown HP. Textbook of Wood Technology: Structure, Identification, Uses, and Properties of the Commercial Woods of the United States. New York: MacGraw-Hill; 1964.
[5] Engelund ET, Garbrecht Thygesen L, Svensson S, Hill CAS. A critical discussion of the physics of wood-water interactions. Wood Sci Technol Berlin: Springer; 2012.
[6] Cox J, McDonald PJ, Gardiner BA. A study of water exchange in wood by means of 2D NMR relaxation correlation and exchange. Holzforschung. 2010; 64:259–266.

8.4 Paper and parchment

8.4.1 Introduction

Papyrus and palm leaves are amongst the earliest examples of writing materials. Parchment was developed in Pergamon about 200 years BC as a substitute for papyrus. *Paper* was developed in China about 2000 years ago, from where paper making technology spread to the Islamic world in about 800 AD. It took another 400 years before paper was used in southern Europe, and by 1400 AD it had arrived in Germany where shortly later in about 1450 AD Gutenberg invented book printing with a mechanical press using movable letters.

8.4.2 Objective

The primary objective in studying paper and parchment by NMR is to learn about the state of degradation and to study the degradation mechanisms for the purpose of developing suitable conservation strategies. While solid-state NMR spectroscopy of paper [2] is destructive because samples have to be taken, stray-field NMR measurements with the NMR-MOUSE are not and can be done on intact and valuable documents at the storage site. Such nondestructive measurements with portable instruments are described in the following.

8.4.3 Further reading

Capitani D, Di Tullio V, Proietti N. Nuclear Magnetic Resonance to characterize and monitor Cultural Heritage. Prog Nuc Magn Reson Spectr. 2012; 64:29–69.

Blümich B, Casanova F, Perlo J. Mobile single-sided NMR. Prog Nucl Magn Reson Spectrosc. 2008; 52:197–269.

8.4.4 Theory

Parchment is made predominately from the skin of calf, goat, sheep and deer. It mostly consists of *collagen*, a protein, which exists as elongated fibrils in skin similar to tendon (Fig. 6.2.1). In the preparation of parchment, the animal skin is de-haired usually in a strong alkaline bath, and the epidermis and the subcutaneous tissues are removed. Depending on the preparation method, residual amounts of lipids are present in the material. New parchment is considered to contain 15% water in bound and unbound states at different sites. The *bound water* plays a central role in the stability of the collagen material. If stored at low humidity, parchment becomes brittle. If stored at high humidity, it gelates, forming translucent areas where the fibrous structure of the triple-collagen helices is destroyed, and becomes vulnerable to fungus attack. *Parchment* also degrades by hydrolysis and oxidation [1].

Paper is made from cellulose and water with small amounts of additives [2]. Early paper was handmade from the long cellulose fibers of hemp, flax or cotton, while most modern, machine-made paper is made from short cellulose fibers mixed with hemicellulose and lignin as well as further additives. In dry paper, water is contained in small pores. Loss of this water irreversibly destroys the material. *Paper* is damaged mostly by biological attack from fungi, bacteria, and insects but also by chemical oxidation and exposure to acids. This changes the content of bound water and the crystallinity of the cellulose. Both bound water and free water can be observed by NMR relaxation measurements and correlated with the state of conservation [2, 3].

8.4.5 Hardware

To observe the signal from bound water, an NMR-MOUSE with a short dead time is needed, for example, a Profile NMR-MOUSE with 3 or 5 mm depth access (Fig. 1.2.2b) or a bar magnet NMR-MOUSE (Fig. 1.2.1, top right). Early measurements were carried out with a simple U-shaped NMR-MOUSE (Fig. 8.4.1).

Figure 8.4.1. An early version of an NMR-MOUSE® operated by a Bruker Minispec spectrometers measuring the CPMG relaxation decay from a 17th century edition of the Kurfürstenbibel.

8.4.6 Pulse sequences and parameters

Like most objects of art and cultural heritage, paper and parchment are also measured with stray-field instruments using multi-echo sequences like the CPMG sequence (Fig. 5.1.5a) and the multi-solid echo sequence (Fig. 5.1.5b). In a depth scan with only a few echoes, the position of the object is determined, and then further measurements are carried out at the selected depth. The measurements need to be executed with a short echo time to detect the signals not only from free water, but also from bound water and possibly the amorphous polymer. Default parameters are given in Tab. 8.4.1. Advanced measurements concern the analysis of longitudinal relaxation with the saturation recovery sequence (Fig. 3.2.2a) and T_1-T_2 correlation experiments (Fig. 3.2.4a).

8.4.7 Beginner's level measurements

Often damage in paper as well as in parchment gives rise to lower longitudinal and transverse relaxation times. This indicates a loss of water, which is supported by increasing brittleness or loss of elasticity of these materials. Two relaxation times were found in the paper of a biologically degraded book from 1605 (Fig. 8.4.2a). They were assigned to bound water and amorphous cellulose. The signal from crystalline cellulose is largely lost in the dead time of the sensor but can be estimated by extrapolation of the signal amplitude to zero time (Fig. 8.4.2b). The data were collected with an early version of the NMR-MOUSE (Fig. 8.4.1), and Hahn echoes were measured instead

Table 8.4.1. Default acquisition parameters for the PM3 NMR-MOUSE to measure paper and parchment

Parameter	CPMG sequence
transmitter frequency ν_{rf}	17.1 MHz
transmitter attenuation for 90° pulse	−8 dB at 300 W
duration t_p of the 90° pulse	2 μs
dwell time Δt	0.5 μs
acquisition time t_{acq}	5 μs
echo time t_E	50 μs
number n_E of echoes	1024
recycle delay t_R	1 s
number n_s of scans	512

of CPMG echo trains to suppress the signal from the adhesive in the sensor fixing the rf coil [3].

In an *artificial aging* study it has been shown that measurements of transverse relaxation decays can reveal the degradation of *paper* earlier than visual inspection, and that nondestructive measurements with the NMR-MOUSE and destructive measurements on samples in a homogeneous field produce results in good agreement with each other [2]. Similar to biological degradation, oxidation of cellulose in $NaIO_4$ shortens the transverse relaxation (Fig. 8.4.2c). Moreover, corrosive effects of iron gall ink on historical paper have been detected [4]. Different inks have different effects on the NMR relaxation times. Even faded inks can in some cases be traced. Also, it has been shown that the relaxation signals from *oil stains* on paper can be correlated with the type and the degree of cross-linking of the oil for the benefit of identifying suitable restoration strategies [5]. Common issues encountered when measuring paper and parchment are summarized in Tab. 8.4.2.

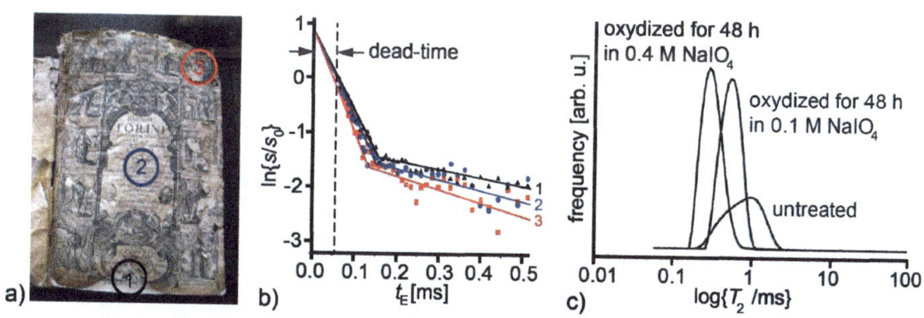

Figure 8.4.2. Transverse relaxation in aged paper. (a) Water-damaged book from 1605 with different paper degradation levels arbitrarily denoted as little degraded (1), medium degraded (2) and strongly degraded (3). (b) Relative transverse relaxation decays measured with Hahn echoes at different echo times t_E at the three spots marked in (a). Lines are drawn to guide the eye [adapted from [3] with permission). (c) Distributions of relaxation times measured with a stray-field sensor on artificially aged Whatman paper (adapted from [2] with permission).

Table 8.4.2. Common issues encountered when measuring paper and parchment

– The sample is bent and the sensitive slice is not fully inside the material
– The material is so brittle that the signal decays in less than ten echoes
– The echo time is too long to detect the rapidly relaxing magnetization components
– The sample is pressed against the NMR-MOUSE with another object that gives a ^{1}H signal
– The number of echoes is high and the recycle delay short so that the rf coil heats up

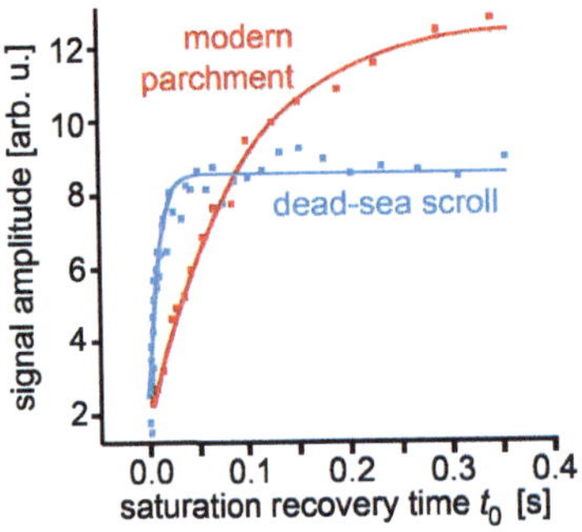

Figure 8.4.3. Longitudinal magnetization build-up curves measured with the saturation recovery sequence for modern parchment and ancient parchment (adapted from [6]).

8.4.8 Advanced level measurements

Like paper, most parchment becomes more brittle with age. However, while with paper primarily the transverse relaxation changes with ongoing degradation, in parchment it is the longitudinal relaxation. A pronounced difference in T_1 has been observed when comparing the relaxation times from modern parchment and the ancient parchment of the Dead Sea Scrolls (Fig. 8.4.3) [6]. This observation is confirmed with measurements of T_1-T_2 *correlation* maps of degraded parchment from a 17th century, water-damaged book cover (Fig. 8.4.4). The change in T_2 is much weaker than that in T_1 with

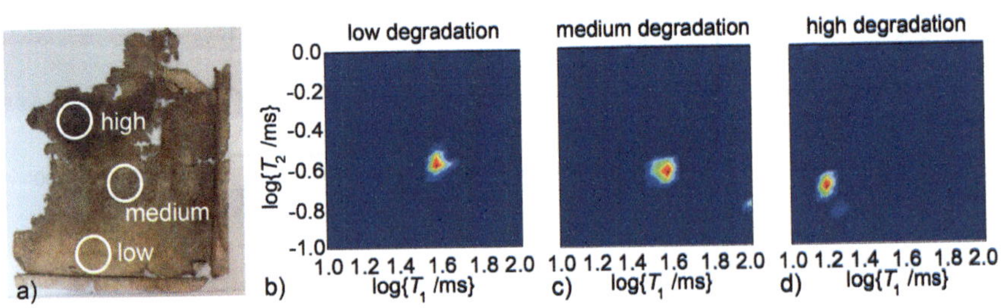

Figure 8.4.4. T_1-T_2 correlation maps of differently degraded regions of the cover from a water-damaged 17th century book. (a) Photo of the parchment fragment. (b) Map from a weakly degraded region. (c) Map from a medium degraded region. (d) Map from a highly degraded region.

degradation increasing from low to medium to high. Average values of T_1 for modern parchments made from the hides of different animals are all about 45±2 ms. However, depending on the *aging* mechanism, more than one T_1 component may be observed and T_1 may also increase with the time of artificial aging. A shortening of T_1 is observed by aging in heat and humidity. It has been attributed to gelatinization of the originally fibrous collagen material [6].

8.4.9 Data processing

NMR relaxation data from paper and parchment are processed in analogy to the data from paintings (Section 8.2.9).

8.4.10 References

[1] Kennedy CJ, Weiss T. The structure of collagen within parchment – A review. Restaurator. 2003; 24:61–80.

[2] Proietti N, Capitani D, Pedemonte E, Blümich B, Segre AL. Monitoring degradation in paper: Non-invasive analysis by unilateral NMR. Part II. J Magn Reson. 2004; 170:113–120.

[3] Blümich B, Anferova S, Sharma S, Segre AL, Federici C. Degradation of historical paper: Nondestructive analysis by the NMR-MOUSE®. J Magn Reson. 2003; 161:204–209.

[4] Viola I, Bubici S, Casieri C, De Luca F. The codex major of the Collectio Altaempsiana: A non-invasive NMR study of paper. J Cult Heritage. 2004; 5: 257–261.

[5] Del Federico E, Centeno SA, Kehlet C, Currier P, Stockman D, Jerschow A. Unilateral NMR applied to the conservation of works of art. Anal Bioanal Chem. 2010; 396:213–220.

[6] Masic A, Chierotti MR, Gobetto R, MartraG, Rabin I, Coluccia S. Solid-state and unilateral NMR study of deterioration of a Dead Sea Scroll fragment. Anal Bioanal Chem. 2012; 402:1551–1557.

8.5 Mummies and bones

8.5.1 Introduction

Mummies are often wrapped in textiles. The NMR signal of *mummies* derives from textiles, skin and bone, depending on the measurement depth. *Bone* is a regenerative rigid connective tissue that supports and protects living organisms. Bone consists of compact bone and of spongy bone with a cellular structure (Fig. 8.5.1). Long bones contain a medullary cavity, which hosts the bone marrow in living bone.

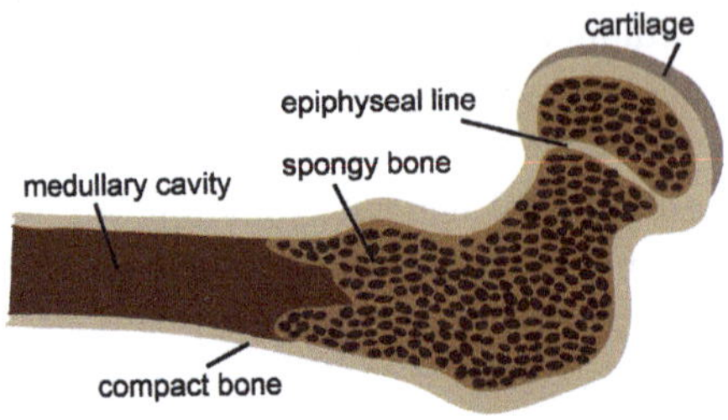

Figure 8.5.1. Bone anatomy. Bone consists of compact bone and spongy bone. The medullary cavity in long bones contains bone marrow.

8.5.2 Objective

Major objectives in measuring mummies and bones are to determine the thickness of different layers and to determine the material density and its variation with depth. The proton density of untreated bone is an important indicator for the state of preservation of the organic collagen matrix of the bone material. High proton bone density may identify a bone as a good candidate for DNA analysis. In comparison with reference data, the presence of foreign organic substrates can be noticed, which may give evidence of moisture or past preservation measures.

8.5.3 Further reading

Capitani D, Di Tullio V, Proietti N. Nuclear Magnetic Resonance to characterize and monitor Cultural Heritage. Prog Nuc Magn Reson Spectr. 2012; 64:29–69.

Blümich B, Casanova F, Perlo J. Mobile single-sided NMR. Prog Nucl Magn Reson Spectrosc. 2008; 52:197–269.

8.5.4 Theory

Bone is a rigid composite material mostly from organic collagen, inorganic hydroxylapatite ($Ca_{10}(PO_4)_6(OH)_2$) and water. The collagen gives it elasticity and the hydroxylapatite mechanical strength. While X-ray radiography serves to measure mineral bone density, NMR can be employed to measure organic *bone density*, unless the bone has undergone conflicting conservation treatment.

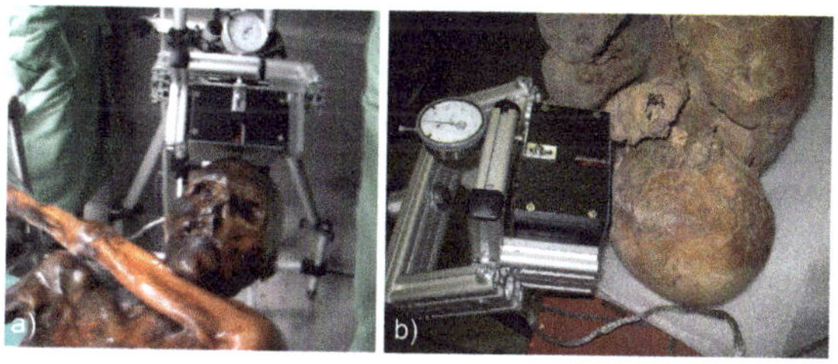

Figure 8.5.2. Setup of the NMR-MOUSE to measure mummies. The sensor is mounted on a manual lift attached to the arm of an aluminum scaffold. (a) The Neolithic iceman Ötzi at the Archeological Museum in Bozen, Italy. (b) A mummy from Peru (courtesy of Frank Rühli).

8.5.5 Hardware

Intact *bones* and *mummies* are preferably studied with an NMR-MOUSE (Fig. 8.5.2). Because these objects are hardly ever planar, the depth range should be chosen sufficiently large, e.g. 10 mm, so that the sensitive slice can be completely inside the material. For the measurement of the Neolithic mummy of the iceman Ötzi and a Peruvian mummy, a PM10 NMR-MOUSE was mounted on a manual lift attached to a scaffold, which could be positioned near the stretcher with the iceman (Fig. 8.5.2). This setup proved to be sufficiently stable so that only a few second long delays were required for the vibrations to settle down that were caused by readjusting the measurement depth by hand when acquiring a depth profile.

8.5.6 Pulse sequences and parameters

Measurements on hard objects are conducted with the CPMG sequence (Fig. 5.1.5a) and the multi-solid echo sequence (Fig. 5.1.5b). Only the initial part of the echo train is acquired if only spin density is needed, e.g. to map bone density. In studies of *mummies* with layers of textile, skin and bone, however, it is advisable to sample complete CPMG decays at each depth so that the contrast in the depth profile can be adjusted even after measurement by fitting the relaxation signal with model functions and extracting fit parameters (Tab. 3.1.2) or varying the integration limits of the w-function (eqn. (3.1.5)). Default parameters for measuring CPMG signals of bone are given in Tab. 8.5.1. The number of echoes varies depending on the sample. Ancient bone may give as few as six echoes only.

8.5.7 Beginner's level measurements

The *organic bone density* of dry bones from humans and animals can readily be determined with the NMR-MOUSE from the proton density and correlated with the state of

Table 8.5.1. Default acquisition parameters for the PM5 NMR-MOUSE to measure bone

Parameter	CPMG sequence
transmitter frequency ν_{rf}	17.1 MHz
transmitter attenuation for 90° pulse	−8 dB at 300 W
duration t_p of the 90° pulse	5 μs
dwell time Δt	0.5 μs
acquisition time t_{acq}	5 μs
echo time t_E	50 μs
number n_E of echoes	64
recycle delay t_R	0.5 s
number n_s of scans	128

preservation. As such the NMR-MOUSE is an instrument of interest for probing bone artifacts and mummies in the museum or at the site of the excavation.

The transverse magnetization from the dry tissue of historic bone and mummies decays rapidly. Consequently CPMG echo trains are short, and are usually fitted with mono-exponential decay functions to extract signal amplitude corresponding to proton density. Because bone is rarely flat and heterogeneous, the signal amplitude corresponds to organic bone density only if the sensitive volume of the sensor is fully inside the bone region of interest. To find these positions, complete depth profiles should be acquired if the structure or the position of the bone are not accurately known. The issues encountered when measuring bone with the NMR-MOUSE are summarized in Tab. 8.5.2.

When bone deteriorates, the *bone density* decreases. In this process, the correlation between organic and inorganic bone densities is highly nonlinear. A combined study with the NMR-MOUSE and X-ray tomography of differently degraded, old tibias from the Dahlheim monastery [1] revealed that the organic collagen matrix decomposes first, which consequently releases the mineral components (Fig. 8.5.3).

Depth profiles of *bones* reveal the variation of bone volume in the sensitive volume of the sensor and the bone density when the sensitive volume is completely inside the bone unless other organic substances are also present. In an attempt to study the degree of conservation of the *tibia* from Charlemagne, which is stored in the treasury of the Aachen Cathedral, NMR depth profiles were taken at different positions of the tibia

Table 8.5.2. Common issues encountered when measuring bone

- Due to the curvature of the object, the sensitive slice is not fully inside the material
- The material is brittle and the signal decays in just a few echoes
- The echo time is too long to detect the rapidly relaxing magnetization components
- The bone is wet or has been treated with a proton-containing conservation agent, so that the echo-train amplitude does not correspond to bone density

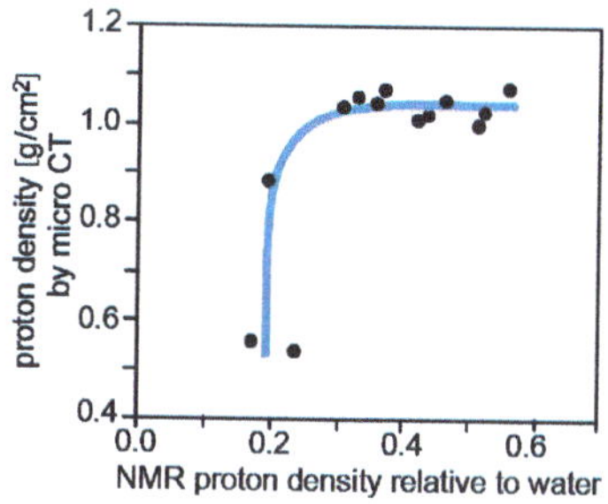

Figure 8.5.3. Correlation of X-ray and NMR bone densities for differently degraded historic tibias. The X-ray bone density was determined by micro computer tomography and refers to mineral bone density. The NMR bone density was determined at the same spots from CPMG echo train amplitudes measured with an NMR-MOUSE. Computer tomography measures mineral density and NMR hydrogen density from organic matter. Both components deplete at different rates during bone decay.

(Fig. 8.5.4a). These depth profiles (Fig. 8.5.4b) show varying thickness of the compact bone, which agrees with that in the CT cross sections taken from the same positions (Fig. 8.5.4c). The signal amplitudes, however, are considerably larger than those ob-

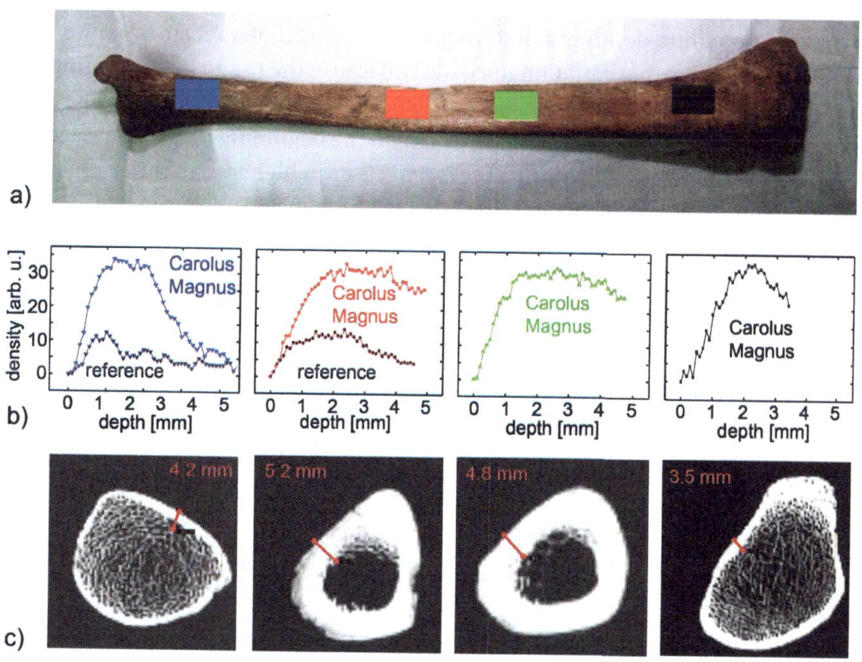

Figure 8.5.4. Study of the left tibia from Charlemagne (Carolus Magnus, 747–814 AD). (a) Photo of the tibia with the positions marked where NMR depth profiles were measured. (b) NMR depth profiles from Charlemagne's tibia and a reference tibia from the Dahlheim monastery. (c) CT cross sections of Charlemagne's tibia at the same positions where the NMR depth profiles were measured.

served at comparable positions in a tibia from the *Dahlheim monastery*, which is of similar age (Fig. 8.5.4b, left two profiles). This gives evidence that the tibia of Charlemagne had been treated with an organic agent in a past conservation effort. From the length of the tibia, the height of Charlemagne has been estimated at 184 cm [2].

NMR signal amplitude as a measure of *bone degradation* has been determined in a study of the Neolithic mummy of the iceman Ötzi [3]. The *iceman* was murdered about 3100 BC and remained frozen since then until he was found in 1991 near Tisenjoch in the Tyrolian Alps. The mummy is on display in the Archeological Museum in Bozen, Italy, where he is kept frozen at about -6.5° C and 97 to 99% humidity. His body is covered with a thin sheet of ice. The NMR investigation of his body was limited to about 60 minutes to avoid thawing from the spectrometer heat and the body heat of the investigators in the cold room (Fig. 8.5.2a).

The depth profile through the iceman's forehead (Fig. 8.5.5a) shows the layer of *amorphous ice*, a signal from skin and deeper down the signal from the bone. Within the bone the signal first peaks and then decreases. This decrease is explained by the anatomy of the skull bone (inset), which shows a compact outer layer, a spongy middle layer and then again a compact bone layer at the inside of the head. Depth profiles from a well-preserved skull from the anatomical collection of the University of Zürich and a degraded skull from the Dahlheim monastery confirm this variation of *bone density*. The corresponding depth profile through the forehead of a recently mummified cadaver does not show this variation most likely because the tissue layer covering the skull bone is too thick and bone signal is not detected, while the profile of an Egyptian mummy head shows the tissue wrapping and then the beginning of the bone signal.

Interestingly, the bone signal from the *iceman* is even higher than the bone signal from the recent *skull*. On account of the fact that the iceman had been exposed to water

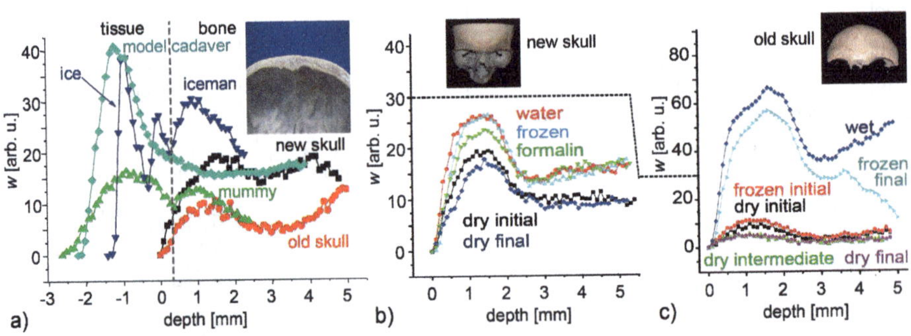

Figure 8.5.5. Depth profiles through foreheads. (a) Depth profiles of the iceman with ice layer and skin, of a model cadaver with skin and tissue, of an Egyptian mummy head, of a deteriorated skull from the Dahlheim monastery, and of a recent skull from the anatomical collection of the University of Zürich. The irregularity of the skull bone and the compact and spongy bone layers are visible in the inset. (b) Profiles dry, wet, and frozen for a section taken from the new skull. (c) Profiles dry, wet, and frozen for a section taken from the old skull.

all the time, this observation gave rise to a study of the effect of water upon soaking, freezing, thawing and drying bone. Small pieces of the *skull bone* were cut from the forehead of the old and the recent skulls, and depth profiles were measured when dry, wet, frozen at -30° C, and dried again. The profiles from the dry skull section before and after soaking in water are slightly different, reflecting the residual *moisture content*, which depends on the humidity and the way of drying. The profiles from the wet and the frozen new skull section overlap (Fig. 8.5.5b) and are slightly lower than the bone profile from the iceman. On the other hand, the dry profile of the degraded old skull (Fig. 8.5.5c) is lower than that of the new skull, indicating lower bone density, and upon soaking is much higher than that of the wet new skull. The loss of collagen thus goes along with larger pores, which can take up more water when the bone is wet. Upon freezing the bone profile amplitude decreases, indicating that some of the water can no longer be observed because it is frozen to crystalline ice in the large pores, where the freezing point depression is low. The small difference between the profile amplitudes of the iceman and the wet/frozen new skull suggest that neither the water in the forehead of the iceman nor that in the cold bone section from the new skull is frozen because the pores are small and the freezing point depression is high. This comparison with the wet and frozen profiles from the degraded skull section indicates that the bone structure of the iceman is about as well preserved as that of the new skull.

8.5.8 Advanced level measurements

Advanced level NMR measurements of bone usually employ high-field MRI and solid-state NMR spectroscopy. Most clinical studies image the soft tissue surrounding the bone [4]. However, new imaging modalities with short echo times are available which are capable of imaging bone directly. They have also been applied to study dry mummy tissue with short transverse relaxation times.

8.5.9 References

[1] Collection of Frank Rühli, University of Zürich, Zürich, Switzerland.
[2] Rühli F, Blümich B, Henneberg M. Charlemagne was very tall, but not robust. Econ Hum Bio. 2010; 8:289–290.
[3] Rühli F, Böni T, Perlo J, Casanova F, Baias M, Egarter E, Blümich B. Non-invasive spatial tissue discrimination in ancient mummies and bones in situ by portable nuclear magnetic resonance. J Cultural Heritage. 2007; 8:257–263.
[4] Wehrli FW. Magnetic resonance of calcified tissue. J Magn Reson. 2013; 229:35–48.

9 Concluding remarks

9.1 Moving ahead

NMR is a simple physical phenomenon with a large number of applications in the field of analysis. This number steadily grows as technology progressively reduces the noise figures and the physical size of the electronic components needed for signal generation and reception. The early NMR magnets were resistive electromagnets (Fig. 9.1a). Their field was swept through resonance, and the wide resonance lines from condensed matter were analyzed in terms of spectral moments to access the transverse relaxation times. Once the fine structure of the NMR spectrum was discovered and understood to be of chemical origin (Fig. 9.1b), the main interest in NMR moved from physics to chemistry. High magnetic fields became the prime target of instrumental development because the frequency range of the chemical shift is proportional to the field strength and the sensitivity approximately grows with the square of the field strength. The advent of superconducting magnets and the development of multidimensional NMR spectroscopy established NMR spectroscopy in the life sciences, because structure, conformation, and function of large molecules can be analyzed with it in their natural environment and in non-crystalline assemblies.

Chemical analysis was the main topic until 1973, when magnetic resonance imaging was discovered. Subsequently the primary use of NMR moved from chemistry to medicine, and the quest for high magnetic fields generated by ever more powerful superconducing magnets remains a strategic development goal (Fig. 9.2a). Desktop NMR with permanent magnets conquered a niche in the food industry with measurements based on relaxation and diffusion, which eventually expanded into other areas of materials science. A genuine push came with the advent of well-logging NMR instruments demonstrating that useful NMR measurements can be conducted in the stray field outside the magnet and the rf coil. This principle is used also by the NMR-MOUSE for non-destructive materials testing but with hardware much smaller than well-logging tools and superconducting magnets.

The major technological advance being witnessed currently is the commercialization of compact permanent magnets which are shimmed and temperature stabilized to a perfection sufficient to resolve the chemical shift in proton spectra of solutions to a degree that is attractive for bench-top NMR spectroscopy in the chemistry laboratory (Fig. 9.1c,d). For reasons of sensitivity at low fields like 1 or 2 T, large sample volumes are preferred, but solutions in sub-molar concentrations contained in standard 5 mm diameter sample tubes produce good spectra provided the field homogeneity affords narrow lines.

Yet the miniaturization continues. A palm-sized NMR relaxometer with the magnet included has already been tested [2], and small magnets for chemical shift resolved spectroscopy are becoming available [3]. But smaller magnets condition smaller sam-

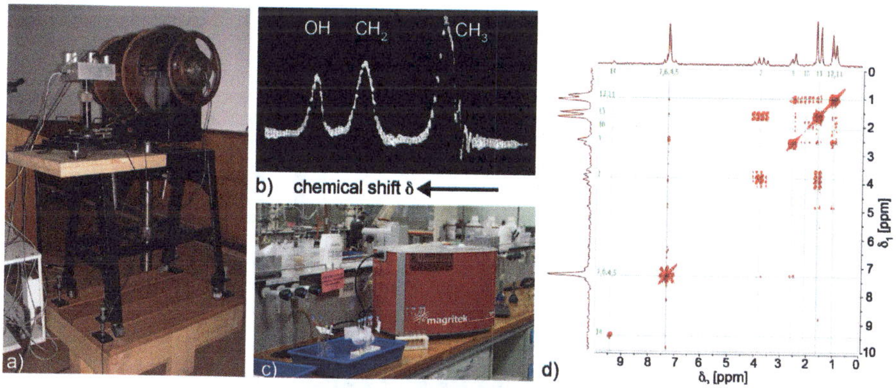

Figure 9.1. NMR hardware and spectra. (a) Magnet from Wilhelm Schütz, student of of Walther Gerlach, which has been moved from München via Königsberg to Jena, where Gerhard Scheler set it up to demonstrate CW NMR experiments. (b) NMR spectrum of ethanol from 1951 measured by sweeping the magnetic field with an electromagnet at 32 MHz in 2 minutes acquisition time (adapted from [1] with permission). (c) Permanent-magnet desktop NMR spectrometer for high-resolution Fourier NMR spectroscopy at 42 MHz. (d) 2D COSY spectrum of ibuprofen acquired in 8 minutes with the spectrometer shown in (c).

ple volumes, and for practical use in quantitative analysis, the sensitivity needs to be boosted. Considerable effort is being spent today in finding ways to improve the sensitivity, and different strategies are becoming apparent. In the context of compact spectrometers, the most successful one is based on targeting specific molecular markers for disease in complex fluids by functionalized relaxation agents, e.g. by ferromagnetic nanoparticles, which can be counted in a relaxation analysis with a small relaxometer [3, 4]. High sensitivity with this approach is gained from the use of microcoils, the sensitive volume of which tightly matches the volume of the concentrated analyte. Other techniques to boost sensitivity are not yet available in compact form. But strategies of current interest concern *hyperpolarization* and *alternative detection* schemes.

Hyperpolarization denotes the generation of magnetization larger than that afforded by thermodynamic equilibrium. To this end, the sample can be prepolarized in a high but inhomogeneous field and subsequently transferred to a low but homogeneous field for spectroscopic analysis; noble gases, in particular ^{129}Xe can be hyperpolarized with lasers and the polarization transferred to other nuclei; magnetic polarization can be transferred from the electrons in nearby radicals to the nuclei by *dynamic nuclear polarization* (*DNP*); and the magnetic order of the two protons in *para*-hydrogen molecules can be transferred to specific target molecules by temporary complexation or by hydrogentation and from there to other nuclei. Apart from the prepolarization method, the other methods are selective in amplifying the polarization of particular chemical groups. While this seems to be a disadvantage at first sight, it is an advantage when analyzing complex fluid mixtures with many resonances at low field. Sensitivity can also be gained when resorting to detection schemes which are alter-

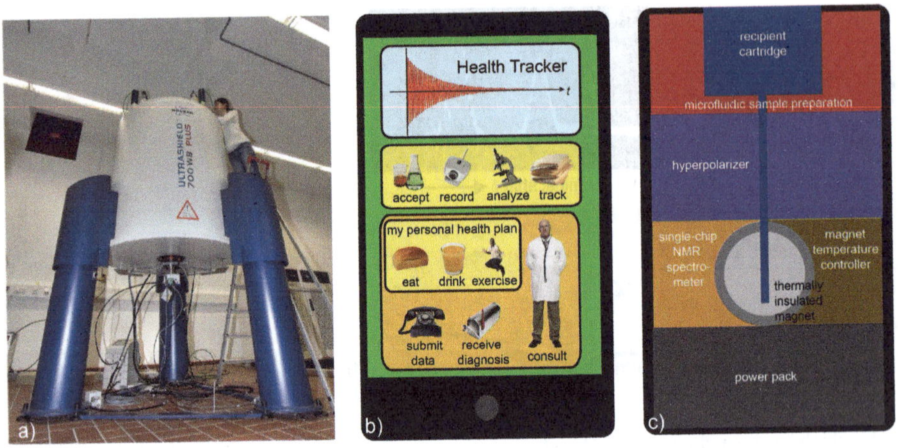

Figure 9.2. NMR hardware now and for the future. (a) 700 MHz NMR magnet for chemical analysis. (b) Conceptual drawing of a smartphone-sized NMR spectrometer for quantitative analysis of body fluids and personal health care. (c) Conceptual drawing of the NMR components of a personal health tracker.

natives to conventional detection with an rf coil by nuclear induction. Here superconducting quantum interference devices (*SQUIDs*), and *optical magnetometers* including nitrogen vacancies in diamond are some of the novel avenues being explored.

So far, none of these technologies has been compacted to a size of a smartphone while working at field strength near 1 T where the chemical shift resolution is still acceptable for chemical analysis. But once solved [5], and techniques have been refined to produce small permanent magnets with magnetic fields sufficiently homogeneous to resolve the chemical shift, microfluidic analyzers can be designed that utilize NMR spectroscopy in addition to other techniques like infrared spectroscopy to screen body fluids for point-of-care medical diagnostics and for home use to track personal health and to optimize one's own nutrition and exercise plan via internet feedback (Fig. 9.2b,c).

With the advent of compact NMR machines today, NMR already begins to be used outside the NMR center or the hospital at the site of the object for materials testing, chemical analysis, quality control of products and process control in the factory. The speculation of eventually finding NMR in point-of-care medical analyzers and in every household is shared by a number of visionaries and even promoted by Qualcomm Inc., a research and development enterprise and one of the largest chip manufacturers, who announced the Qualcomm Tricorder X-Prize in 2012 for development of a tool capable of capturing key health metrics and diagnosing a set of 15 diseases [6]. Chemical analysis of body fluids and tissue by NMR techniques is expected to be an important component of this device.

9.2 Further Reading

Zalesskiy SS, Danieli , Blümich B, Ananikov VP. Miniaturization of NMR systems: Desktop spectrometers, microcoil spectroscopy and "NMR on a Chip" for chemistry, biochemistry and industry. Chem Rev. 2014; in press.

Webb AG. Radiofrequency microcoils for magnetic resonance imaging and spectroscopy. J Magn Reson. 2013; 229:55–66.

Qian C, Zabow G, Koretsky A. Engineering novel detectors and sensors for MRI. J Magn Reson. 2013; 229:67–74.

Espy E, Matlashov A, Volegov P. SQUID-detected ultra-low field MRI. J Magn Reson. 2013; 229:127–141.

Lilburn DML, Pavlovskaya GE, Meersmann T. Perspectives of hyperpolarized noble gas MRI beyond ^{3}He. J Magn Reson. 2013; 229:173–186.

Green RA, Adams RW, Duckett SB, Mewis RE, Williamson DC, Green CGR. The theory and practice of hyperpolarization in magnetic resonance using parahydrogen. Prog Nucl Magn Reson Spectr. 2012: 67:1–48.

Acosta RH, Blümler P, Münnemann K, Spiess HW. Mixture and dissolution of laser polarized noble gases: Spectroscopic and imaging applications. Prog Nucl Magn Reson Spectr. 2012; 66:40–69.

Utz M, Landers J. Magnetic resonance and microfluidics. Science. 2010; 330:1056–1058.

Harel E. Lab-on-a-chip detection by magnetic resonance methods. Prog Nuc Magn Reson Spectr. 2010; 57:293–305.

Blümich B, Casanova F, Appelt S. NMR at low magnetic fields. Chem Phys Let. 2009; 477:231–240.

Budker D, Romalis M. Optical magnetometry. Nature Physics. 2007; 3:227–234.

Beckonert O, Keun HC, Ebbels TMD, Bundy J, Holmes E, et al. Metabolic profiling, metabolomic and metabonomic procedures for NMR spectroscopy of urine, plasma, serum and tissue extracts. Nature Protocols. 2007; 2:2692–2703.

Jelezko F, Wrachtrup J. Single defect centres in diamond: A review. Physica Status Solidi A. 2006; 203:3207–3225.

9.3 References

[1] Arnold JT, Dharmatti SS, Packard ME. Chemical effects on nuclear induction signal from organic compounds. J Chem Phys. 1951; 19:507.

[2] Sun N, Yoon TJ, Lee H, Andress W, Weissleder R, Ham D. Member IEEE Palm NMR and 1-Chip NMR. IEEE J Solid-State Circuits. 2011; 46:342–352.

[3] Sillerud LO, McDowell AF, Adolphi NL, Serda RE, Adams DP, Vasile MJ, Alam TM. ^{1}H NMR detection of superparamagnetic nanoparticles at 1 T using a microcoil and novel tuning circuit. J Magn Reson. 2006; 181:181–190.

[4] Haun JB, Castro CM, Wang R, Peterson VM, Marinelli BS, Lee H, Weissleder R. Micro-NMR for rapid molecular analysis of human tumor samples. Sci Transl Med. 2011; 3:71ra16.

[5] Kennedy DJ, Jimenez-Martinez R, Donley EA, Knappe S, Seltzer SJ, Ring HL et al. A Microfabricated Xenon Hyperpolarizer. Abstract ENC Asilomar Cal. 2013.

[6] http://www.qualcommtricorderxprize.org (23 August 2013)

Index

Printed and bound by CPI Group (UK) Ltd, Croydon, CR0 4YY
13/09/2026
14960854-0001